Genetic and Epigenetic Modulation of Cell Functions by Physical Exercise

Genetic and Epigenetic Modulation of Cell Functions by Physical Exercise

Special Issue Editor

Italia Di Liegro

MDPI • Basel • Beijing • Wuhan • Barcelona • Belgrade

Special Issue Editor
Italia Di Liegro
Universita degli Studi di
Palermo, Department of
Experimental Biomedicine and
Clinical Neurosciences
(BioNeC), Palermo, Italy

Editorial Office
MDPI
St. Alban-Anlage 66
4052 Basel, Switzerland

This is a reprint of articles from the Special Issue published online in the open access journal *Genes* (ISSN 2073-4425) in 2019 (available at: https://www.mdpi.com/journal/genes/special_issues/ Genetics_Epigenetics_Cell_Exercise).

For citation purposes, cite each article independently as indicated on the article page online and as indicated below:

LastName, A.A.; LastName, B.B.; LastName, C.C. Article Title. *Journal Name* **Year**, *Article Number*, Page Range.

ISBN 978-3-03928-480-1 (Pbk)
ISBN 978-3-03928-481-8 (PDF)

Cover image courtesy of Italia Di Liegro.

© 2020 by the authors. Articles in this book are Open Access and distributed under the Creative Commons Attribution (CC BY) license, which allows users to download, copy and build upon published articles, as long as the author and publisher are properly credited, which ensures maximum dissemination and a wider impact of our publications.

The book as a whole is distributed by MDPI under the terms and conditions of the Creative Commons license CC BY-NC-ND.

Contents

About the Special Issue Editor

Italia Di Liegro get a Master Degree in Biology in 1975 at the University of Palermo, Italy, with a thesis on sea urchin chromatin proteins. In 1981 she began working as University Researcher at a project on chromatin organization in sea urchin embryos. Afterwards, she turned to the rat developing central nervous system (CNS). In 2000 she was appointed as Associated Professor in Biochemistry, and in 2005 as Full Professor in Biochemistry, at the University of Palermo, Faculty of Medicine and Surgery. The studies on rat CNS allowed demonstration that regulation of histone variant expression is mainly post-transcriptional and involves both specific and general RNA-binding factors. In particular, at least two cDNAs were cloned in Di Liegro's laboratory which encoded novel proteins (CSD-C2/PIPPin and LPI), possibly involved in this regulation. In parallel, IDL was involved in setting and analyzing an in vitro model of blood-brain barrier (BBB). In the last few years, Di Liegro's group demonstrated that both neurons and glial cells shed extracellular vesicles (EVs) that transport angiogenic factors, and is now involved with studies aimed at ascertaining the role of EVs in both physiological and pathological conditions.

Editorial

Genetic and Epigenetic Modulation of Cell Functions by Physical Exercise

Italia Di Liegro

Department of Biomedicine, Neurosciences and Advanced Diagnostics (Dipartimento di Biomedicina, Neuroscienze e Diagnostica avanzata) (Bi.N.D.), University of Palermo, 90128 Palermo, Italy; italia.diliegro@unipa.it; Tel.: +39-091-238-97415/446

Received: 12 December 2019; Accepted: 13 December 2019; Published: 16 December 2019

Abstract: Since ancient times, the importance of physical activity (PA) and of a wholesome diet for human health has been clearly recognized. However, only recently, it has been acknowledged that PA can reverse at least some of the unwanted effects of a sedentary lifestyle, contributing to the treatment of pathologies such as hypertension and diabetes, to the delay of aging and neurodegeneration, and even to the improvement of immunity and cognitive processes. At the same time, the cellular and molecular bases of these effects are beginning to be uncovered. The original research articles and reviews published in this Special Issue on "Genetic and Epigenetic Modulation of Cell Functions by Physical Exercise" focus on different aspects of the genetics and molecular biology of PA effects on health and, in addition, on the effects of different genotypes on the ability to perform PA. All authors have read and agreed to the published version of the manuscript.

Keywords: physical activity; physical exercise; aerobic exercise; exercise and health

---

1. Introduction

The importance of exercise for health had already been recognized by ancient physicians [1]: for example, Hippocrates (c460–c370 BC) underlined that: "eating alone will not keep a man well, he must also take exercise" [1–3]. Moreover, Hippocrates considered exercise like a medicine and was the first recorded physician to provide a detailed written exercise prescription [Tipton, 2014]. Similarly, Herophilus (c335–c280 BC.) believed that exercise and a healthy diet were fundamental for maintaining a healthy body and a healthy mind [4]. Over the centuries, this belief has been expressed and clarified in detail many times [1,5]. Only recently, however, we have been starting to understand the cellular and molecular bases of the detrimental effects on health of a sedentary lifestyle, as well as the reasons why physical activity (PA) can function as a medicine to counteract these effects.

This Special Issue was destined to discuss how cell functions can be modulated at both the genetic and the epigenetic level by physical exercise. Some of the original papers also deal with the effects of gene variants on the ability to perform PA.

2. Effects of Physical Exercise on Cardiovascular Disease, Inflammation, and Immunity

The fact of the matter is that PA can counteract the development of hypertension in a dose-dependent manner [6]. This finding is of high relevance when we consider that hypertension is a widespread disorder as well as the most common risk factor for cardiovascular disease (CVD), which, in turn, is one of the leading causes of death [7–9]. In spite of these clear correlations, however, only a few patients with hypertension actually exercise. As discussed in one of the reviews published in this Special Issue, the reasons for the lack of interest in PA of most patients are many; one of them, however, can be found in the genetically determined efficacy of exercise, as shown in a systematic review of studies published in Pubmed and other data banks, describing the association between candidate genes and the antihypertensive effects of exercise [10]. These data have been used to derive a signature of genes related to blood pressure and exercise effects; the corresponding gene exons have been deep-sequenced. From this study, the approach based on deep-sequencing of systematically assembled signature of genes emerges as a cost- and time-efficient way to predict the antihypertensive effects of exercise. The conclusions are relevant for the long-term goal of the described work, that is to develop personalized exercise prescriptions based upon genetic predispositions and other clinical features of the patients [10].

A condition that seemingly links PA and CVD is inflammation, evaluated as the number of circulating white blood cells [11]. In their study, published in this Special Issue, Prins and co-workers [12] investigated the association between the genetic substrate and the ability to perform exercise, together with the association between systemic inflammation and PA. Briefly, 68 nucleotide polymorphisms, objectively associated with specific levels of PA, were analyzed, focusing on circulating blood cells. Notably, the authors report that increased PA ability correlates with decreased inflammation and fewer circulating lymphocytes and eosinophils [12].

Modifications induced by acute bouts of exercise were also evidenced in natural killer (NK) cells of the innate immune system, purified from peripheral blood. In particular, in a pilot study described in this Special Issue, Schenk and co-workers found that the DNA methylation level of 25 genes with different regulatory roles was changed in five women after an incremental step test done using a bicycle ergometer [13]. Although this finding is preliminary and new studies are necessary to correlate the methylation of these genes with functional effects, it is anyway of note, since it suggests that exercise can induce epigenetic adaptations, possibly responsible for the improvement of immune system functions.

3. Impact of Physical Exercise on Body Mass Index and Lipid Metabolism

Obesity is a leading risk factor for many chronic pathologies, such as CVD and diabetes. The main causes of obesity lie in a positive balance between intake and expenditure of calories, in the consumption of high-energy food [14], and in a sedentary lifestyle, very often also linked to socio-economic conditions [15]. As described in one of the papers published in this Special Issue, a fundamental role for human health and disease is also played by the gut microbial communities (microbiota), initially established at birth but further enriched in the first 3–4 years of life. Importantly, quality and vitality of these communities are influenced by environmental and nutritional cues [16].

Obviously, the genetic background also has an impact on obesity, since it influences the susceptibility to metabolic adaptations [17]. Among the genes that have been linked to obesity, that encoding the Fat mass and obesity-associated protein (FTO) is one of the best known [17]. In mice models, it has been shown that some FTO deletions and point mutations cause an increase of metabolic rate, while FTO overexpression induces dyslipidemia [17]. On the other hand, some FTO variants have been associated with obesity in humans, although the underlying molecular mechanisms have not been completely clarified. One interesting observation concerns the fact that FTO is an N6-methyl-adenosine RNA demethylase that might act as a post-transcriptional epigenetic effector [18]. In an original study reported in the present issue, the effect on body mass index (BMI) of the obesity-inducing FTO rs3751812 variant has been analyzed in Taiwanese adults who exercise in comparison with individuals who do not exercise [19]. As expected, individuals with the variant gene both in heterozygosity and homozygosity had a higher BMI than wild-type individuals. However, obesity susceptibility, measured as BMI increase, induced by the FTO variant, was found to be reduced by PA [19]. In another paper of this Special Issue, the authors studied the possible correlation between aerobic exercise and high-density lipoprotein-associated cholesterol (HDL-C) in a large group of 30–70-year-old individuals with different BMI and waist/hip ratio (WHR). The impact of a single nucleotide polymorphism on hepatic lipase (rs1800588 variant) was also analyzed [20]. The results reported in the paper show that, as expected, individuals with abnormal BMI and/or abnormal WHR had significantly lower HDL-C levels with respect to individuals with normal parameters. Notably, aerobic exercise significantly induced higher values of HDL-C in comparison with no exercise [20]. On the other hand, the authors found that in individuals with a CC genotype, HDL-C was not modified by aerobic exercise in any of the BMI and WHR classes. This finding confirms the idea that the effect of PA can be different in different genetic (and perhaps epigenetic) backgrounds (Figure 1).

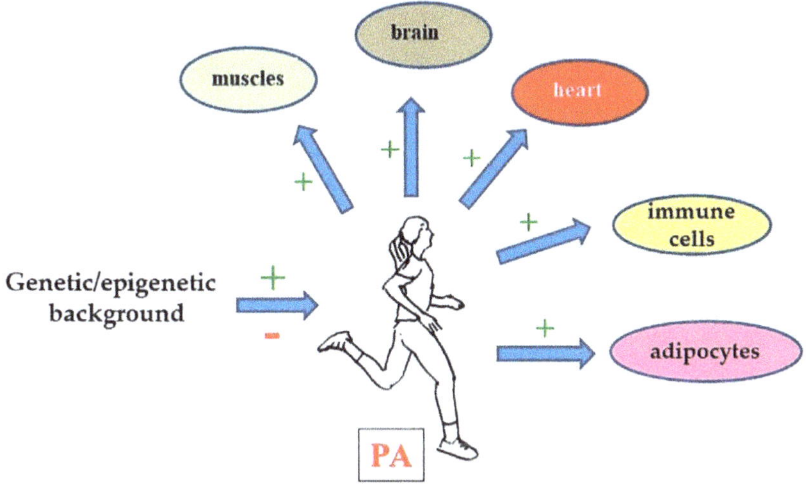

Figure 1. As described in this Special Issue, physical activity (PA) has a beneficial effect on the functions of most tissues in the body, from the heart and circulation to the immune and nervous systems. It can also help in reducing or at least in contrasting obesity, diabetes, and other chronic pathologies, by itself or in association with conventional drugs. However, it is increasingly clear that the genetic (and probably the epigenetic) background of a patient can have both positive and negative effects on the efficacy of exercise. Thus, an important goal for the future should be to identify panels of genes that control the responses to PA.

4. Impact of the Genetic Background on PA

As discussed in paragraph 1, the efficacy of exercise can be genetically determined. On the other hand, the ability to perform PA can be influenced by the genotype. In one of the original papers published in this Special Issue, Del Coso and co-workers [21] analyzed the exercise phenotypes of recreational marathon runners endowed with different variants of the gene encoding α-actinin-3 (ACTN3). Normally, the expression of this protein is limited to fast-type muscle fibers. Individuals homozygotes for a null allele (X) of the gene partially compensate by expressing a higher amount of α-actinin-2 in fast-type fibers, a process that, however, does not prevent increased susceptibility to contraction-induced damage [22]. It was reported that the ACTN3 genotype might influence exercise phenotypes in recreational marathoners [22]. In particular, deficiency of α-actinin-3 seems to correlate with higher body fatness, lower muscle strength, and higher muscle flexibility [21].

The role of genetic background in exercise aptitude is also the topic of another original article of this Special Issue. In this work, 451 subjects were genotyped for gene variants correlated with different functional and metabolic parameters (among which inflammation, vascular function, carbohydrate metabolism, and lipid metabolism). Specific questionnaires were also used to evaluate both quantitatively and qualitatively the average physical exercise level [23]. Notably, the authors found that the carriers of a minor allele of the gene encoding glucokinase regulator (GCKR) tended to show a greater frequency of physical exercise in comparison to the major homozygous genotype carriers [23].

5. Other PA Effects

As a consequence of the hypoxia that often accompanies endurance exercise, oxygen transportation has to be increased in both well-trained and amateur athletes. An increase of oxygen requirement means, in turn, an increase in hemoglobin and myoglobin and thus affects iron concentration and availability, as well as the synthesis of proteins (enzymes, transporters, and regulators) involved in iron and heme metabolism. The levels of mRNAs encoding some of these proteins have been studied by Grzybkowska and co-workers [24] in the leukocytes of 24 amateur runners, before a marathon and 3 h and 24 h after running [24]. The results show a significant modification of gene expression and serum iron and C-reactive protein levels, while ferritin concentration remained unchanged. Notably, the authors found that the running pace also influence the levels of the studied mRNAs [24].

Finally, this Special Issue also includes a review paper that summarizes bibliographic data suggesting that PA improves cognitive processes and memory, together with exerting analgesic and antidepressant effects [25]. The potential mechanisms underlying the effects of PA on brain health are discussed, taking into account hormones, neurotrophines, and neurotransmitters, the release of which is modulated by PA. When already discovered, intra- and extracellular pathways that regulate the expression of some of the genes involved in the neuroprotective and anti-aging effects of PA on brain health are also discussed Notably, experiments are reported that suggest that PA can also function as a medicine in the therapy of neurodegeneration [25].

6. Conclusions

In conclusion, this Special Issue has tried to open a forum on the effects of PA on different functions of our body. The results collected concern different aspects of exercise impact, from the effects on the heart and circulation to those on the immune and nervous systems. The results reported in both the reviews and the original papers show that, in most cases, PA can be considered as a real therapy, to be used both alone and in association with conventional drugs.

Notably, however, in most articles, it is clearly shown that the genetic (and probably the epigenetic) background can affect all the PA effects (Figure 1). Thus, probably, as a long-term goal, the studies on the therapeutic effects of PA should also take into account the specific genome of each athlete or patient considered. Actually, this concept is not new: it is becoming increasingly clear, indeed, that these considerations apply to any therapy, even if we are not yet ready to integrate them into treatment designs.

Funding: This research received no external funding

Conflicts of Interest: The author declares no conflicts of interest

References

1. Tipton, C.M. The history of "Exercise Is Medicine" in ancient civilizations. *Adv. Physiol. Educ.* **2014**, *38*, 109–117. [CrossRef] [PubMed]
2. Sallis, R.E. Exercise is medicine and physicians need to prescribe it! *Br. J. Sports Med.* **2009**, *43*, 3–4. [CrossRef] [PubMed]
3. Berryman, J.W. Exercise is medicine: A historical perspective. *Curr. Sports Med. Rep.* **2010**, *9*, 195–201. [CrossRef] [PubMed]
4. Russo, L. *The Forgotten Revolution*; Springer-Verlag: Berlin/Heidelberg, Germany, 2003; ISBN 3-540-20068-1.
5. Booth, F.W.; Roberts, C.K.; Thyfault, J.P.; Ruegsegger, G.N.; Toedebusch, R.G. Role of Inactivity in Chronic Diseases: Evolutionary Insight and Pathophysiological Mechanisms. *Physiol. Rev.* **2017**, *97*, 1351–1402. [CrossRef] [PubMed]
6. Diaz, K.M.; Shimbo, D. Physical activity and the prevention of hypertension. *Curr. Hypertens. Rep.* **2013**, *15*, 659–668. [CrossRef] [PubMed]

7. Pescatello, L.S.; MacDonald, H.V.; Lamberti, L.; Johnson, B.T. Exercise for Hypertension: A Prescription Update Integrating Existing Recommendations with Emerging Research. *Curr. Hypertens. Rep.* **2015**, *17*, 87. [CrossRef] [PubMed]
8. Schultz, M.G.; La Gerche, A.; Sharman, J.E. Blood Pressure Response to Exercise and Cardiovascular Disease. *Curr. Hypertens. Rep.* **2017**, *19*, 89. [CrossRef]
9. Lachman, S.; Boekholdt, S.M.; Luben, R.N.; Sharp, S.J.; Brage, S.; Khaw, K.-T.; Peters, R.J.; Wareham, N.J. Impact of physical activity on the risk of cardiovascular disease in middle-aged and older adults: EPIC Norfolk prospective population study. *Eur. J. Prev. Cardiol.* **2018**, *25*, 200–208. [CrossRef]
10. Pescatello, L.S.; Parducci, P.; Livingston, J.; Taylor, B.A. A Systematically Assembled Signature of Genes to be Deep-Sequenced for Their Associations with the Blood Pressure Response to Exercise. *Genes* **2019**, *10*, 295. [CrossRef]
11. Willis, E.A.; Shearer, J.J.; Matthews, C.E.; Hofmann, J.N. Association of physical activity and sedentary time with blood cell counts: National Health and Nutrition Survey 2003-2006. *PLoS ONE* **2018**, *13*, e0204277. [CrossRef]
12. Prins, F.M.; Said, M.A.; van de Vegte, Y.J.; Verweij, N.; Groot, H.E.; van der Harst, P. Genetically Determined Physical Activity and Its Association with Circulating Blood Cells. *Genes* **2019**, *10*, 908. [CrossRef] [PubMed]
13. Schenk, A.; Koliamitra, C.; Bauer, C.J.; Schier, R.; Schweiger, M.R.; Bloch, W.; Zimmer, P. Impact of Acute Aerobic Exercise on Genome-Wide DNA-Methylation in Natural Killer Cells-A Pilot Study. *Genes* **2019**, *10*, 380. [CrossRef] [PubMed]
14. Spiegelman, B.M.; Flier, J.S. Obesity and the regulation of energy balance. *Cell* **2001**, *104*, 531–543. [CrossRef]
15. Blüher, M. Obesity: Global epidemiology and pathogenesis. *Nat. Rev. Endocrinol.* **2019**, *15*, 288–298. [CrossRef] [PubMed]
16. Paoli, A.; Mancin, L.; Bianco, A.; Thomas, E.; Mota, J.F.; Piccini, F. Ketogenic Diet and Microbiota: Friends or Enemies? *Genes* **2019**, *10*, 534. [CrossRef] [PubMed]
17. Chang, J.Y.; Park, J.H.; Park, S.E.; Shon, J.; Park, Y.J. The Fat Mass- and Obesity-Associated (FTO) Gene to Obesity: Lessons from Mouse Models. *Obesity (Silver Spring)* **2018**, *26*, 1674–1686. [CrossRef]
18. Zhao, X.; Yang, Y.; Sun, B.-F.; Zhao, Y.-L.; Yang, Y.-G. FTO and Obesity: Mechanisms of Association. *Curr. Diabetes Rep.* **2014**, *14*, 486. [CrossRef]
19. Liaw, Y.C.; Liaw, Y.P.; Lan, T.H. Physical Activity Might Reduce the Adverse Impacts of the FTO Gene Variant rs3751812 on the Body Mass Index of Adults in Taiwan. *Genes* **2019**, *10*, 354. [CrossRef]
20. Nassef, Y.; Nfor, O.N.; Lee, K.J.; Chou, M.C.; Liaw, Y.P. Association between Aerobic Exercise and High-Density Lipoprotein Cholesterol Levels across Various Ranges of Body Mass Index and Waist-Hip Ratio and the Modulating Role of the Hepatic Lipase rs1800588 Variant. *Genes* **2019**, *10*, 440. [CrossRef]
21. Del Coso, J.; Moreno, V.; Gutiérrez-Hellín, J.; Baltazar-Martins, G.; Ruíz-Moreno, C.; Aguilar-Navarro, M.; Lara, B.; Lucía, A. ACTN3 R577X Genotype and Exercise Phenotypes in Recreational Marathon Runners. *Genes* **2019**, *10*, 413. [CrossRef]
22. Seto, J.T.; Lek, M.; Quinlan, K.G.R.; Houweling, P.J.; Zheng, X.F.; Garton, F.; MacArthur, D.G.; Raftery, J.M.; Garvey, S.M.; Hauser, M.A.; et al. Deficiency of a-actinin-3 is associated with increased susceptibility to contraction-induced damage and skeletal muscle remodeling. *Hum. Mol. Genet.* **2011**, *20*, 2914–2927. [CrossRef] [PubMed]
23. Espinosa-Salinas, I.; de la Iglesia, R.; Colmenarejo, G.; Molina, S.; Reglero, G.; Martinez, J.A.; Loria-Kohen, V.; Ramirez de Molina, A. GCKR rs780094 Polymorphism as A Genetic Variant Involved in Physical Exercise. *Genes* **2019**, *10*, 570. [CrossRef] [PubMed]
24. Grzybkowska, A.; Anczykowska, K.; Ratkowski, W.; Aschenbrenner, P.; Antosiewicz, J.; Bonisławska, I.; Żychowska, M. Changes in Serum Iron and Leukocyte mRNA Levels of Genes Involved in Iron Metabolism in Amateur Marathon Runners-Effect of the Running Pace. *Genes* **2019**, *10*, 460. [CrossRef] [PubMed]
25. Di Liegro, C.M.; Schiera, G.; Proia, P.; Di Liegro, I. Physical Activity and Brain Health. *Genes* **2019**, *10*, 720. [CrossRef] [PubMed]

© 2019 by the author. Licensee MDPI, Basel, Switzerland. This article is an open access article distributed under the terms and conditions of the Creative Commons Attribution (CC BY) license (http://creativecommons.org/licenses/by/4.0/).

 *genes* 

Review

A Systematically Assembled Signature of Genes to be Deep-Sequenced for Their Associations with the Blood Pressure Response to Exercise

Linda S. Pescatello [1,2,*], Paul Parducci [1], Jill Livingston [3] and Beth A. Taylor [1,2,4]

[1] Department of Kinesiology, University of Connecticut, Storrs, CT 06269, USA;
Paul.Parducci@uconn.edu (P.P.); Beth.Taylor@uconn.edu (B.A.T.)
[2] Institute for Systems Genomics, University of Connecticut, Storrs, CT 06269, USA
[3] Homer Babbidge Library, Health Sciences, University of Connecticut, Storrs, CT 06269, USA;
Jill.Livingston@uconn.edu
[4] Preventive Cardiology, Hartford Hospital, Hartford, CT 06269, USA
* Correspondence: Linda.Pescatello@uconn.edu; Tel.: +1-860-486-0008; Fax: +1-860-486-1123

Received: 9 March 2019; Accepted: 4 April 2019; Published: 11 April 2019

Abstract: *Background*: Exercise is one of the best nonpharmacologic therapies to treat hypertension. The blood pressure (BP) response to exercise is heritable. Yet, the genetic basis for the antihypertensive effects of exercise remains elusive. *Methods*: To assemble a prioritized gene signature, we performed a systematic review with a series of Boolean searches in PubMed (including Medline) from earliest coverage. The inclusion criteria were human genes in major BP regulatory pathways reported to be associated with: (1) the BP response to exercise; (2) hypertension in genome-wide association studies (GWAS); (3) the BP response to pharmacotherapy; (4a) physical activity and/or obesity in GWAS; and (4b) BP, physical activity, and/or obesity in non-GWAS. Included GWAS reports disclosed the statistically significant thresholds used for multiple testing. *Results*: The search yielded 1422 reports. Of these, 57 trials qualified from which we extracted 11 genes under criteria 1, 18 genes under criteria 2, 28 genes under criteria 3, 27 genes under criteria 4a, and 29 genes under criteria 4b. We also included 41 genes identified from our previous work. *Conclusions*: Deep-sequencing the exons of this systematically assembled signature of genes represents a cost and time efficient approach to investigate the genomic basis for the antihypertensive effects of exercise.

Keywords: antihypertensive therapy; healthy lifestyle; physical exercise

---

1. Introduction

Cardiovascular disease (CVD) is the leading cause of death worldwide, claiming 31% (17.5 million) of deaths globally [1,2]. Hypertension is the most common, costly, and preventable CVD risk factor [1]. The American College of Cardiology (ACC)/American Heart Association (AHA) recently redefined hypertension to a lower blood pressure (BP) threshold of 130 mmHg for systolic BP (SBP) or 80 mmHg for diastolic BP (DBP) [3]. This change has been met with some opposition. Nonetheless, the lower BP thresholds now classify 50% of adults in the United States with hypertension [1], underscoring the importance of hypertension as a public health problem.

The authors of the AHA/ACC report rated exercise as one of the best nonpharmacologic therapies to treat hypertension because aerobic exercise training lowers BP 5–8 mmHg among adults with hypertension [3]. The magnitude of these BP reductions rival those that result from taking antihypertensive medication [4,5], may lower the risk of CVD risk by 4–22% and stroke by 6–41% [6–8], and reduce the resting BP of some adults with hypertension into normal ranges [9,10]. Accordingly, professional organizations from around the world recommend adults with hypertension participate in 30–60 min/day of aerobic exercise, such as walking or jogging, on most days of the week [5,11].

Despite the well-documented antihypertensive benefits of exercise, many adults with hypertension do not exercise to lower their BP [12,13]. Reasons for this non-adherence are varied; however, one reason relevant to the topic of this systematic review is that there is significant inter-individual variability in the BP response to exercise partially attributed to genetic predispositions that led some to believe exercise does not work as antihypertensive therapy [14,15]. Indeed, investigators from the *HE*alth, *RI*sk Factors Exercise *TrA*ining and *GE*netics or *HERITAGE* Family Study involving over 700 subjects established the BP response to aerobic exercise is heritable (h = 0.13–0.42) [14,16–19]. Nearly 20 years ago this discovery prompted our laboratory group [20–24] and others [16–19,25] to conduct candidate genes association studies to identify genetic variants that account for a clinically meaningful proportion of the variability in the BP response to exercise. However, these efforts have met with little success [17,25,26].

More recently, we exploited advances in genomic technology that emerged since our discovery phase candidate gene association studies as well as other strategies to bolster our statistical power to detect genetic variants associated with the BP response to exercise in a replication cohort of subjects with hypertension whose characteristics resembled those from our earlier studies. In this series of studies, we deep-sequenced the exons of genes using the Illumina TruSeq Custom Amplicon kit (Catalog# FC-130-1001, Illumina, San Diego, CA, USA) on a prioritized signature of 41 BP and exercise genes that contained genes identified from our prior work in addition to those obtained from a systematic review of the literature of genes reported to be associated with hypertension or the BP response to exercise or pharmacotherapy [27–29]. After adjustment for multiple testing, despite the small sample size, we found variants in 61% of the 41 genes in the prioritized panel associated with the BP response to exercise. We attributed the high proportion of the significant BP-genotype associations that we found to a focused inquiry of variants with a gene signature obtained from a systematic review of the literature that reduced the search space within the genome; and the use of high throughput exon sequencing to concentrate on functional gene regions and standardized BP and exercise protocols that were well controlled and closely supervised.

The purpose of this systematic review is to update and expand our original systematic review to assemble a prioritized signature of BP and exercise genes whose exons can then be deep-sequenced among a larger, more ethnically and gender diverse sample of adults that were reported in the qualifying studies to have hypertension to better inform the genomic basis for the antihypertensive effects of exercise. The long-term goal of this work is to develop personalized exercise prescriptions based upon genetic predispositions and other clinical characteristics to optimize the BP benefits of exercise.

2. Systematic Review Methods

This systematic review followed the specifications of the Preferred Reporting Items for Systematic Reviews and Meta-Analyses (PRISMA) statement [30,31]. A series of comprehensive Boolean searches were run in PubMed (including Medline). The first search replicated the original search done by Bruneau et al. [25] in the four databases of Medline, Biosis, Scopus, and Web of Science and included contiguous dates. The additional three searches were run from earliest coverage to 11 May 2017 to locate trials that met our predetermined inclusion criteria. These *a priori* criteria were human genes in major BP regulatory pathways reported to be associated with: (1) the BP response to exercise; (2) hypertension in genome-wide association studies (GWAS); (3) the BP response to pharmacotherapy; (4a) physical activity and/or obesity, a major risk factor for hypertension as 80% of adults with hypertension are overweight to obese [32] in GWAS; and (4b) BP, physical activity, and/or obesity in non-GWAS from relevant reference searches of the authors' files. We also included the genes from the prioritized signature of BP and exercise genes from our previous work [27–29].

The major regulatory BP pathways that a qualifying gene could be in were the renin angiotensin aldosterone system, endothelial nitric oxide synthase pathway, and/or the sympathetic nervous system, and/or pathways involved with vascular function and structure, fluid and electrolyte control, inflammation, insulin sensitivity, lipid metabolism, and obesity. Of note, many of the GWAS we initially located as indicating they involved genes associated with hypertension upon closer scrutiny involved genes associated with resting BP rather than hypertension *per se* so that they were excluded. In all GWAS for the gene to be included on the panel, the reported associations with the BP response to exercise must have met preestablished statistically significant thresholds for multiple testing which is commonly set at $p = 5 \times 10^{-8}$. The full search strategy for each of the four *a priori* criteria are depicted in Tables 1–4.

Table 1. Criteria 1: Genes (n = 11) in major blood pressure regulatory pathways associated with the blood pressure response to exercise. Two searches were conducted on 05/11/2017 and a total of 321 records were identified.

Search 1:	(2006/01:2017/06 [edat]) AND ("mean arterial" OR "blood pressure"[mesh] OR "blood pressure" OR "blood pressures" OR "arterial pressure" OR "arterial pressures" OR hypertension OR hypotension OR normotension OR hypertensive OR hypotensive OR normotensive OR "systolic pressure" OR "diastolic pressure" OR "pulse pressure" OR "venous pressure" OR "pressure monitor" OR hypotension OR "pre hypertension" OR "bp response" OR "bp decrease" OR "bp reduction" OR "bp monitor" OR "bp monitors" OR "bp measurement") AND ("exercise"[mesh] OR exercise OR exercises OR running[mesh] OR "bicycle" OR "bicycles" OR "bicycling" OR walking[mesh] OR treadmill* OR "weight lifting" OR "weight training" OR "weight bearing" OR "resistance training" OR "strength training" OR "endurance training" OR "speed training" OR "training duration" OR "training frequency" OR "training intensity" OR "aerobic endurance") AND ("randomized controlled trial"[pt] OR "nonrandomized controlled" OR "nonrandomized control" OR controlled clinical trial[pt] OR "randomized controlled trial"[publication type] OR random allocation[mh] OR clinical trial[pt] OR "comparative study" OR "comparative studies" OR clinical trials[mh] OR "clinical trial"[tw] OR "latin square"[tw] OR random*[tw] OR research design[mh:noexp] OR "comparative study"[publication type] OR "evaluation studies"[publication type] OR "prospective studies"[mh] OR "cross-over studies"[mh] OR "control"[tw] OR "controlled"[tw]) AND ("gene" OR "genes" OR "genotype" OR "genotypes" OR "snp" OR polymorphism* OR "DNA" OR "minor allele" OR "minor alleles" OR "single nucleotide polymorphism" OR "single nucleotides polymorphisms" OR genetic*) NOT ("DASH"[tiab] OR "cancer" OR "neoplasms" OR "review"[pt] OR "fibromyalgia" OR "alzheimers" OR "alzheimer" OR "pregnant" OR "pregnancy" OR "obesity/drug therapy"[mesh] OR "diet therapy"[mesh] OR "diet therapy"[subheading] OR "caffeine" OR "eating change" OR "activities of daily living" OR "dehydration" OR "dehydrate" OR "dehydrated" OR "dietary salt" OR "epilepsy" OR "influenza" OR "flu" OR "pneumonia" OR "septicemia" OR "hiv" OR "Acquired Immunodeficiency Syndrome" OR "meningitis" OR "substance abuse" OR "alcoholism" OR "drug abuse" OR "Cross-Sectional Studies"[MeSH Terms] OR "Prospective Studies"[MeSH Terms] OR "epidemiology"[Subheading]).
Search 2:	("mean arterial" OR "blood pressure"[mesh] OR "blood pressure" OR "blood pressures" OR "arterial pressure" OR "arterial pressures" OR hypertension OR hypotension OR normotension OR hypertensive OR hypotensive OR normotensive OR "systolic pressure" OR "diastolic pressure" OR "pulse pressure" OR "venous pressure" OR "pressure monitor" OR hypotension OR "pre hypertension" OR "bp response" OR "bp decrease" OR "bp reduction" OR "bp monitor" OR "bp monitors" OR "bp measurement") AND ("ambulatory blood pressure" OR "exercise"[mesh] OR exercise[ti] OR exercises OR running[mesh] OR running[ti] OR "bicycle" OR "bicycles" OR "bicycling" OR walking[mesh] OR walking[ti] OR treadmill* OR "weight lifting" OR "weight training" OR "weight bearing" OR "resistance training" OR "strength training" OR "endurance training" OR "speed training" OR "training duration" OR "training frequency" OR "training intensity" OR "aerobic endurance" OR "aerobic training") AND (gwa[ti] OR gwas[ti] OR genome[ti] OR "gene"[ti] OR "genes"[ti] OR "genotype"[ti] OR "genotypes"[ti] OR "genotyping"[ti] OR "snp"[ti] OR "snps"[ti] OR polymorphism*[ti] OR "DNA"[ti] OR allele[ti] OR alleles[ti] OR "minor allele" OR "minor alleles" OR "single nucleotide polymorphism" OR "single nucleotides polymorphisms" OR genetic*[ti] OR "trait locus"[ti] OR "loci"[ti] OR "Genetic Predisposition to Disease"[MeSH] OR "Genotype"[MeSH] OR "Gene Frequency"[MeSH] OR "Polymorphism, Single Nucleotide/genetics"[MESH] OR "Polymorphism, Single Nucleotide"[MAJR] OR "Genetic Loci"[Mesh] OR "Genetic Association Studies"[Mesh] OR "Genetic Variation"[Mesh]) NOT ("DASH"[tiab] OR "cancer" OR "neoplasms" OR "review"[pt] OR "fibromyalgia" OR "alzheimers" OR "alzheimer" OR "pregnant" OR "pregnancy" OR "obesity/drug therapy"[mesh] OR "diet therapy"[mesh] OR "diet therapy"[subheading] OR "caffeine" OR "eating change" OR "activities of daily living" OR "dehydration" OR "dehydrate" OR "dehydrated" OR "dietary salt" OR "epilepsy" OR "influenza" OR "flu" OR "pneumonia" OR "septicemia" OR "hiv" OR "Acquired Immunodeficiency Syndrome" OR "meningitis" OR "substance abuse" OR "alcoholism" OR "drug abuse" OR "Cross-Sectional Studies"[MeSH Terms] OR "Prospective Studies"[MeSH Terms] OR "epidemiology"[Subheading]).

Table 2. Criteria 2: Genes (n = 18) in major blood pressure regulatory pathways associated with hypertension in genome wide association studies. The search was conducted on 05/11/2017 and a total of 188 records were identified.

("mean arterial" OR "blood pressure"[mesh] OR "blood pressure" OR "blood pressures" OR "arterial pressure" OR "arterial pressures" OR hypertension OR hypotension OR normotension OR hypertensive OR hypotensive OR normotensive OR "systolic pressure" OR "diastolic pressure" OR "pulse pressure" OR "venous pressure" OR "pressure monitor" OR hypotension OR "pre hypertension" OR "bp response" OR "bp decrease" OR "bp reduction" OR "bp monitor" OR "bp monitors" OR "bp measurement" OR "Blood Pressure/genetics"[MeSH]) AND ("resting blood pressure" OR "resting BP" OR "exercise"[mesh] OR exercise[ti] OR exercises OR running[mesh] OR running[ti] OR "bicycle" OR "bicycles" OR "bicycling" OR walking OR walking[mesh] OR walking[ti] OR treadmill* OR "weight lifting" OR "weight training" OR "weight bearing" OR "resistance training" OR "strength training" OR "endurance training" OR "speed training" OR "training duration" OR "training frequency" OR "training intensity" OR "aerobic endurance" OR "aerobic training" OR "physical activity" OR "motor activity"[mesh] OR Overweight OR "Overweight"[Mesh] OR BMI OR "body mass" OR "Body Mass Index"[MeSH] OR "Waist Circumference"[MeSH] OR obesity OR "Obesity"[Mesh] OR obese) AND (gwa OR gwas OR "Genetic Association Studies"[Mesh] OR genomewide OR "genome wide" OR "genomewide association" OR "genome wide association" OR "genome-wide interaction") NOT ("review"[pt] OR "Cross-Sectional Studies"[MeSH Terms] OR Comment[pt] OR Editorial[pt] OR Letter[pt] OR "Case Reports"[pt] OR "case control"[ti] OR "case report"[ti] OR "case study"[ti] OR "case series"[ti] OR "Case-Control Studies"[Mesh] OR "Follow-Up Studies"[Mesh] OR "observational study"[ti] OR "prospective cohort"[ti] OR "cohort studies" [Mesh:NoExp] OR "cohort study"[ti] OR "Longitudinal Studies" [Mesh:NoExp] OR "Follow-Up Studies"[mesh] OR "Retrospective Studies"[mesh] OR "non-randomized"[ti] OR "follow up study"[ti] OR "Cross-Sectional Studies"[MeSH Terms] OR "Prospective Studies"[MeSH Terms] OR "epidemiology"[Subheading] OR "pulmonary hypertension" OR "pulmonary arterial hypertension" OR "heart transplant" OR "heart failure" OR "cystic fibrosis" OR "cancer" OR "neoplasms" OR "fibromyalgia" OR "alzheimers" OR "alzheimer" OR "pregnant" OR "pregnancy" OR "obesity/drug therapy"[mesh] OR "diet therapy"[mesh] OR "diet therapy"[subheading] OR "DASH"[tiab] OR meal[ti] OR "nutritional intervention" OR "dietary intervention" OR "nutritional counseling" OR "dietary counseling" OR "caffeine" OR "eating change" OR "activities of daily living" OR "dehydration" OR "dehydrate" OR "dehydrated" OR "dietary salt" OR sodium OR "epilepsy" OR "influenza" OR "flu" OR "pneumonia" OR "septicemia" OR arthritis OR "hiv" OR "Acquired Immunodeficiency Syndrome" OR "meningitis" OR "substance abuse" OR "alcoholism" OR "drug abuse" OR "spinal cord"[ti] OR "Sleep"[Majr] OR "Sleep Apnea Syndromes"[Majr] OR sleep[ti] OR contraceptive*[ti] OR (animals[mesh] NOT humans[mesh]) OR rat[ti] OR rats[ti] OR mouse[ti] OR mice[ti] OR pig[ti] OR pigs[ti] OR dog[ti] OR dogs[ti] OR canine[ti] OR cow[ti] OR cows[ti] OR bovine[ti]).

Table 3. Criteria 3. Genes (n = 28) in major blood pressure regulatory pathways associated with the blood pressure response to pharmacotherapy. The search was conducted on 05/11/2017 and a total of 711 records were identified.

("mean arterial" OR "blood pressure"[mesh] OR "blood pressure" OR "blood pressures" OR "arterial pressure" OR "arterial pressures" OR hypertension OR hypotension OR normotension OR hypertensive OR hypotensive OR normotensive OR "systolic pressure" OR "diastolic pressure" OR "pulse pressure" OR "venous pressure" OR "pressure monitor" OR hypotension OR "pre hypertension" OR "bp response" OR "bp decrease" OR "bp reduction" OR "bp monitor" OR "bp monitors" OR "bp measurement" OR "Blood Pressure/genetics"[Mesh]) AND ("Antihypertensive Agents"[Mesh] OR "Antihypertensive Agents" [Pharmacological Action] OR "anti-hypertensive agent" OR "antihypertensive agent" OR "anti hypertensive agent" OR "anti-hypertensive agents" OR "antihypertensive agents" OR "anti hypertensive agents" OR "anti-hypertensive drug" OR "antihypertensive drug" OR "anti hypertensive drug" OR "anti-hypertensive drugs" OR "antihypertensive drugs" OR "anti hypertensive drugs" OR "anti-hypertensive medication" OR "antihypertensive medication" OR "anti hypertensive medication" OR "anti-hypertensive medications" OR "antihypertensive medications" OR "anti hypertensive medications" OR "anti-hypertensives" OR antihypertensives OR "anti hypertensives" OR diuretic OR diuretics OR "Diuretics"[Mesh] OR acebutolol OR aliskiren OR Ambrisentan OR amlodipine OR atenolol OR "azilsartan medoxomil" OR benazepril OR betaxolol OR bisoprolol OR bosentan OR "candesartan cilexetil" OR captopril OR carteolol OR carvedilol OR chlorthalidone OR clonidine OR cilazapril OR clevidipine OR deserpidine OR diazoxide OR diltiazem OR doxazosin OR enalapril OR enalaprilat OR "eprosartan mesylate" OR hydrochlorothiazide OR felodipine OR fenoldopam OR fosinopril OR guanabenz OR guanadrel OR guanethidine OR guanfacine OR hydralazine OR irbesartan OR isradipine OR labetalol OR lisinopril OR "losartan potassium" OR macitentan OR mecamylamine OR methyldopa OR metoprolol OR metyrosine OR mibefradil OR minoxidil OR moexipril OR moxonidine OR nadolol OR nebivolol OR nicardipine OR nifedipine OR nisoldipine OR nitroprusside OR "olmesartan medoxomil" OR omapatrilat OR penbutolol OR perindopril OR phentolamine OR pindolol OR prazosin OR propranolol OR quinapril OR ramipril OR rescinnamine OR reserpine OR sildenafil OR "sodium nitroprusside" OR tadalafil OR telmisartan OR terazosin OR timolol OR trandolapril OR treprostinil OR trimethaphan OR valsartan OR verapamil OR diuretic OR diuretics OR thiazide OR "adrenergic beta-antagonist" OR "adrenergic beta-antagonists" OR "adrenergic alpha-antagonist" OR "adrenergic alpha-antagonists" OR "Angiotensin-Converting Enzyme Inhibitors"[Mesh] OR "ace inhibitor" OR "ace inhibitors" OR "angiotensin-converting enzyme inhibitor" OR "angiotensin-converting enzyme inhibitors" OR "angiotensin II Receptor Blockers" OR "angiotensin II Receptor Blockers" OR "Angiotensin II Type 2 Receptor Blockers"[Mesh] OR "calcium channel blocker" OR "calcium channel blockers" OR "ganglionic blocker" OR "ganglionic blockers" OR "vasodilator agent" OR "vasodilator agents" OR "Vasodilator Agents"[Mesh] OR nitrates OR "Nitrates"[Mesh] OR nitrites OR "Nitrites"[Mesh]) AND (gwa[ti] OR gwas[ti] OR genome[ti] OR "gene"[ti] OR "genes"[ti] OR "genotype"[ti] OR "genotypes"[ti] OR "genotyping"[ti] OR "snp"[ti] OR "snps"[ti] OR polymorphism*[ti] OR "DNA"[ti] OR allele[ti] OR alleles[ti] OR "minor allele" OR "minor alleles" OR "single nucleotide polymorphism" OR "single nucleotides polymorphisms" OR genetic[ti] OR "trait locus"[ti] OR "loci"[ti] OR "Genetic Predisposition to Disease"[MeSH] OR "Genotype"[MeSH] OR "Gene Frequency"[MeSH] OR "Polymorphism, Single Nucleotide/genetics"[MESH] OR "Polymorphism, Single Nucleotide"[MAJR] OR "Genetic Loci"[Mesh] OR "Genetic Association Studies"[Mesh] OR "Genetic Variation"[Mesh] OR "Blood Pressure/genetics"[Mesh]) NOT ("review"[pt] OR "Cross-Sectional Studies"[MeSH Terms] OR Comment[pt] OR Editorial[pt] OR Letter[pt] OR "Case Reports"[pt] OR "case control"[ti] OR "case report"[ti] OR "case study"[ti] OR "case series"[ti] OR "Case-Control Studies"[Mesh] OR "Follow-Up Studies"[Mesh] OR "observational study"[ti] OR "prospective cohort"[ti] OR "cohort studies" [Mesh:NoExp] OR "cohort study"[ti] OR "Longitudinal Studies" [Mesh:NoExp] OR "Follow-Up Studies"[mesh] OR "Retrospective Studies"[mesh] OR "non-randomized"[ti] OR "follow up study"[ti] OR "Cross-Sectional Studies"[MeSH Terms] OR "Prospective Studies"[MeSH Terms] OR "epidemiology"[Subheading] OR "pulmonary hypertension" OR "pulmonary arterial hypertension" OR "heart transplant" OR "heart failure" OR "cystic fibrosis" OR "cancer" OR "neoplasms" OR "fibromyalgia" OR "alzheimers" OR "alzheimer" OR "pregnant" OR "pregnancy" OR "obesity/drug therapy"[mesh] OR "diet therapy"[mesh] OR "diet therapy"[subheading] OR "DASH"[tiab] OR meal[ti] OR "nutritional intervention" OR "dietary intervention" OR "nutritional counseling" OR "dietary counseling" OR "caffeine" OR "eating change" OR "activities of daily living" OR "dehydration" OR "dehydrate" OR "dehydrated" OR "dietary salt" OR sodium OR "epilepsy" OR "influenza" OR "flu" OR "pneumonia" OR "septicemia" OR arthritis OR "hiv" OR "Acquired Immunodeficiency Syndrome" OR "meningitis" OR "substance abuse" OR "alcoholism" OR "drug abuse" OR "spinal cord"[ti] OR "Sleep"[Majr] OR "Sleep Apnea Syndromes"[Majr] OR sleep[ti] OR contraceptive*[ti] OR (animals[mesh] NOT humans[mesh]) OR rat[ti] OR rats[ti] OR mouse[ti] OR mice[ti] OR pig[ti] OR pigs[ti] OR dog[ti] OR dogs[ti] OR canine[ti] OR cow[ti] OR cows[ti] OR bovine[ti]).

Table 4. Criteria 4a: Genes (n = 27) in major blood pressure regulatory pathways associated with physical activity and/or obesity in genome wide association studies. The search was conducted on 05/11/2017 and a total of 188 records were identified.

("mean arterial" OR "blood pressure"[mesh] OR "blood pressure" OR "blood pressures" OR "arterial pressure" OR "arterial pressures" OR hypertension OR hypotension OR normotension OR hypertensive OR hypotensive OR normotensive OR "systolic pressure" OR "diastolic pressure" OR "pulse pressure" OR "venous pressure" OR "pressure monitor" OR hypotension OR "pre hypertension" OR "bp response" OR "bp decrease" OR "bp reduction" OR "bp monitor" OR "bp monitors" OR "bp measurement" OR "Blood Pressure/genetics"[MeSH]) AND ("resting blood pressure" OR "resting BP" OR "exercise"[mesh] OR exercise[ti] OR exercises OR running[mesh] OR running[ti] OR "bicycle" OR "bicycles" OR "bicycling" OR walking OR walking[mesh] OR walking[ti] OR treadmill* OR "weight lifting" OR "weight training" OR "weight bearing" OR "resistance training" OR "strength training" OR "endurance training" OR "speed training" OR "training duration" OR "training frequency" OR "training intensity" OR "aerobic endurance" OR "aerobic training" OR "physical activity" OR "motor activity"[mesh] OR Overweight OR "Overweight"[Mesh] OR BMI OR "body mass" OR "Body Mass Index"[MeSH] OR "Waist Circumference"[MeSH] OR obesity OR "Obesity"[Mesh] OR obese) AND (gwa OR gwas OR "Genetic Association Studies"[Mesh] OR genomewide OR "genome wide" OR "genomewide association" OR "genome wide association" OR "genome-wide interaction") NOT ("review"[pt] OR "Cross-Sectional Studies"[MeSH Terms] OR Comment[pt] OR Editorial[pt] OR Letter[pt] OR "Case Reports"[pt] OR "case control"[ti] OR "case report"[ti] OR "case study"[ti] OR "case series"[ti] OR "Case-Control Studies"[Mesh] OR "Follow-Up Studies"[Mesh] OR "observational study"[ti] OR "prospective cohort"[ti] OR "cohort studies" [Mesh:NoExp] OR "cohort study"[ti] OR "Longitudinal Studies" [Mesh:NoExp] OR "Follow-Up Studies"[mesh] OR "Retrospective Studies"[mesh] OR "non-randomized"[ti] OR "follow up study"[ti] OR "Cross-Sectional Studies"[MeSH Terms] OR "Prospective Studies"[MeSH Terms] OR "epidemiology"[Subheading] OR "pulmonary hypertension" OR "pulmonary arterial hypertension" OR "heart transplant" OR "heart failure" OR "cystic fibrosis" OR "cancer" OR "neoplasms" OR "fibromyalgia" OR "alzheimers" OR "alzheimer" OR "pregnant" OR "pregnancy" OR "obesity/drug therapy"[mesh] OR "diet therapy"[mesh] OR "diet therapy"[subheading] OR "DASH"[tiab] OR meal[ti] OR "nutritional intervention" OR "dietary intervention" OR "nutritional counseling" OR "dietary counseling" OR "caffeine" OR "eating change" OR "activities of daily living" OR "dehydration" OR "dehydrate" OR "dehydrated" OR "dietary salt" OR sodium OR "epilepsy" OR "influenza" OR "flu" OR "pneumonia" OR "septicemia" OR arthritis OR "hiv" OR "Acquired Immunodeficiency Syndrome" OR "meningitis" OR "substance abuse" OR "alcoholism" OR "drug abuse" OR "spinal cord"[ti] OR "Sleep"[Majr] OR "Sleep Apnea Syndromes"[Majr] OR sleep[ti] OR contraceptive*[ti] OR (animals[mesh] NOT humans[mesh]) OR rat[ti] OR rats[ti] OR mouse[ti] OR mice[ti] OR pig[ti] OR pigs[ti] OR dog[ti] OR dogs[ti] OR canine[ti] OR cow[ti] OR cows[ti] OR bovine[ti]).

Potential reports were screened by PP for title and abstract, and by PP and LSP for full text review to determine if they qualified. In addition, the authors performed manual searches of reference lists from relevant reports in their files for possible inclusion. Any questions regarding qualification were resolved by discussion by PP and LSP.

3. Results

The search yielded 1408 reports plus 14 records obtained from manual searches of the authors' files. Of these, 57 trials qualified. Please see Figure 1 for the flowchart detailing the systematic search of potential reports and selection process for the reports that qualified. Of the qualifying trials containing human genes in major BP regulatory pathways, we extracted 11 genes associated with the BP response to exercise under criteria 1 [33–40]; 18 genes associated with hypertension in GWAS under criteria 2 [41–48]; 28 genes associated with the BP response to pharmacotherapy under criteria 3 [49–70]; 27 genes associated with physical activity and/or obesity in GWAS under criteria 4a [71–85]; and 29 genes associated with BP, physical activity, and/or obesity in non-GWAS under criteria 4b [86–88] displayed in Supplemental Digital Content (SDC) 1 Table S1. In addition, we included the 41 genes from our preliminary work that were reported to be associated with hypertension or the BP response to exercise or pharmacotherapy [27–29].

See SDC 1 Table S1 for the complete signature of 154 prioritized BP and exercise genes by our predetermined inclusion criteria.

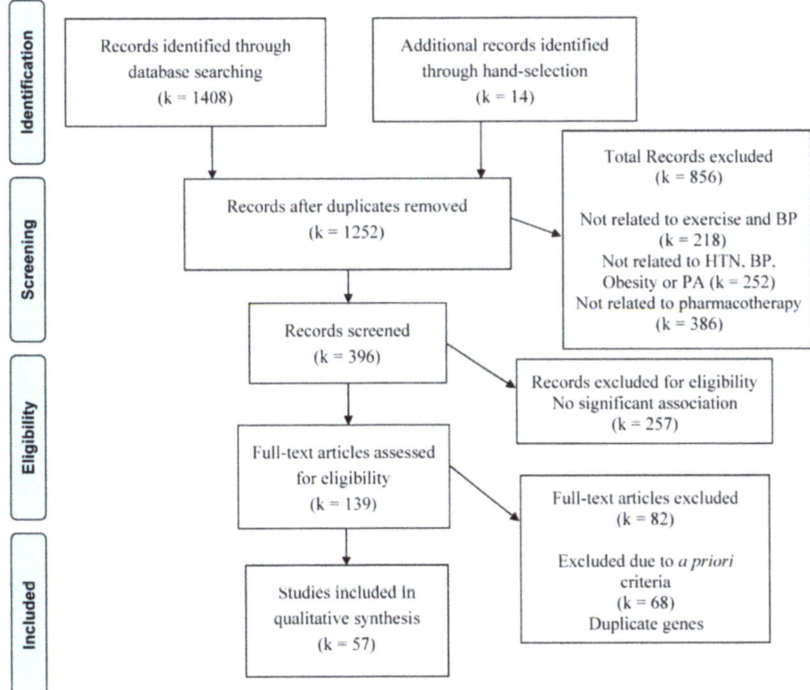

Figure 1. The systematic search trial selection process. BP: blood pressure; HTN: hypertension; PA: physical activity.

4. Discussion

The AHA/ACC rated exercise as one of the best lifestyle therapeutic approaches to prevent and treat hypertension because aerobic exercise training lowers BP 5-8 mmHg among adults with hypertension [3]. Nearly 20 years ago, the HERITAGE study investigators established that the BP response to exercise was heritable. This finding set off investigator-initiated studies examining the genetic basis for the antihypertensive effects of exercise using a candidate gene approach, as this was the technology available at that time. Despite the best intentions and efforts of these research teams, these candidate gene studies were met with little success [17,25,26]. The failure of the candidate gene approach is best illustrated by a systematic review we recently completed on the influence of the angiotensin converting enzyme (*ACE*) insertion/deletion polymorphism rs4340 on human endurance exercise performance and cardiovascular health [89]. *ACE* rs4340 is the most widely investigated genetic variant in the exercise genomic literature. Despite the extensive volume of literature on *ACE* rs4340, we concluded that due to disparate findings no definitive conclusions could be made regarding the role of *ACE* rs4340 on endurance exercise performance or the BP response to exercise.

More recently, using newer genomic technology that emerged since our discovery phase candidate gene association studies, in a replication cohort we deep-sequenced the exons of genes contained on a prioritized signature of 41 genes identified from our earlier work combined with those from a systematic review of the literature of genes reported to be associated with hypertension or the BP response to exercise or pharmacotherapy using the Illumina TruSeq Custom Amplicon kit [27–29]. Even after adjustment for multiple testing, we found that 61% of the genes in the prioritized panel associated with the BP response to exercise. Clinical features such as resting BP, age, gender, and cardiometabolic biomarkers explained 66%–92% of the variation in the BP response to exercise. Yet, the genetic variants that emerged from these analyses explained 2%–15% of the variance in the BP response to exercise, a magnitude that is larger than typically reported in exercise genomic studies [25,90].

We attributed the high proportion of significant BP-genotype associations that we found to the following methodological strategies we instituted [27–29]: a randomized control repeated measure design with subjects who were their own control; a focused inquiry of variants with a prioritized panel of genes obtained from a systematic review of the literature that reduced the search space within the genome; high throughput exon sequencing on functional gene regions; standardized protocols that included a closely monitored, well-controlled exercise exposure; and adjustment for multiple testing based on genetic variants exhibiting variability in the number of minor alleles and with unique genotypic values. This series of studies are proof of concept that a focused genomic inquiry on functional portions of genes based upon a systematic review of the literature is a time and cost-efficient way to investigate the genomic basis for the antihypertensive effects of exercise. See Figure 2 for a conceptual overview of this systematic review approach to investigate the genomics of the antihypertensive effects of exercise. However, future investigator-initiated studies remain to be done expanding upon this approach among a large ethnically and gender diverse sample of adults with hypertension to continue to gain information about the genomic basis for the antihypertensive effects of exercise with the long-term goal of fine-tuning exercise prescriptions to optimize the BP benefits of exercise.

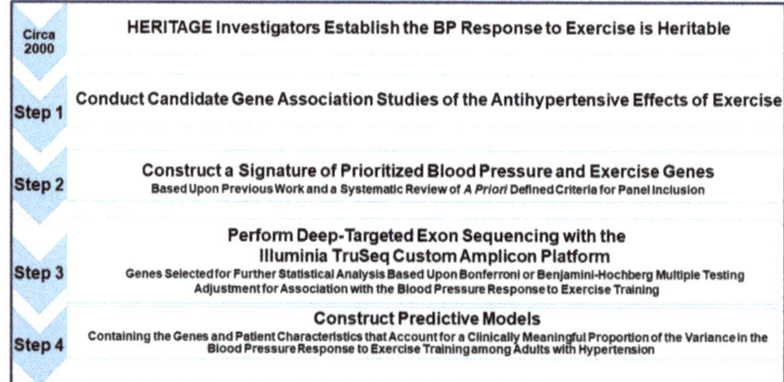

Figure 2. A Systematic review approach to assemble a targeted gene panel to study the genomics of the antihypertensive effects of exercise. HERITAGE: *HE*alth, *RI*sk Factors Exercise *TrA*ining and *Ge*netics; BP: blood pressure.

We acknowledge a limitation of our approach is that there are more sophisticated system analyses and bioinformatic approaches now being used with GWAS, human disease, and pathway data sets such as ENCODE and NHGRI GWAS that integrate omic high throughput technology of the genome, transcriptome, proteome, and epigenome to elucidate the genomic basis for hypertension (See SDC 1 Table 1 criteria 4a). In fact, we used such data sets in our most recent work to derive regulatory elements for the polymorphisms that passed multiple testing threshold for significant associations with the BP to exercise [27–29]. Another limitation is that genes have transcription-factor and post-transcriptional regulators so that RNA expression levels could also be used to find signatures associated with the antihypertensive effects of exercise as opposed to BP and exercise genes *per se*. Genes have epistatic and pleiotropic effects further complicating the identification of genetic variants that explain a clinically meaningful proportion of the BP response to exercise. Presently no data set repositories exist for genes that have been reported to associated with the BP response to exercise that have passed preestablished thresholds for multiple testing. Considering the importance of hypertension as a public health problem and the critical role exercise has in the treatment of hypertension, we posit that the methods we have presented in this systematic review represent a time and cost-efficient approach to construct targeted gene signatures whose exons can be deep-sequenced to gain insight into the genomic basis for the antihypertensive effects of exercise. Furthermore, these methods could easily be adapted to design prioritized signatures of the transcriptome, proteome, and epigenome as they regulate the BP response to exercise.

Supplementary Materials: The following are available online at http://www.mdpi.com/2073-4425/10/4/295/s1, Supplemental Digital Content 1. Table S1. The Prioritized Panel of 154 Blood Pressure Genes from a Systematic Review of the Literature by the Four *A Priori* Criteria Combined with Our Preliminary Work.

Funding: The University of Connecticut (UConn) Research Foundation; UConn Institute for Health, Intervention, and Policy; Connecticut Institute for Clinical and Translational Science.

Acknowledgments: The Jackson Laboratory's Scientific Services (Genome Technologies) for performing the deep-targeting exon sequencing and preparation of the variant calling files. L.S.P. contributed to study design conceptualization, data collection, statistical analyses, results interpretation, manuscript writing and reviewing, and acquisition of funding; Elizabeth D. Schifano contributed to statistical analyses, results interpretation, and manuscript writing and reviewing; Garrett I. Ash contributed to study design conceptualization, data collection, statistical analyses, results interpretation, manuscript reviewing, and acquisition of funding; Gregory A. Panza contributed to data collection, statistical analyses, results interpretation, and manuscript writing and reviewing; Lauren ML Corso contributed to data collection and manuscript writing and reviewing; Ming-Hui Chen contributed to statistical analyses, results interpretation, and manuscript reviewing; Ved Deshpande contributed to statistical analyses, results interpretation, and manuscript reviewing; Amanda Zaleski contributed to data collection, manuscript writing and reviewing, and acquisition of funding; Burak Cilhoroz contributed to statistical analyses, results interpretation, and manuscript writing and reviewing; Paulo Farinatti contributed to data collection, manuscript writing and reviewing, and acquisition of funding; B.A.T. contributed to data collection, manuscript writing and reviewing, and acquisition of funding; Rachel J. O'Neill contributed to results interpretation and manuscript writing and reviewing; and Paul D. Thompson contributed to data collection and manuscript writing and reviewing.

Conflicts of Interest: The authors declare no conflict of interest.

References

1. Benjamin, E.J.; Muntner, P.; Alonso, A.; Bittencourt, M.S.; Callaway, C.W.; Carson, A.P.; Chamberlain, A.M.; Chang, A.R.; Cheng, S.; Das, S.R.; et al. Heart Disease and Stroke Statistics-2019 Update: A Report from the American Heart Association. *Circulation* **2019**, *139*, e56–e528. [CrossRef] [PubMed]

2. Fisher, N.D.L.; Curfman, G. Hypertension-A Public Health Challenge of Global Proportions. *JAMA* **2018**, *320*, 1757–1759. [CrossRef] [PubMed]

3. Whelton, P.K.; Carey, R.M.; Aronow, W.S.; Casey, D.E., Jr.; Collins, K.J.; Dennison Himmelfarb, C.; DePalma, S.M.; Gidding, S.; Jamerson, K.A.; Jones, D.W.; et al. 2017 ACC/AHA/AAPA/ABC/ACPM/AGS/ APhA/ASH/ASPC/NMA/PCNA Guideline for the Prevention, Detection, Evaluation, and Management of High Blood Pressure in Adults: A Report of the American College of Cardiology/American Heart Association Task Force on Clinical Practice Guidelines. *Hypertension* **2018**, *71*, e127–e248.

4. Naci, H.; Salcher-Konrad, M.; Dias, S.; Blum, M.R.; Anova, S.; Nunan, D.; Ioannidis, J. How Does Exercise Treatment Compare with Antihypertensive Medications? A Network Meta-Analysis of 391 RCTs Assessing Exercise and Medication Effects on Systolic Blood Pressure. *Br. J. Sports Med.* **2018**. [CrossRef] [PubMed]

5. Pescatello, L.S. Exercise Measures Up to Medication as Antihypertensive Therapy: Its Value has Long been Underestimated. *Br. J. Sports Med.* **2018**. [CrossRef]

6. Chobanian, A.V.; Bakris, G.L.; Black, H.R.; Cushman, W.C.; Green, L.A.; Izzo, J.L., Jr.; Jones, D.W.; Materson, B.J.; Oparil, S.; Wright, J.T., Jr.; et al. Seventh Report of the Joint National Committee on Prevention, Detection, Evaluation, and Treatment of High Blood Pressure. *Hypertension* **2003**, *42*, 1206–1252. [CrossRef] [PubMed]

7. Law, M.R.; Morris, J.K.; Wald, N.J. Use of Blood Pressure Lowering Drugs in the Prevention of Cardiovascular Disease: Meta-Analysis of 147 Randomised Trials in the Context of Expectations from Prospective Epidemiological Studies. *BMJ* **2009**, *338*, b1665. [CrossRef] [PubMed]

8. Whelton, P.K.; He, J.; Appel, L.J.; Cutler, J.A.; Havas, S.; Kotchen, T.A.; Roccella, E.J.; Stout, R.; Vallbona, C.; Winston, M.C.; et al. Primary Prevention of Hypertension: Clinical and Public Health Advisory from the National High Blood Pressure Education Program. *JAMA* **2002**, *288*, 1882–1888. [CrossRef] [PubMed]

9. Physical Activity Guidelines Advisory Committee. *2018 Physical Activity Guidelines Advisory Committee Scientific Report*; Department of Health and Human Services: Washington, DC, USA, 2018.

10. Powell, K.E.; King, A.C.; Buchner, D.M.; Campbell, W.W.; DiPietro, L.; Erickson, K.I.; Hillman, C.H.; Jakicic, J.M.; Janz, K.F.; Katzmarzyk, P.T.; et al. The Scientific Foundation for the Physical Activity Guidelines for Americans, 2nd Edition. *J. Phys. Act. Health* **2018**, *16*, 1–11. [CrossRef] [PubMed]

11. Pescatello, L.S.; MacDonald, H.V.; Ash, G.I.; Lamberti, L.M.; Farquhar, W.B.; Arena, R.; Johnson, B.T. Assessing the Existing Professional Exercise Recommendations for Hypertension: A Review and Recommendations for Future Research Priorities. *Mayo Clin. Proc.* **2015**, *90*, 801–812. [CrossRef] [PubMed]

12. Kim, H.; Andrade, F.C.D. Diagnostic Status and Age at Diagnosis of Hypertension on Adherence to Lifestyle Recommendations. *Prev. Med. Rep.* **2018**, *13*, 52–56. [CrossRef] [PubMed]

13. Mu, L.; Cohen, A.J.; Mukamal, K.J. Prevalence and Predictors of Resistance and Aerobic Exercise among Hypertensive Adults in the United States. *J. Hum. Hypertens.* **2015**, *29*, 394–395. [CrossRef] [PubMed]

14. Bouchard, C.; Rankinen, T. Individual Differences in Response to Regular Physical Activity. *Med. Sci. Sports Exerc.* **2001**, *33*, S446–S451. [CrossRef] [PubMed]

15. Bouchard, C.; Blair, S.N.; Church, T.S.; Earnest, C.P.; Hagberg, J.M.; Hakkinen, K.; Jenkins, N.T.; Karavirta, L.; Kraus, W.E.; Leon, A.S.; et al. Adverse Metabolic Response to Regular Exercise: Is it a Rare Or Common Occurrence? *PLoS ONE* **2012**, *7*, e37887. [CrossRef] [PubMed]

16. Rankinen, T.; Gagnon, J.; Perusse, L.; Chagnon, Y.C.; Rice, T.; Leon, A.S.; Skinner, J.S.; Wilmore, J.H.; Rao, D.C.; Bouchard, C. AGT M235T and ACE ID Polymorphisms and Exercise Blood Pressure in the HERITAGE Family Study. *Am. J. Physiol. Heart Circ. Physiol.* **2000**, *279*, H368–H374. [CrossRef] [PubMed]

17. Rankinen, T.; Roth, S.M.; Bray, M.S.; Loos, R.; Perusse, L.; Wolfarth, B.; Hagberg, J.M.; Bouchard, C. Advances in Exercise, Fitness, and Performance Genomics. *Med. Sci. Sports Exerc.* **2010**, *42*, 835–846. [CrossRef]

18. Rankinen, T.; Rice, T.; Perusse, L.; Chagnon, Y.C.; Gagnon, J.; Leon, A.S.; Skinner, J.S.; Wilmore, J.H.; Rao, D.C.; Bouchard, C. NOS3 Glu298Asp Genotype and Blood Pressure Response to Endurance Training: The HERITAGE Family Study. *Hypertension* **2000**, *36*, 885–889. [CrossRef] [PubMed]

19. Rankinen, T.; Church, T.; Rice, T.; Markward, N.; Leon, A.S.; Rao, D.C.; Skinner, J.S.; Blair, S.N.; Bouchard, C. Effect of Endothelin 1 Genotype on Blood Pressure is Dependent on Physical Activity Or Fitness Levels. *Hypertension* **2007**, *50*, 1120–1125. [CrossRef]

20. Augeri, A.L.; Tsongalis, G.J.; van Heest, J.L.; Maresh, C.M.; Thompson, P.D.; Pescatello, L.S. The Endothelial Nitric Oxide Synthase -786 T>C Polymorphism and the Exercise-Induced Blood Pressure and Nitric Oxide Responses among Men with Elevated Blood Pressure. *Atherosclerosis* **2009**, *204*, e28–e34. [CrossRef]

21. Ash, G.I.; Eicher, J.D.; Pescatello, L.S. The Promises and Challenges of the use of Genomics in the Prescription of Exercise for Hypertension: The 2013 Update. *Curr. Hypertens. Rev.* **2013**, *9*, 130–147. [CrossRef]

22. Blanchard, B.E.; Tsongalis, G.J.; Guidry, M.A.; LaBelle, L.A.; Poulin, M.; Taylor, A.L.; Maresh, C.M.; Devaney, J.; Thompson, P.D.; Pescatello, L.S. RAAS Polymorphisms Alter the Acute Blood Pressure Response to Aerobic Exercise among Men with Hypertension. *Eur. J. Appl. Physiol.* **2006**, *97*, 26–33. [CrossRef]

23. Pescatello, L.S.; Blanchard, B.E.; Tsongalis, G.J.; Maresh, C.M.; O'Connell, A.; Thompson, P.D. The α-Adducin Gly460Trp Polymorphism and the Antihypertensive Effects of Exercise among Men with High Blood Pressure. *Clin. Sci. (Lond.)* **2007**, *113*, 251–258. [CrossRef] [PubMed]

24. Pescatello, L.S.; Blanchard, B.E.; van Heest, J.L.; Maresh, C.M.; Gordish-Dressman, H.; Thompson, P.D. The Metabolic Syndrome and the Immediate Antihypertensive Effects of Aerobic Exercise: A Randomized Control Design. *BMC Cardiovasc. Disord.* **2008**, *8*, 12. [CrossRef] [PubMed]

25. Bruneau, M.L., Jr.; Johnson, B.T.; Huedo-Medina, T.B.; Larson, K.A.; Ash, G.I.; Pescatello, L.S. The Blood Pressure Response to Acute and Chronic Aerobic Exercise: A Meta-Analysis of Candidate Gene Association Studies. *J. Sci. Med. Sport* **2016**, *19*, 424–431. [CrossRef] [PubMed]

26. Bouchard, C. Overcoming Barriers to Progress in Exercise Genomics. *Exerc. Sport Sci. Rev.* **2011**, *39*, 212–217. [CrossRef]

27. Cilhoroz, B.T.; Schifano, E.D.; Panza, G.A.; Ash, G.I.; Corso, L.; Chen, M.H.; Deshpande, V.; Zaleski, A.; Farinatti, P.; Santos, L.P.; et al. FURIN Variant Associations with Postexercise Hypotension are Intensity and Race Dependent. *Physiol. Rep.* **2019**, *7*, e13952. [CrossRef]

28. Pescatello, L.S.; Shifano, E.D.; Ash, G.I.; Panza, G.A.; Lamberti, L.; Chen, M.; Deshpande, V.; Zaleski, A.; Farinatti, P.; Taylor, B.A.; et al. Deep-Targeted Exon Sequencing Reveals Renal Polymorphisms Associate with Postexercise Hypotension among African Americans. *Physiol. Rep.* **2016**, *4*, e12992. [CrossRef]

29. Pescatello, L.S.; Schifano, E.D.; Ash, G.I.; Panza, G.A.; Corso, L.M.L.; Chen, M.H.; Deshpande, V.; Zaleski, A.; Cilhoroz, B.; Farinatti, P.; et al. Deep-Targeted Sequencing of Endothelial Nitric Oxide Synthase Gene Exons Uncovers Exercise Intensity and Ethnicity-Dependent Associations with Post-Exercise Hypotension. *Physiol. Rep.* **2017**, *5*. [CrossRef]

30. Moher, D.; Liberati, A.; Tetzlaff, J.; Altman, D.G. PRISMA Group. Preferred Reporting Items for Systematic Reviews and Meta-Analyses: The PRISMA Statement. *J. Clin. Epidemiol.* **2009**, *62*, 1006–1012. [CrossRef]

31. Moher, D.; Shamseer, L.; Clarke, M.; Ghersi, D.; Liberati, A.; Petticrew, M.; Shekelle, P.; Stewart, L.A.; PRISMA-P Group. Preferred Reporting Items for Systematic Review and Meta-Analysis Protocols (PRISMA-P) 2015 Statement. *Syst. Rev.* **2015**, *4*, 1. [CrossRef] [PubMed]

32. Lavie, C.J.; de Schutter, A.; Parto, P.; Jahangir, E.; Kokkinos, P.; Ortega, F.B.; Arena, R.; Milani, R.V. Obesity and Prevalence of Cardiovascular Diseases and Prognosis-the Obesity Paradox Updated. *Prog. Cardiovasc. Dis.* **2016**, *58*, 537–547. [CrossRef] [PubMed]

33. Deschamps, C.L.; Connors, K.E.; Klein, M.S.; Johnsen, V.L.; Shearer, J.; Vogel, H.J.; Devaney, J.M.; Gordish-Dressman, H.; Many, G.M.; Barfield, W. The *ACTN3* R577X Polymorphism is Associated with Cardiometabolic Fitness in Healthy Young Adults. *PLoS ONE* **2015**, *10*, e0130644. [CrossRef] [PubMed]

34. Franks, P.W.; Bhattacharyya, S.; Luan, J.; Montague, C.; Brennand, J.; Challis, B.; Brage, S.; Ekelund, U.; Middelberg, R.P.; O'Rahilly, S.; et al. Association between Physical Activity and Blood Pressure is Modified by Variants in the G-Protein Coupled Receptor 10. *Hypertension* **2004**, *43*, 224–228. [CrossRef]

35. Heradien, M.; Revera, M.; van der Merwe, L.; Goosen, A.; Corfield, V.A.; Brink, P.A.; Mayosi, B.M.; Moolman-Smook, J.C. Abnormal Blood Pressure Response to Exercise Occurs More Frequently in Hypertrophic Cardiomyopathy Patients with the R92W Troponin T Mutation than in those with Myosin Mutations. *Heart Rhythm* **2009**, *6*, S18–S24. [CrossRef] [PubMed]

36. Jayewardene, A.F.; Mavros, Y.; Gwinn, T.; Hancock, D.P.; Rooney, K.B. Associations between CD36 Gene Polymorphisms and Metabolic Response to a Short-Term Endurance-Training Program in a Young-Adult Population. *Appl. Physiol. Nutr. Metab.* **2015**, *41*, 157–167. [CrossRef]

37. Masuki, S.; Mori, M.; Tabara, Y.; Miki, T.; Sakurai, A.; Morikawa, M.; Miyagawa, K.; Higuchi, K.; Nose, H.; Shinshu University Genetic Research Consortium. Vasopressin V1a Receptor Polymorphism and Interval Walking Training Effects in Middle-Aged and Older People. *Hypertension* **2010**, *55*, 747–754. [CrossRef]

38. Nunes, R.A.B.; Barroso, L.P.; da Costa Pereira, A.; Krieger, J.E.; Mansur, A.J. Gender-Related Associations of Genetic Polymorphisms of α-Adrenergic Receptors, Endothelial Nitric Oxide Synthase and Bradykinin B2 Receptor with Treadmill Exercise Test Responses. *Open Heart* **2014**, *1*, e000132. [CrossRef] [PubMed]

39. Rivera, M.A.; Echegaray, M.; Rankinen, T.; Pérusse, L.; Rice, T.; Gagnon, J.; Leon, A.S.; Skinner, J.S.; Wilmore, J.H.; Rao, D. Angiogenin Gene-Race Interaction for Resting and Exercise BP Phenotypes: The HERITAGE Family Study. *J. Appl. Physiol.* **2001**, *90*, 1232–1238. [CrossRef] [PubMed]

40. Shiwaku, K.; Nogi, A.; Anuurad, E.; Kitajima, K.; Enkhmaa, B.; Shimono, K.; Yamane, Y. Difficulty in Losing Weight by Behavioral Intervention for Women with Trp64Arg Polymorphism of the β 3-Adrenergic Receptor Gene. *Int. J. Obes.* **2003**, *27*, 1028. [CrossRef] [PubMed]

41. Fujimaki, T.; Oguri, M.; Horibe, H.; Kato, K.; Matsuoka, R.; Abe, S.; Tokoro, F.; Arai, M.; Noda, T.; Watanabe, S. Association of a Transcription Factor 21 Gene Polymorphism with Hypertension. *Biomed. Rep.* **2015**, *3*, 118–122. [CrossRef] [PubMed]

42. Huan, T.; Esko, T.; Peters, M.J.; Pilling, L.C.; Schramm, K.; Schurmann, C.; Chen, B.H.; Liu, C.; Joehanes, R.; Johnson, A.D. A Meta-Analysis of Gene Expression Signatures of Blood Pressure and Hypertension. *PLoS Genet.* **2015**, *11*, e1005035. [CrossRef]

43. Lu, X.; Wang, L.; Lin, X.; Huang, J.; Charles Gu, C.; He, M.; Shen, H.; He, J.; Zhu, J.; Li, H. Genome-Wide Association Study in Chinese Identifies Novel Loci for Blood Pressure and Hypertension. *Hum. Mol. Genet.* **2014**, *24*, 865–874. [CrossRef] [PubMed]

44. Murakata, Y.; Fujimaki, T.; Yamada, Y. Association of a Butyrophilin, Subfamily 2, Member A1 Gene Polymorphism with Hypertension. *Biomed. Rep.* **2014**, *2*, 818–822. [CrossRef] [PubMed]

45. Ong, K.L.; Li, M.; Tso, A.W.; Xu, A.; Cherny, S.S.; Sham, P.C.; Tse, H.F.; Lam, T.H.; Cheung, B.M.; Lam, K.S.; et al. Association of Genetic Variants in the Adiponectin Gene with Adiponectin Level and Hypertension in Hong Kong Chinese. *Eur. J. Endocrinol.* **2010**, *163*, 251–257. [CrossRef]

46. Rong, S.; Zhou, X.; Wang, Z.; Wang, X.; Wang, Y.; Xue, C.; Li, B. Glutathione S-Transferase M1 and T1 Polymorphisms and Hypertension Risk: An Updated Meta-Analysis. *J. Hum. Hypertens.* **2018**, 1. [CrossRef]

47. Zhang, K.; Rao, F.; Miramontes-Gonzalez, J.P.; Hightower, C.M.; Vaught, B.; Chen, Y.; Greenwood, T.A.; Schork, A.J.; Wang, L.; Mahata, M. Neuropeptide Y (NPY): Genetic Variation in the Human Promoter Alters Glucocorticoid Signaling, Yielding Increased NPY Secretion and Stress Responses. *J. Am. Coll. Cardiol.* **2012**, *60*, 1678–1689. [CrossRef]

48. Zhang, W.; Wang, H.; Guan, X.; Niu, Q.; Li, W. Variant rs2237892 of KCNQ1 is Potentially Associated with Hypertension and Macrovascular Complications in Type 2 Diabetes Mellitus in a Chinese Han Population. *Genom. Proteom. Bioinform.* **2015**, *13*, 364–370. [CrossRef]

49. Chen, Y.; Liu, D.; Zhang, P.; Zhong, J.; Zhang, C.; Wu, S.; Zhang, Y.; Liu, G.; He, M.; Jin, L. Impact of ACE2 Gene Polymorphism on Antihypertensive Efficacy of ACE Inhibitors. *J. Hum. Hypertens.* **2016**, *30*, 766. [CrossRef]

50. Bruck, H.; Schwerdtfeger, T.; Toliat, M.; Leineweber, K.; Heusch, G.; Philipp, T.; Nürnberg, P.; Brodde, O. Presynaptic α-2C Adrenoceptor-mediated Control of Noradrenaline Release in Humans: Genotype-or Age-Dependent? *Clin. Pharmacol. Ther.* **2007**, *82*, 525–530. [CrossRef]

51. Chen, G.; Jiang, S.; Mao, G.; Zhang, S.; Hong, X.; Tang, G.; Li, Z.; Liu, X.; Zhang, Y.; Xing, H. CYP2C9 Ile359Leu Polymorphism, Plasma Irbesartan Concentration and Acute Blood Pressure Reductions in Response to Irbesartan Treatment in Chinese Hypertensive Patients. Methods Find. *Exp. Clin. Pharmacol.* **2006**, *28*, 19–24.

52. Chen, W.; Shu, Y.; Li, Q.; Xu, L.; Roederer, M.W.; Fan, L.; Wu, L.; He, F.; Luo, J.; Tan, Z. Polymorphism of *ORM1* is Associated with the Pharmacokinetics of Telmisartan. *PLoS ONE* **2013**, *8*, e70341. [CrossRef]

53. Anthony, E.G.; Richard, E.; Lipkowitz, M.S.; Kelley, S.T.; Alcaraz, J.E.; Shaffer, R.A.; Bhatnagar, V. Association of Phosphodiesterase 4 Polymorphism (rs702553) with Blood Pressure in the African American Study of Kidney Disease and Hypertension Genomics Study. *Pharm. Genom.* **2013**, *23*, 442–444. [CrossRef] [PubMed]

54. Donner, K.M.; Hiltunen, T.P.; Hannila-Handelberg, T.; Suonsyrjä, T.; Kontula, K. STK39 Variation Predicts the Ambulatory Blood Pressure Response to Losartan in Hypertensive Men. *Hypertens. Res.* **2012**, *35*, 107. [CrossRef] [PubMed]

55. Duarte, J.D.; Turner, S.T.; Tran, B.; Chapman, A.B.; Bailey, K.R.; Gong, Y.; Gums, J.G.; Langaee, T.Y.; Beitelshees, A.L.; Cooper-Dehoff, R.M. Association of Chromosome 12 Locus with Antihypertensive Response to Hydrochlorothiazide may Involve Differential YEATS4 Expression. *Pharm. J.* **2013**, *13*, 257. [CrossRef] [PubMed]

56. Frau, F.; Zaninello, R.; Salvi, E.; Ortu, M.F.; Braga, D.; Velayutham, D.; Argiolas, G.; Fresu, G.; Troffa, C.; Bulla, E. Genome-Wide Association Study Identifies CAMKID Variants Involved in Blood Pressure Response to Losartan: The SOPHIA Study. *Pharmacogenomics* **2014**, *15*, 1643–1652. [CrossRef] [PubMed]

57. Duan, R.; Cui, W.; Wang, H. Association of the Antihypertensive Response of Iptakalim with KCNJ11 (Kir6. 2 Gene) Polymorphisms in Chinese Han Hypertensive Patients. *Acta Pharmacol. Sin.* **2011**, *32*, 1078. [CrossRef]

58. Bhatnagar, V.; O'connor, D.T.; Brophy, V.H.; Schork, N.J.; Richard, E.; Salem, R.M.; Nievergelt, C.M.; Bakris, G.L.; Middleton, J.P.; Norris, K.C. G-Protein-Coupled Receptor Kinase 4 Polymorphisms and Blood Pressure Response to Metoprolol among African Americans: Sex-Specificity and Interactions. *Am. J. Hypertens.* **2009**, *22*, 332–338. [CrossRef]

59. Kamide, K.; Asayama, K.; Katsuya, T.; Ohkubo, T.; Hirose, T.; Inoue, R.; Metoki, H.; Kikuya, M.; Obara, T.; Hanada, H. Genome-Wide Response to Antihypertensive Medication using Home Blood Pressure Measurements: A Pilot Study Nested within the HOMED-BP Study. *Pharmacogenomics* **2013**, *14*, 1709–1721. [CrossRef]

60. Kamide, K.; Yang, J.; Matayoshi, T.; Takiuchi, S.; Horio, T.; Yoshii, M.; Miwa, Y.; Yasuda, H.; Yoshihara, F.; Nakamura, S. Genetic Polymorphisms of L-Type Calcium Channel α1C and α1D Subunit Genes are Associated with Sensitivity to the Antihypertensive Effects of L-Type Dihydropyridine Calcium-Channel Blockers. *Circ. J.* **2009**, *73*, 732–740. [CrossRef]

61. Konoshita, T.; Kato, N.; Fuchs, S.; Mizuno, S.; Aoyama, C.; Motomura, M.; Makino, Y.; Wakahara, S.; Inoki, I.; Miyamori, I.; et al. Genetic Variant of the Renin-Angiotensin System and Diabetes Influences Blood Pressure Response to Angiotensin Receptor Blockers. *Diabetes Care* **2009**, *32*, 1485–1490. [CrossRef]

62. He, F.; Luo, J.; Luo, Z.; Fan, L.; He, Y.; Zhu, D.; Gao, J.; Deng, S.; Wang, Y.; Qian, Y. The KCNH2 Genetic Polymorphism (1956, C> T) is a Novel Biomarker that is Associated with CCB and α, β-ADR Blocker Response in EH Patients in China. *PLoS ONE* **2013**, *8*, e61317. [CrossRef]

63. He, F.; Luo, J.; Zhang, Z.; Luo, Z.; Fan, L.; He, Y.; Wen, J.; Zhu, D.; Gao, J.; Wang, Y. The RGS2 (-391, C> G) Genetic Variation Correlates to Antihypertensive Drug Responses in Chinese Patients with Essential Hypertension. *PLoS ONE* **2015**, *10*, e0121483. [CrossRef]

64. Hong, X.; Xing, H.; Yu, Y.; Wen, Y.; Zhang, Y.; Zhang, S.; Tang, G.; Xu, X. Genetic Polymorphisms of the Urea Transporter Gene are Associated with Antihypertensive Response to Nifedipine GITS. *Methods Find. Exp. Clin. Pharmacol.* **2007**, *29*, 3–10. [CrossRef]

65. Guo, R.; Chen, L.; Li, L.; Guo, X.; Sun, J.; Xiong, X.; Cheng, Z.; Li, Y.; Chen, X. Association of GSTM1 Null Polymorphism with Isosorbide-5-Mononitrate Cardiovascular Response and Involvement of CGRP in Healthy Chinese Male Volunteers. *Pharm. Genom.* **2011**, *21*, 142–151. [CrossRef]

66. Geshi, E.; Kimura, T.; Yoshimura, M.; Suzuki, H.; Koba, S.; Sakai, T.; Saito, T.; Koga, A.; Muramatsu, M.; Katagiri, T. A Single Nucleotide Polymorphism in the Carboxylesterase Gene is Associated with the Responsiveness to Imidapril Medication and the Promoter Activity. *Hypertens. Res.* **2005**, *28*, 719. [CrossRef]

67. Jia, J.; Men, C.; Tang, K.; Zhan, Y. Apelin Polymorphism Predicts Blood Pressure Response to Losartan in Older Chinese Women with Essential Hypertension. *Genet. Mol. Res.* **2015**, *14*, 6561–6568. [CrossRef] [PubMed]

68. Lanzani, C.; Citterio, L.; Glorioso, N.; Manunta, P.; Tripodi, G.; Salvi, E.; Carpini, S.D.; Ferrandi, M.; Messaggio, E.; Staessen, J.A.; et al. Adducin- and Ouabain-Related Gene Variants Predict the Antihypertensive Activity of Rostafuroxin, Part 2: Clinical Studies. *Sci. Transl. Med.* **2010**, *2*, 59ra87. [CrossRef] [PubMed]

69. Zhang, Y.; Zhang, M.; Niu, T.; Xu, X.; Zhu, G.; Huo, Y.; Chen, C.; Wang, X.; Xing, H.; Peng, S. D919G Polymorphism of Methionine Synthase Gene is Associated with Blood Pressure Response to Benazepril in Chinese Hypertensive Patients. *J. Hum. Genet.* **2004**, *49*, 296. [CrossRef]

70. Zhang, Y.; Hong, X.; Xing, H.; Li, J.; Yong, H.; Xu, X. E112D Polymorphism in the Prolylcarboxypeptidase Gene is Associated with Blood Pressure Response to Benazepril in Chinese Hypertensive Patients. *Chin. Med. J.* **2009**, *122*, 2461–2465. [PubMed]

71. Andreassen, O.A.; Djurovic, S.; Thompson, W.K.; Schork, A.J.; Kendler, K.S.; O'Donovan, M.C.; Rujescu, D.; Werge, T.; van de Bunt, M.; Morris, A.P. Improved Detection of Common Variants Associated with Schizophrenia by Leveraging Pleiotropy with Cardiovascular-Disease Risk Factors. *Am. J. Hum. Genet.* **2013**, *92*, 197–209. [CrossRef]

72. Flaquer, A.; Baumbach, C.; Kriebel, J.; Meitinger, T.; Peters, A.; Waldenberger, M.; Grallert, H.; Strauch, K. Mitochondrial Genetic Variants Identified to be Associated with BMI in Adults. *PLoS ONE* **2014**, *9*, e105116. [CrossRef] [PubMed]

73. Flores-Alfaro, E.; Fernández-Tilapa, G.; Salazar-Martínez, E.; Cruz, M.; Illades-Aguiar, B.; Parra-Rojas, I. Common Variants in the CRP Gene are Associated with Serum C-Reactive Protein Levels and Body Mass Index in Healthy Individuals in Mexico. *Genet. Mol. Res.* **2012**, *11*, 2258. [CrossRef]

74. Hotta, K.; Kitamoto, A.; Kitamoto, T.; Mizusawa, S.; Teranishi, H.; Matsuo, T.; Nakata, Y.; Hyogo, H.; Ochi, H.; Nakamura, T.; et al. Genetic Variations in the *CYP17A1* and *NT5C2* Genes are Associated with a Reduction in Visceral and Subcutaneous Fat Areas in Japanese Women. *J. Hum. Genet.* **2012**, *57*, 46–51. [CrossRef]

75. Hager, J.; Dina, C.; Francke, S.; Dubois, S.; Houari, M.; Vatin, V.; Vaillant, E.; Lorentz, N.; Basdevant, A.; Clement, K. A Genome-Wide Scan for Human Obesity Genes Reveals a Major Susceptibility Locus on Chromosome 10. *Nat. Genet.* **1998**, *20*, 304. [CrossRef]

76. Kaess, B.M.; Barnes, T.A.; Stark, K.; Charchar, F.J.; Waterworth, D.; Song, K.; Wang, W.Y.; Vollenweider, P.; Waeber, G.; Mooser, V. FGF21 Signalling Pathway and Metabolic traits–genetic Association Analysis. *Eur. J. Hum. Genet.* **2010**, *18*, 1344. [CrossRef] [PubMed]

77. Kong, X.; Zhang, X.; Xing, X.; Zhang, B.; Hong, J.; Yang, W. The Association of Type 2 Diabetes Loci Identified in Genome-Wide Association Studies with Metabolic Syndrome and its Components in a Chinese Population with Type 2 Diabetes. *PLoS ONE* **2015**, *10*, e0143607. [CrossRef] [PubMed]

78. Kim, Y.K.; Kim, Y.; Hwang, M.Y.; Shimokawa, K.; Won, S.; Kato, N.; Tabara, Y.; Yokota, M.; Han, B.; Lee, J.H. Identification of a Genetic Variant at 2q12.1 Associated with Blood Pressure in East-Asians by Genome-Wide Scan Including Gene-Environment Interactions. *BMC Med. Genet.* **2014**, *15*, 65. [CrossRef] [PubMed]

79. Muiya, N.P.; Wakil, S.; Al-Najai, M.; Tahir, A.I.; Baz, B.; Andres, E.; Al-Boudari, O.; Al-Tassan, N.; Al-Shahid, M.; Meyer, B.F. A Study of the Role of *GATA2* Gene Polymorphism in Coronary Artery Disease Risk Traits. *Gene* **2014**, *544*, 152–158. [CrossRef] [PubMed]

80. Sull, J.W.; Yang, S.J.; Kim, S.; Jee, S.H. The *ABCG2* Polymorphism rs2725220 is Associated with Hyperuricemia in the Korean Population. *Genom. Inform.* **2014**, *12*, 231–235. [CrossRef] [PubMed]
81. Teixeira, A.A.; Quinto, B.M.R.; Dalboni, M.A.; Rodrigues, C.J.D.O.; Batista, M.C. Association of IL-6 Polymorphism-174G/C and Metabolic Syndrome in Hypertensive Patients. *BioMed Res. Int.* **2015**, *2015*, 927589. [CrossRef] [PubMed]
82. Wang, M.; Li, J.; Yeung, V.; Zee, B.; Yu, R.; Ho, S.; Waye, M. Four Pairs of gene–gene Interactions Associated with Increased Risk for Type 2 Diabetes (CDKN2BAS–KCNJ11), Obesity (SLC2A9–IGF2BP2, FTO–APOA5), and Hypertension (MC4R–IGF2BP2) in Chinese Women. *Meta Gene* **2014**, *2*, 384–391. [CrossRef]
83. Wang, Y.; Bos, S.D.; Harbo, H.F.; Thompson, W.K.; Schork, A.J.; Bettella, F.; Witoelar, A.; Lie, B.A.; Li, W.; McEvoy, L.K. Genetic Overlap between Multiple Sclerosis and several Cardiovascular Disease Risk Factors. *Mult. Scler. J.* **2016**, *22*, 1783–1793. [CrossRef] [PubMed]
84. Wei, F.; Cai, C.; Yu, P.; Lv, J.; Ling, C.; Shi, W.; Jiao, H.; Chang, B.; Yang, F.; Tian, Y. Quantitative Candidate Gene Association Studies of Metabolic Traits in Han Chinese Type 2 Diabetes Patients. *Genet. Mol. Res.* **2015**, *14*, 15471–15481. [CrossRef] [PubMed]
85. Zeller, T.; Wild, P.; Szymczak, S.; Rotival, M.; Schillert, A.; Castagne, R.; Maouche, S.; Germain, M.; Lackner, K.; Rossmann, H.; et al. Genetics and Beyond–the Transcriptome of Human Monocytes and Disease Susceptibility. *PLoS ONE* **2010**, *5*, e10693. [CrossRef] [PubMed]
86. Akintunde, A.; Nondi, J.; Gogo, K.; Jones, E.S.; Rayner, B.L.; Hackam, D.G.; Spence, J.D. Physiological Phenotyping for Personalized Therapy of Uncontrolled Hypertension in Africa. *Am. J. Hypertens.* **2017**, *30*, 923–930. [CrossRef]
87. Ehret, G.B.; Ferreira, T.; Chasman, D.I.; Jackson, A.U.; Schmidt, E.M.; Johnson, T.; Thorleifsson, G.; Luan, J.; Donnelly, L.A.; Kanoni, S.; et al. The Genetics of Blood Pressure Regulation and its Target Organs from Association Studies in 342,415 Individuals. *Nat. Genet.* **2016**, *48*, 1171–1184. [CrossRef]
88. Wahl, S.; Drong, A.; Lehne, B.; Loh, M.; Scott, W.R.; Kunze, S.; Tsai, P.; Ried, J.S.; Zhang, W.; Yang, Y. Epigenome-Wide Association Study of Body Mass Index, and the Adverse Outcomes of Adiposity. *Nature* **2017**, *541*, 81. [CrossRef] [PubMed]
89. Pescatello, L.S.; Corso, L.M.L.; Santos, L.P.; Livingston, J.; Taylor, B.A. Chapter 21 Angiotensin Converting Enzyme and the Genomics of Endurance Performance. In *Routledge Handbook of Sport and Exercise Systems Genetics*, 1st ed.; Lightfoot, J.T., Hubal, M.J., Roth, S.M., Eds.; Routledge Taylor and France Group: London, UK, 2019.
90. Pescatello, L.S.; Devaney, J.M.; Hubal, M.J.; Thompson, P.D.; Hoffman, E.P. Highlights from the Functional Single Nucleotide Polymorphisms Associated with Human Muscle Size and Strength Or FAMuSS Study. *Biomed. Res. Int.* **2013**, *2013*, 643575. [CrossRef] [PubMed]

© 2019 by the authors. Licensee MDPI, Basel, Switzerland. This article is an open access article distributed under the terms and conditions of the Creative Commons Attribution (CC BY) license (http://creativecommons.org/licenses/by/4.0/).

 *genes*

Article

Genetically Determined Physical Activity and Its Association with Circulating Blood Cells

Femke M. Prins, M. Abdullah Said, Yordi J. van de Vegte, Niek Verweij, Hilde E. Groot and Pim van der Harst *

Department of Cardiology, University of Groningen, University Medical Center Groningen, 9713 GZ Groningen, The Netherlands; f.m.prins@umcg.nl (F.M.P.); m.a.said@umcg.nl (M.A.S.); y.j.van.de.vegte@umcg.nl (Y.J.v.d.V.); n.verweij@umcg.nl (N.V.); h.e.groot@umcg.nl (H.E.G.)
* Correspondence: p.van.der.harst@umcg.nl; Tel.: +31-50-36-12234; Fax: +31-50-361-48-84

Received: 18 September 2019; Accepted: 5 November 2019; Published: 7 November 2019

Abstract: Lower levels of physical activity (PA) have been associated with increased risk of cardiovascular disease. Worldwide, there is a shift towards a lifestyle with less PA, posing a serious threat to public health. One of the suggested mechanisms behind the association between PA and disease development is through systemic inflammation, in which circulating blood cells play a pivotal role. In this study we investigated the relationship between genetically determined PA and circulating blood cells. We used 68 single nucleotide polymorphisms associated with objectively measured PA levels to perform a Mendelian randomization analysis on circulating blood cells in 222,645 participants of the UK Biobank. For inverse variance fixed effects Mendelian randomization analyses, $p < 1.85 \times 10^{-3}$ (Bonferroni-adjusted p-value of 0.05/27 tests) was considered statistically significant. Genetically determined increased PA was associated with decreased lymphocytes ($\beta = -0.03$, SE = 0.008, $p = 1.35 \times 10^{-3}$) and decreased eosinophils ($\beta = -0.008$, SE = 0.002, $p = 1.36 \times 10^{-3}$). Although further mechanistic studies are warranted, these findings suggest increased physical activity is associated with an improved inflammatory state with fewer lymphocytes and eosinophils.

Keywords: physical activity; blood cell counts; single nucleotide polymorphisms; inflammation

---

1. Introduction

Reduced physical activity (PA) poses a serious threat for public health. Accumulating evidence shows that lower levels of PA are associated with an increased risk of cardiovascular disease (CVD) and all-cause mortality [1–3]. Although the World Health Organization (WHO) recommends at least 150 minutes of moderate-intensity aerobic PA throughout the week, the proportion of Europeans reported to not meet these recommendations has increased in recent years to 46% [4,5]. This trend is not limited to Europeans but is occurring worldwide [6]. Although health care cost estimates related to physical inactivity vary across studies (e.g., 2.4–11.1% of the healthcare expenditure in the United States of America), it is generally believed that physical inactivity is a costly pandemic and associated with a substantial disease burden [7].

However, the exact mechanisms underlying the associations of PA the development of disease are incompletely understood. It has been suggested that systemic inflammation plays a pivotal role in the association between PA and CVD, possibly through changes in circulating (inflammatory) blood cells [8,9]. It is therefore important to investigate whether the effects of PA on CVD risk could be linked through changes in circulating blood cells. However, the association between PA activity and circulating blood cells has only been investigated using traditional observational analyses, which are prone to suffering from confounding effects [10,11]. PA is determined by both genetic and environmental factors [6,12]. A recent genome-wide association study (GWAS) using objectively measured data from wrist-worn accelerometers of PA in a large sub-cohort of the UK Biobank identified newly associated single nucleotide polymorphisms (SNPs) and studied whether activity might contribute causally to disease outcomes [12]. In this study, we aimed to investigate the relationship between genetically determined PA levels, based on the previously reported SNPs, and circulating blood cells using a Mendelian randomization (MR) strategy to minimize confounding effects. We hypothesize a genetically determined higher level of PA is associated with a lower inflammatory state with fewer circulating inflammatory blood cells.

2. Material and Methods

2.1. UK Biobank Participants

The UK Biobank study design and population have been described in detail elsewhere [13]. In brief, the UK Biobank is a large community-based prospective study in the United Kingdom that recruited over 500,000 participants aged 40–69 years aiming to improve the prevention, diagnosis, and treatment of a plethora of diseases. All participants gave informed consent for the study [13]. At the baseline visit, vital signs and biological samples were collected, together with data of self-completed questionnaires, interviews, and physical measurements. The present study was conducted under application number 12,006 of the UK Biobank resource.

2.2. Genotyping and Imputation

The genotyping process and arrays used in the UK Biobank study have been described elsewhere in more detail [14]. Briefly, participants were genotyped using the custom UK Biobank lung exome variant evaluation axiom (Affymetrix: Santa Clara, CA, United States; n = 49,949), which includes 807,411 SNPs or the custom UK Biobank axiom array (Affymetrix; n = 452,713), which includes 820,967 SNPs [13]. The arrays have insertion and deletion markers with more than 95% common content [14,15]. Imputed genotype data were provided by UK Biobank, based on merged UK10K and 1000 Genomes phase 3 panels [16]. Figure 1 shows a flowchart of the study sample selection and is further described below. Participants were excluded if there was no genetic data available or if there was a mismatch between genetic and reported sex (n = 378). Furthermore, participants with high missingness or excess heterozygosity were excluded (n = 963). Participants with familial relatedness or who were not of white British descent were excluded as well (n = 64,535). In addition, participants included in the GWAS on physical activity (n = 90,277) [12] and participants without lab measurements were excluded (n = 18,498). Lastly, participants with diseases or medication affecting the immune response were excluded as well (n = 105,263). We created a set of 222,645 individuals for the present analyses.

For the definitions of diseases, we used hospital episode statistics data in combination with self-reported diagnoses and medication, as described previously [17]. Further information on the definitions of diseases is presented in Supplementary Table S1.

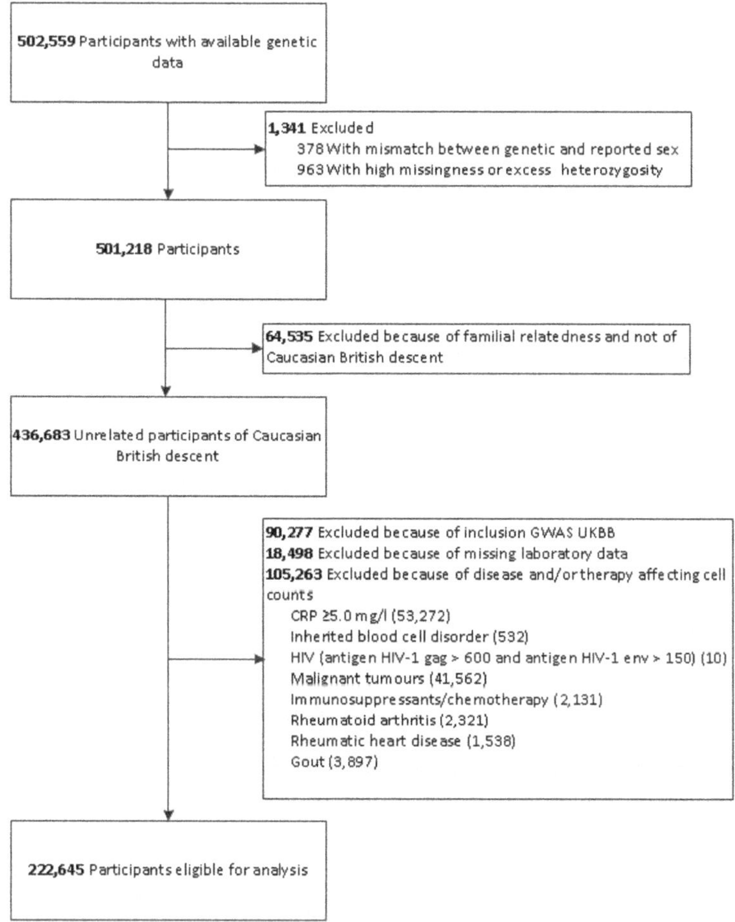

Figure 1. Flowchart for the selection of the analyzed study sample from the UK Biobank Study. GWAS: Genome-wide association study; UKBB: UK Biobank; CRP: C-reactive protein; HIV: human immunodeficiency virus.

2.3. Single Nucleotide Polymorphisms

For our analyses between genetically determined PA and circulating blood cells, we used a set of 68 SNPs identified in the GWAS on physical activity by Doherty et al. [12]. Similar to Doherty et al., we used 68 SNPs that were associated with physical activity with $p < 5 \times 10^{-6}$ (Doherty et al., Supplementary Figure S8) to explain more phenotypic variance than the three SNPs at $p < 5 \times 10^{-8}$. Supplementary Table S2 contains a detailed list of the extracted SNPs. SNPs associated with sleep duration in the study by Doherty et al. were not present in this list of SNPs.

2.4. Statistical Analyses

Normally distributed continuous variables were summarized as mean ± standard deviation (SD) and skewed variables as median and interquartile range (IQR). Linear regression analyses were performed to assess the association between PA and blood cell counts. Regression analyses between SNPs and circulating cells were adjusted for age at the baseline visit, sex, genotyping chip, and the first 30 principal components provided by UK Biobank (to adjust for population structure). Because the PA

SNPs were identified in a GWAS on objectively measured PA, the associations between the SNPs with self-reported PA were not tested as these were considered separate entities. Linear regression analyses were performed using Stata 15 (StataCorp, College Station, TX, United States).

The association between genetically determined increased physical activity with outcomes was first assessed using a fixed-effects inverse-variance weighted (IVW) meta-analysis method, combining the Mendelian randomization (MR) estimates for each SNP with the outcome. To adjust for multiple testing, we applied a Bonferroni correction (significance level divided by number of independent tests) and considered a two-sided *p*-value of less than $0.05/27 = 1.85 \times 10^{-3}$ as statistically significant for the main analyses using the MR-IVW fixed effects model. For the MR-IVW effects model to be valid, the assumption of absence of pleiotropy needs to be fulfilled. Pleiotropy occurs when genetic variants associated with the exposure of interest, exert their effect on the same outcome through multiple pathways. Heterogeneity tests are an easy way to evaluate possible pleiotropy, since low heterogeneity indicates that estimates between the genetic variants' association with the outcome should vary by chance only, which is only possible in case of absence of pleiotropic effects. The Rücker framework was adopted to differentiate between preferred models. In case Cochran's Q in the MR-IVW fixed effects analyses was significant ($p < 0.05$), suggesting heterogeneity and thus non-random error, the MR-IVW random effects model was adopted. Rücker's Q was calculated to evaluate the heterogeneity within the MR-Egger analyses [18]. We then calculated whether the Rücker's Q' statistic was different ($p < 0.05$) from Cochran's Q statistic (Q-Q') [18]. A statistically significant difference indicates the MR-Egger test to be the best approach in case Q-Q' is large and positive [18]. MR-Egger assumes pleiotropic effects of the SNPs on the outcome are independent of their association with PA, and therefore allows for a non-zero intercept [11]. For this, the MR-Egger must not violate the InSIDE assumption, which assumes the association of the SNPs with the exposure are independent of their direct pleiotropic effects on the outcome. The MR-Egger provides additional information on pleiotropy as a non-significant different intercept from 0 ($p > 0.05$) indicates evidence for absent pleiotropic bias. MR-Steiger filtering was performed to remove variants with stronger associations (R^2) with the outcome than the exposure [18]. Beta values (β) and standard errors (SE) are provided for the MR outcomes. Lastly, we performed a weighted median analysis, which allows up to 50% of the information from variants to violate the MR assumptions [19]. For sensitivity analyses, we adopted a *p*-value of <0.05.

MR analyses were performed using the R (version 3.5.1) and the package TwoSampleMR version 0.4.22 (https://mrcieu.github.io/TwoSampleMR/).

3. Results

3.1. Population Characteristics

Baseline characteristics are provided in Table 1. Of the 222,645 participants included in the UK-Biobank, 105,970 (47.6%) were male, and the mean age was 56 ± 8 years. The population was slightly overweight (body mass index ≥25 kg/m^2) with a mean body mass index of 27.0 kg/m^2. More than half of the population never smoked or smoked <100 cigarettes. On average, participants spent 4.88 (inter quartile range (IQR): 1.5–11.3) hours per week doing moderate PA and 0.75 (IQR 0.0–2.7) hours per week doing vigorous PA, based on self-reported data using questionnaires.

Table 1. Baseline characteristics.

Characteristic	No. (%)
Total, No	222,645
Age, mean (SD), years	56 (8)
Sex, male (%)	105,970 (47.6%)
Body Mass Index, mean (SD), kg/m^2	27.0 (4.4)

Table 1. *Cont.*

Characteristic	No. (%)
Smoking behavior, No (%)	
Never or <100 cigarettes	125,777 (58.1%)
Ex-smokers	66,895 (30.9%)
Current	23,849 (11.0%)
Hypertension, No (%)	60,900 (27.4%)
Hyperlipidemia, No (%)	40,099 (18.0%)
Diabetes Mellitus type 2, No (%)	6,984 (3.1%)
PA phenotypes	
Moderate PA, median (IQR), h/week	4.9 (1.5–11.3)
Vigorous PA, median (IQR), h/week	0.8 (0.0–2.7)
Blood cell counts	
Leukocytes, median (IQR), 10^9/L	6.57 (5.60–7.70)
Erythrocytes, median (IQR), 10^{12}/L	4.54 (4.29–4.82)
Neutrophils, median (IQR), 10^9/L	3.95 (3.20–4.82)
Lymphocytes, median (IQR), 10^9/L	0.92 (0.60–1.20)
Monocytes, median (IQR), 10^9/L	0.44 (0.36–0.55)
Eosinophils, median (IQR), 10^9/L	0.13 (0.10–0.21)
Basophils, median (IQR), 10^9/L	0.02 (0.00–0.04)
Reticulocytes, median (IQR), 10^{12}/L	0.06 (0.04–0.07)
Thrombocytes, median (IQR), 10^9/L	246.8 (212.9–285.0)

Data is shown as number (%), as mean with standard deviation (SD) or as median with inter-quartile range (IQR). Units of measurements are indicated. PA: physical activity.

3.2. Genetically Determined Physical Activity and Circulating Blood Cells

Detailed information on the SNPs and their estimates on circulating blood cells is provided in Supplementary Table S3. The association between all 68 SNPs associated with PA with circulating blood cells was assessed using MR analyses. In MR-IVW fixed-effects analyses, genetically determined increased duration of PA was associated with decreased lymphocytes (β = −0.026, SE = 0.008, $p = 1.35 \times 10^{-3}$), decreased eosinophils (β = −0.008, SE = 0.002, $p = 1.36 \times 10^{-3}$), and increased platelet distribution width (β = 0.04, SE = 0.009, $p = 9.07 \times 10^{-5}$) (Table 2). Figures 2–4 displays the individual SNP forest plots for these three outcomes. Supplementary Figure S1–S3 displays the corresponding scatter plots. PA was not associated with any of the other outcomes in the MR-IVW fixed-effects analyses (Table 2), which will therefore not be discussed any further.

Cochran's Q of the association between PA and lymphocytes (Q = 160, DF = 67, $p = 6.50 \times 10^{-10}$) and eosinophils (Q = 180, DF = 67, $p = 6.10 \times 10^{-12}$) indicated the MR-random effects model to be the preferred approach. Using this approach, the association with eosinophils remained significant (β = −0.0078, SE = 0.009, $p = 0.049$), but the association with lymphocytes was attenuated (β = −0.026, SE = 0.014, $p = 0.072$). However, the loss of statistical significance was probably attributable to the larger standard error due to loss of power in the random effects model. In the weighted median MR analyses, the associations between PA and leukocytes (β = −0.030, SE = 0.0134, $p = 0.022$) and eosinophil count (β = −0.0036, SE = 0.0036, $p = 0.3229$) were lost, of which the latter could be attributed to wider standard errors. MR-Steiger filtering indicated all SNP's were more strongly associated with the PA behavior than with lymphocytes and eosinophils. The MR Egger intercept indicated little evidence for pleiotropy for the analyses on lymphocytes and eosinophils.

Table 2. Associations between increased physical activity and circulating blood cells.

	N snps	Inverse Variance Weighted (Fixed Effects)			Inverse Variance Weighted (Multiplicative Random Effects)			MR Egger Fixed Effects			Weighted Median		
		Beta	SE	*p*-Value	Beta	SE	*p*-Value	Beta	SE	*p*-Value	Beta	SE	*p*-Value
Leukocyte count (10^{-9} cells/L)	68	−0.0406	0.0307	1.86×10^{-1}	−0.0406	0.0465	3.83×10^{-1}	0.0106	0.1455	9.42×10^{-1}	0.0024	0.0487	9.61×10^{-1}
Erythrocyte count (10^{-12} cells/L)	68	0.0001	0.0063	9.85×10^{-1}	0.0001	0.0137	9.93×10^{-1}	0.0811	0.0417	5.64×10^{-2}	−0.0126	0.0098	1.99×10^{-1}
Erythrocyte distribution width (%)	68	0.0078	0.0166	6.40×10^{-1}	0.0078	0.0281	7.82×10^{-1}	0.0562	0.0878	5.25×10^{-1}	−0.0047	0.0251	8.53×10^{-1}
Neutrophil count (10^{-9} cells/L)	68	0.0055	0.0243	8.20×10^{-1}	0.0055	0.0357	8.77×10^{-1}	0.0768	0.1118	4.95×10^{-1}	−0.0099	0.0369	7.88×10^{-1}
Neutrophils (%)	68	0.4236	0.1502	4.80×10^{-3}	0.4236	0.2288	6.41×10^{-2}	0.8990	0.7164	2.14×10^{-1}	0.0012	0.2359	9.96×10^{-1}
Lymphocyte count (10^{-9} cells/L)	68	−0.0259	0.0081	1.40×10^{-3}	−0.0259	0.0144	7.21×10^{-2}	−0.0317	0.0454	4.87×10^{-1}	−0.0305	0.0134	2.24×10^{-2}
Lymphocyte (%)	68	−0.3606	0.1294	5.30×10^{-3}	−0.3606	0.2016	7.37×10^{-2}	−0.8313	0.6300	1.92×10^{-1}	0.0836	0.1995	6.75×10^{-1}
Monocyte count (10^{-9} cells/L)	68	−0.0010	0.0035	7.66×10^{-1}	−0.0010	0.0045	8.21×10^{-1}	0.0057	0.0140	6.86×10^{-1}	−0.0013	0.0052	8.09×10^{-1}
Monocytes (%)	68	0.0158	0.0462	7.33×10^{-1}	0.0158	0.0613	7.97×10^{-1}	0.0549	0.1919	7.76×10^{-1}	0.0116	0.0739	8.75×10^{-1}
Eosinophil count (10^{-9} cells/L)	68	−0.0078	0.0024	1.40×10^{-3}	−0.0078	0.0039	4.90×10^{-2}	−0.0071	0.0123	5.67×10^{-1}	−0.0036	0.0036	3.23×10^{-1}
Eosinophils (%)	68	−0.0687	0.0340	4.36×10^{-2}	−0.0687	0.0619	2.67×10^{-1}	−0.1215	0.1946	5.35×10^{-1}	0.0021	0.0534	9.69×10^{-1}
Basophil count (10^{-9} cells/L)	68	−0.0010	0.0009	2.89×10^{-1}	−0.0010	0.0009	2.71×10^{-1}	−0.0006	0.0028	8.32×10^{-1}	−0.0027	0.0013	4.54×10^{-2}
Basophils (%)	68	−0.0141	0.0110	1.99×10^{-1}	−0.0141	0.0110	1.99×10^{-1}	−0.0131	0.0332	6.94×10^{-1}	−0.0262	0.0161	1.04×10^{-1}
Reticulocyte count (10^{-12} cells/L)	68	−0.0256	0.0111	2.04×10^{-2}	−0.0256	0.0188	1.73×10^{-1}	0.0260	0.0586	6.59×10^{-1}	−0.0455	0.0173	8.40×10^{-3}
Reticulocytes (%)	68	−0.0242	0.0108	2.50×10^{-2}	−0.0242	0.0165	1.41×10^{-1}	−0.0001	0.0517	9.99×10^{-1}	−0.0479	0.0161	3.00×10^{-3}
Reticulocyte volume (femtolitres)	68	−0.3151	0.1397	2.41×10^{-2}	−0.3151	0.2491	2.06×10^{-1}	−0.9755	0.7778	2.14×10^{-1}	0.0022	0.2097	9.92×10^{-1}
Immature reticuloyctes fraction	68	−0.0027	0.0011	1.42×10^{-2}	−0.0027	0.0016	1.01×10^{-1}	0.0026	0.0051	6.15×10^{-1}	−0.0055	0.0017	1.30×10^{-3}
Platelet count (10^{-9} cells/L)	68	−2.5671	1.0319	1.29×10^{-2}	−2.5671	1.8209	1.59×10^{-1}	2.7242	5.6655	6.32×10^{-1}	1.4266	1.6611	3.90×10^{-1}
Platelet volume (femtolitres)	68	−0.0001	0.0199	9.96×10^{-1}	−0.0001	0.0352	9.98×10^{-1}	−0.1221	0.1092	2.68×10^{-1}	0.0083	0.0334	8.04×10^{-1}
Platelet packed cell volume (%)	68	−0.0025	0.0008	2.50×10^{-3}	−0.0025	0.0014	6.65×10^{-2}	−0.0003	0.0043	9.41×10^{-1}	0.0000	0.0013	9.76×10^{-1}

Table 2. *Cont.*

	N snps	Inverse Variance Weighted (Fixed Effects)			Inverse Variance Weighted (Multiplicative Random Effects)			MR Egger Fixed Effects			Weighted Median		
		Beta	SE	*p*-Value	Beta	SE	*p*-Value	Beta	SE	*p*-Value	Beta	SE	*p*-Value
Platelet distribution width (%)	68	0.0369	0.0094	1.00×10^{-4}	0.0369	0.0146	1.14×10^{-2}	−0.0499	0.0443	2.64×10^{-1}	0.0106	0.0144	4.64×10^{-1}
Hemoglobin (g/cL)	68	0.0009	0.0177	9.60×10^{-1}	0.0009	0.0315	9.78×10^{-1}	0.1597	0.0974	1.06×10^{-1}	−0.0216	0.0274	4.30×10^{-1}
Hematocrit (%)	68	−0.0500	0.0521	3.37×10^{-1}	−0.0500	0.0948	5.98×10^{-1}	0.5543	0.2887	5.92×10^{-2}	−0.0150	0.0816	8.54×10^{-1}
Mean corpuscular volume (femtoliters)	68	−0.1090	0.0783	1.64×10^{-1}	−0.1090	0.1126	3.33×10^{-1}	−0.3913	0.3527	2.71×10^{-1}	0.0492	0.1203	6.83×10^{-1}
Mean corpuscular hemoglobin (picograms)	68	−0.0078	0.0321	8.08×10^{-1}	−0.0078	0.0447	8.61×10^{-1}	−0.2389	0.1367	8.52×10^{-2}	0.0500	0.0488	3.05×10^{-1}
Mean corpuscular hemoglobin concentration (grams/dL)	68	0.0402	0.0187	3.19×10^{-2}	0.0402	0.0237	8.90×10^{-2}	−0.0970	0.0719	1.82×10^{-1}	0.0304	0.0284	2.84×10^{-1}
Mean sphered cells volume (femtoliters)	68	−0.0564	0.0955	5.55×10^{-1}	−0.0564	0.1836	7.59×10^{-1}	−0.5939	0.5733	3.04×10^{-1}	0.2760	0.1451	5.72×10^{-2}

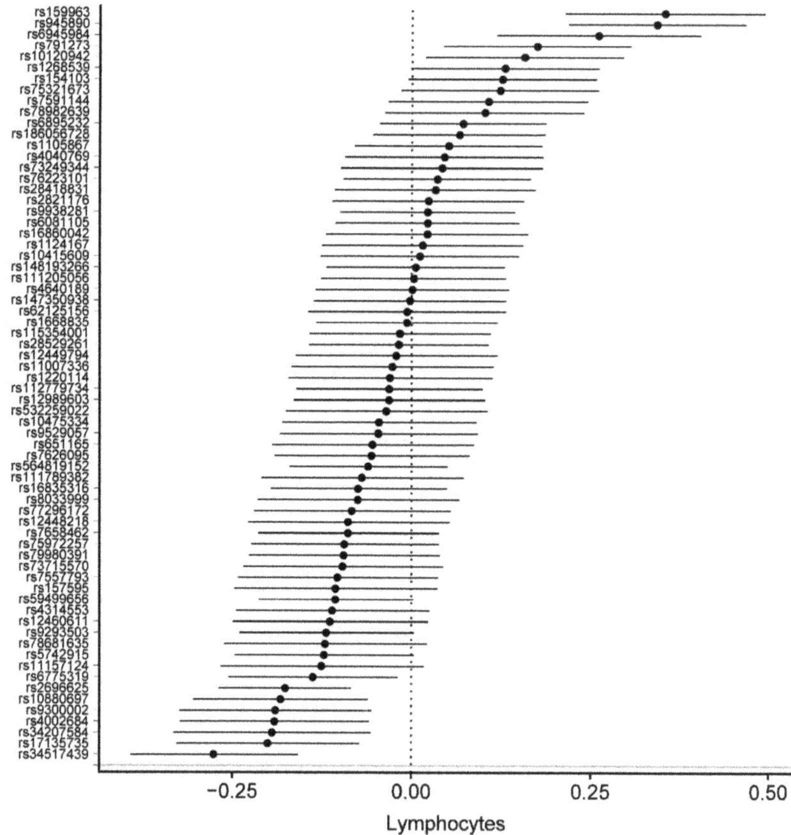

Figure 2. Forest plot of physical activity single nucleotide polymorphisms on lymphocyte levels.

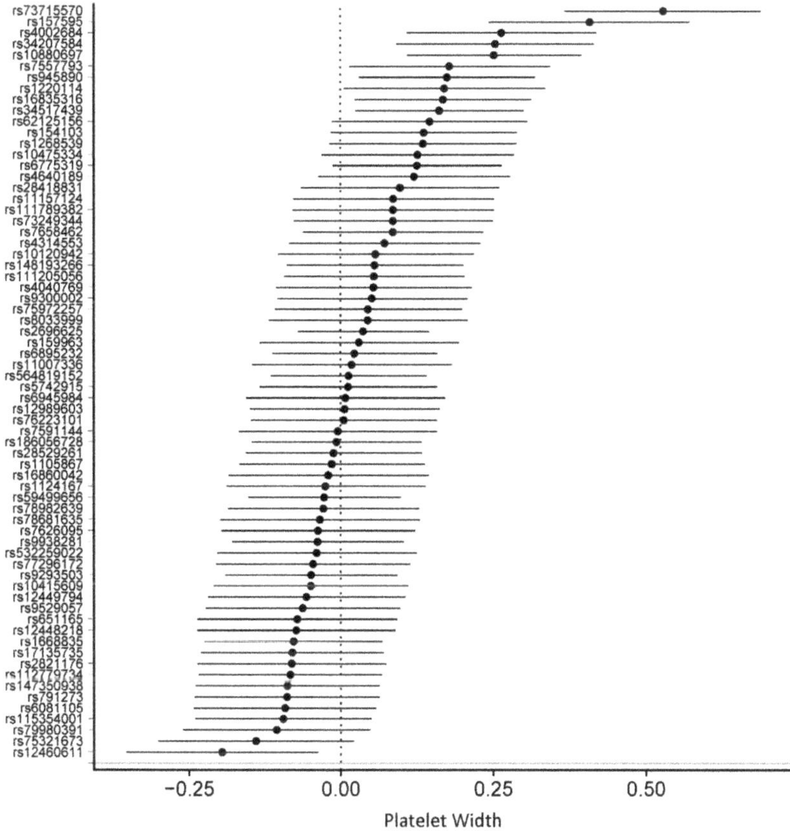

Figure 3. Forest plot of physical activity single nucleotide polymorphisms on platelet width.

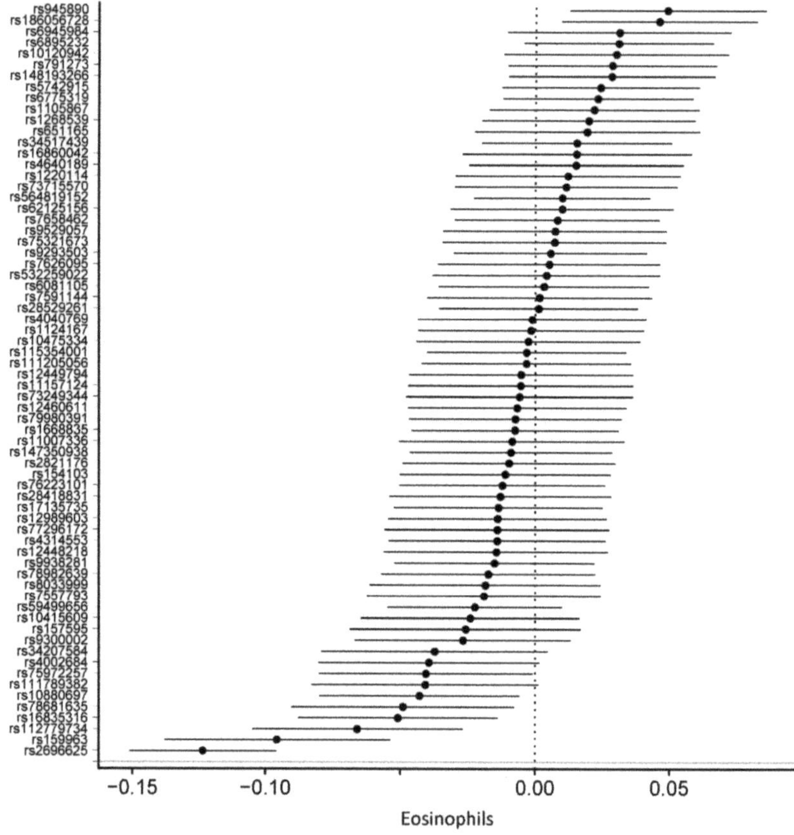

Figure 4. Forest plot of physical activity single nucleotide polymorphisms on eosinophil level.

As an additional analysis, we investigated whether the relationship between genetically determined PA with circulating lymphocytes or eosinophils differed between levels of self-reported PA. However, genetically determined PA was not significantly associated with either lymphocytes or eosinophils amongst individuals with no (n = 16,307), only moderate (n = 52,273), or only vigorous (n = 4536) self-reported PA.

In the heterogeneity analyses, Q-Q' was statistically significant for platelet distribution width (Q = 160 DF = −67 p = 1.80 × 10^{-3}), indicating the MR-Egger analysis to be the best approach. Using this approach, the association between genetically determined PA with platelet distribution width was reversed in effect and no longer significant (β −0.05, SE 0.04, p = 0.26). Furthermore, we performed a look-up in MR-Base to explore whether the 68 genetic variants were associated with other traits than PA. This information can be found in Supplementary Table S5.

4. Discussion

In the present study, we provide evidence for the association between genetically determined PA and circulating blood cells. Genetically determined increased duration of PA was cautiously associated with decreased lymphocytes and decreased eosinophils, suggesting increased levels of PA may improve the inflammatory state. This is in line with our hypothesis.

The present study is the first to report associations between genetically determined PA and circulating blood cells. The association between genetically determined increased PA and decreased lymphocyte levels is partly in line with previous research which studied the association between

PA and total leukocyte count [8]. In 4,857 individuals with a mean age of 43 ± 1 year and 43% females, participating in the National Health and Nutrition Examination Survey, the association between increased PA and a decreased leukocyte count has been observed, suggesting that active individuals might maintain a lower inflammatory state and might be less prone to future chronic disease development [8]. We did not assess the correlations between self-reported PA and circulating blood cells, and this study can therefore not be directly compared with these previous studies. However, our study is of additive value, since we were able to study the association between PA and a broader range of cell types (i.e., monocytes, lymphocytes, neutrophils, eosinophils, and basophils) instead of total leukocyte count solely. We did not observe an association between genetically determined PA and total leukocyte count, but we did observe an association between increased genetically determined PA and decreased lymphocytes levels in the MR-IVW analyses. For a long period, a decrease in lymphocytes has been considered as a suppression of the human immune system and therefore as detrimental [20]. However, recent evidence indicates that this reduction in peripheral blood does not reflect immune suppression, but represents a heightened state of immune surveillance and immune regulation, which is driven by a transfer of cells to peripheral tissues, such as the gut or the lungs [21]. Within the cardiovascular field, T-lymphocytes are known to stimulate macrophages expressing collagen-degrading enzymes and thereby increasing the risk of plaque rupture. Lower lymphocyte levels in the blood stream, might therefore also be beneficial for the risk of CVD [22,23]. The association between lymphocyte count and PA in our study was lost in the sensitivity analyses, although the strength of the effect remained similar. This loss of association may be due to a larger standard error reducing the statistical power. Similarly, no association was observed between genetically determined PA and lymphocytes or eosinophils across levels of self-reported PA, which is likely due to the small sample sizes in these groups. Further MR studies using variants which explain a larger variance in PA and also in larger groups of self-reported PA levels are warranted to further investigate these associations.

There is limited data on the association between PA and eosinophils. Earlier research in 11 athletes running an ultramarathon (90 kilometers), showed a non-allergic activation of eosinophils, reflected by an increase of eosinophil cationic protein [24]. Our study population of community dwelling middle-aged men and women substantially differed from those athletes (e.g., age, ethnicity, and athletic condition), and therefore, it is not possible to formulate any hypotheses or to draw any comparisons [24]. Recently, in 5287 patients who underwent coronary angiography, a negative association between peripheral eosinophil count and the severity of coronary artery disease (CAD) has been observed [25]. Furthermore, previous studies have suggested that eosinophils play a key role in the initiation, progression, and rupture of thrombotic plaques, which was confirmed by tissue samples obtained through thrombus aspiration in patients with myocardial infarction [26,27]. In these samples, a large amount of eosinophils was present [26]. However, these findings (low peripheral eosinophil count associated with increased CAD severity and high eosinophil counts observed in atherosclerotic plaques) were observed in CAD patients, whereas our study was performed in a general population of which only 3.7% had a medical history of CAD at baseline. The present findings indicate increased PA is associated with decreased eosinophil levels. Possibly, less PA leading to higher eosinophil levels could play a role in the development of plaques eventually leading to CAD. In this case, increased PA leading to lower eosinophil levels might be protective for plaques and CAD. Further research is needed to investigate this hypothesis.

Contrary to previous cross-sectional observational studies, we did not observe associations between genetically determined PA and red blood cell indices [28–30]. This might imply that changes in red blood cells during or after PA are bystanders instead of a consequence of PA, although mechanistic studies are necessary to unravel these associations.

Of interest is the specific function of three leading genetic polymorphisms rs564819152, rs2696625, and rs59499656, which may provide more insights in the mechanisms underlying genetically determined PA and circulating immune cells. rs564819152 is located near the SKIDA1 gene. This gene is located on chromosome 10 and encodes the Ski-Dach domain-containing protein 1, which is associated with different types of cancer [31,32]. Furthermore, this gene was found to be associated with lung function [33]. These functions might affect the relation between PA and blood cells, although our heterogeneity tests did not indicate pleiotropy of the genetic variants.

rs2696625 and rss59499656 are located near the genes KANSL1-AS1 and SYT4, respectively. KANSL1-AS1 has been described in the context of Alzheimer's and Parkinson's diseases [34]. SYT4 (synaptotagmin 4) is a protein coding gene that has been associated with various traits, i.e., body mass index, lung function, and body fat percentage [33,35,36]. As obesity is associated with an increased inflammatory state, this route could be involved in the association between genetically determined PA and circulating immune cells [37]. Aside from the leading variants, the association with obesity-related traits was also observed in the MR-base look-up with the other genetic variants. However, further mechanistic studies are warranted to provide more insights into these possible additional pathways.

A major strength of our study is the large cohort size of 222,645 participants. Second, this is the first study to assess genetic variants of PA using SNPs that were found in a GWAS performed with objectively measured PA data using wrist-worn accelerometer data. Thirdly, we used strict exclusion criteria to confine factors affecting immune cells such as infections. Furthermore, we used a stringent threshold *p*-value to reduce false positive rates and increase reproducibility.

As a future perspective, it would be of additional value to investigate the association with the functionality and activity of cells, for example by examining circulating cytokine levels. Circulating cytokines were not measured in the UK Biobank and therefore not tested in the present study. This study could, however, serve as a starting point, providing new insights into where to focus future studies on the association between PA and circulating blood cells.

5. Conclusions

In conclusion, this study shows that genetically determined PA is associated with changes in circulating blood cells. Increased genetically determined PA is associated with decreased lymphocyte and eosinophil levels. Although further mechanistic studies are warranted, these findings cautiously suggest lifestyle changes that include more PA should be encouraged to improve the inflammatory state.

Supplementary Materials: The following are available online at http://www.mdpi.com/2073-4425/10/11/908/s1, Supplementary Table S1: definitions of phenotypes UK Biobank; Supplementary Table S2: Single nucleotide polymorphisms previously associated with physical activity; Supplementary Table S3: Estimates of the associations between the single nucleotide polymorphisms and circulating blood cells; Supplementary Table S4: Mendelian randomization sensitivity analyses; Supplementary Table S5: MR-base look-up of associations between the 68 physical activity SNPs and other traits beside circulating immune cells, observed in the UK Biobank; Supplementary Figure S1: Scatter plot of physical activity single nucleotide polymorphisms on lymphocyte levels; Supplementary Figure S2: Scatter plot of physical activity single nucleotide polymorphisms on platelet width; Supplementary Figure S3: Scatter plot of physical activity single nucleotide polymorphisms on eosinophil level.

Author Contributions: Conceptualization, H.E.G., M.A.S. and Y.J.v.d.V.; formal analysis, M.A.S. and Y.J.v.d.V.; analysis and interpretation of data, all authors; writing—original draft preparation, F.M.P.; writing—review and editing, all authors; visualization, M.A.S. and Y.J.v.d.V.; supervision, N.V. and P.v.d.H.

Funding: N.V. is supported by NWO VENI grant (016.186.125). The funding agencies had no role in the study design, analysis, or interpretation of data; the writing of the manuscript; or in the decision to submit the article for publication.

Acknowledgments: We would like to thank the Centre for Information Technology of the University of Groningen for their support and for providing access to the Peregrine high-performance computing cluster. We thank Ruben N. Eppinga, MD, Tom Hendriks, MD, M. Yldau van der Ende, BSc, Yanick Hagemeijer, MSc, and Jan-Walter Benjamins, BEng (Department of Cardiology, University of Groningen, University Medical Center Groningen, Groningen, the Netherlands) for their contributions to the extraction and processing of data in the UK Biobank. None of the contributors received compensation except for their employment at the University Medical Center Groningen.

Conflicts of Interest: Author N.V. is an employee of Genomics plc. The funders had no role in the design of the study; in the collection, analyses, or interpretation of data; in the writing of the manuscript, or in the decision to publish the results.

References

1. Lachman, S.; Boekholdt, S.M.; Luben, R.N.; Sharp, S.J.; Brage, S.; Khaw, K.-T.; Peters, R.J.; Wareham, N.J. Impact of physical activity on the risk of cardiovascular disease in middle-aged and older adults: EPIC Norfolk prospective population study. *Eur. J. Prev. Cardiol.* **2018**, *25*, 200–208. [CrossRef] [PubMed]
2. Patterson, R.; Mcnamara, E.; Tainio, M.; Hé, T.; De Sá, R.; Smith, A.D.; Sharp, S.J.; Edwards, P.; Woodcock, J.; Brage, S.; et al. Sedentary behaviour and risk of all-cause, cardiovascular and cancer mortality, and incident type 2 diabetes: A systematic review and dose response meta-analysis. *Eur. J. Epidemiol.* **2018**, *33*, 811–829. [CrossRef] [PubMed]
3. Lee, I.M.; Shiroma, E.J.; Lobelo, F.; Puska, P.; Blair, S.N.; Katzmarzyk, P.T.; Alkandari, J.R.; Andersen, L.B.; Bauman, A.E.; Brownson, R.C.; et al. Effect of physical inactivity on major non-communicable diseases worldwide: An analysis of burden of disease and life expectancy. *Lancet* **2012**, *380*, 219–229. [CrossRef]
4. WHO Physical Activity and Adults. Available online: https://www.who.int/dietphysicalactivity/factsheet_adults/en/ (accessed on 1 July 2019).
5. European Union. *Special Eurobarometer 472. "Sport and Physical Activity"*; European Union: Brussels, Belgium, 2018.
6. Althoff, T.; Sosič, R.; Hicks, J.L.; King, A.C.; Delp, S.L.; Leskovec, J. Large-scale physical activity data reveal worldwide activity inequality. *Nature* **2017**, *547*, 336–339. [CrossRef] [PubMed]
7. Ding, D.; Kolbe-Alexander, T.; Nguyen, B.; Katzmarzyk, P.T.; Pratt, M.; Lawson, K.D. The economic burden of physical inactivity: A systematic review and critical appraisal. *Br. J. Sports Med.* **2017**, *51*, 1392–1409. [CrossRef] [PubMed]
8. Willis, E.A.; Shearer, J.J.; Matthews, C.E.; Hofmann, J.N. Association of physical activity and sedentary time with blood cell counts: National Health and Nutrition Survey 2003–2006. *PLoS ONE* **2018**, *13*, e0204277. [CrossRef] [PubMed]
9. Silverman, M.N.; Deuster, P.A. Biological mechanisms underlying the role of physical fitness in health and resilience. *Interface Focus* **2014**, *4*, 20140040. [CrossRef] [PubMed]
10. Davey Smith, G.; Ebrahim, S. 'Mendelian randomization': Can genetic epidemiology contribute to understanding environmental determinants of disease? *Int. J. Epidemiol.* **2003**, *32*, 1–22. [CrossRef]
11. Bowden, J.; Davey Smith, G.; Burgess, S. Mendelian randomization with invalid instruments: Effect estimation and bias detection through Egger regression. *Int. J. Epidemiol.* **2015**, *44*, 512–525. [CrossRef]
12. Doherty, A.; Smith-Byrne, K.; Ferreira, T.; Holmes, M.V.; Holmes, C.; Pulit, S.L.; Lindgren, C.M. GWAS identifies 14 loci for device-measured physical activity and sleep duration. *Nat. Commun.* **2018**, *9*, 5257. [CrossRef]
13. UK Biobank. *Coordinating Centre UK Biobank: Protocol for a Large-Scale Prospective Epidemiological Resource*; UKBB-PROT-09-06 (Main Phase); UK Biobank Coordinating Centre: Stockport, UK, 2007; Volume 6, pp. 1–112.
14. Bycroft, C.; Freeman, C.; Petkova, D.; Band, G.; Elliott, L.T.; Sharp, K.; Motyer, A.; Vukcevic, D.; Delaneau, O.; Connell, J.; et al. Genome-wide genetic data on ~500,000 UK Biobank participants. *bioRxiv* **2017**, 166298.
15. UK Biobank. *Genotyping and Quality Control of UK Biobank, a Large-Scale, Extensively Phenotyped Prospective Resource*; UK Biobank: Stockport, UK, 2018.
16. Marchini, J. *UK Biobank Phasing and Imputation Documentation*; UK Biobank: Stockport, UK, 2018.
17. Abdullah Said, M.; Eppinga, R.N.; Lipsic, E.; Verweij, N.; van der Harst, P. Relationship of arterial stiffness index and pulse pressure with cardiovascular disease and mortality. *J. Am. Heart Assoc.* **2018**, *7*, e007621.
18. Bowden, J.; Del Greco, M.F.; Minelli, C.; Davey Smith, G.; Sheehan, N.; Thompson, J. A framework for the investigation of pleiotropy in two-sample summary data Mendelian randomization. *Stat. Med.* **2017**, *36*, 1783–1802. [CrossRef] [PubMed]
19. Bowden, J.; Davey Smith, G.; Haycock, P.C.; Burgess, S. Consistent Estimation in Mendelian Randomization with Some Invalid Instruments Using a Weighted Median Estimator. *Genet. Epidemiol.* **2016**, *40*, 304–314. [CrossRef] [PubMed]
20. Campbell, J.P.; Turner, J.E. Debunking the Myth of Exercise-Induced Immune Suppression: Redefining the Impact of Exercise on Immunological Health across the Lifespan. *Front. Immunol.* **2018**, *9*, 648. [CrossRef]
21. Dhabhar, F.S. Effects of stress on immune function: The good, the bad, and the beautiful. *Immunol. Res.* **2014**, *58*, 193–210. [CrossRef]
22. Libby, P.; Theroux, P. Pathophysiology of coronary artery disease. *Circulation* **2005**, *111*, 3481–3488. [CrossRef]

23. Rana, J.S.; Boekholdt, S.M.; Ridker, P.M.; Jukema, J.W.; Luben, R.; Bingham, S.A.; Day, N.E.; Wareham, N.J.; Kastelein, J.J.P.; Khaw, K.-T. Differential leucocyte count and the risk of future coronary artery disease in healthy men and women: The EPIC-Norfolk Prospective Population Study. *J. Intern. Med.* **2007**, *262*, 678–689. [CrossRef]

24. McKune, A.J.; Smith, L.L.; Semple, S.J.; Wadee, A.A. Non-allergic activation of eosinophils after strenuous endurance exercise. *S. Afr. J. Sport. Med.* **2004**, *16*, 12–16. [CrossRef]

25. Gao, S.; Deng, Y.; Wu, J.; Zhang, L.; Deng, F.; Zhou, J.; Yuan, Z.; Wang, L. Eosinophils count in peripheral circulation is associated with coronary artery disease. *Atherosclerosis* **2019**, *286*, 128–134. [CrossRef]

26. Sakai, T.; Inoue, S.; Matsuyama, T.-A.; Takei, M.; Ota, H.; Katagiri, T.; Koboyashi, Y. Eosinophils may be involved in thrombus growth in acute coronary syndrome. *Int. Heart J.* **2009**, *50*, 267–277. [CrossRef] [PubMed]

27. Deng, X.; Wang, X.; Shen, L.; Yao, K.; Ge, L.; Ma, J.; Zhang, F.; Qian, J.; Ge, J. Association of eosinophil-to-monocyte ratio with 1-month and long-term all-cause mortality in patients with ST-elevation myocardial infarction undergoing primary percutaneous coronary intervention. *J. Thorac. Dis.* **2018**, *10*, 5449–5458. [CrossRef] [PubMed]

28. Loprinzi, P.D.; Hall, M.E. Physical activity and dietary behavior with red blood cell distribution width. *Physiol. Behav.* **2015**, *149*, 35–38. [CrossRef] [PubMed]

29. Hammam, N.; Ezeugwu, V.E.; Manns, P.J.; Pritchard-Wiart, L. Relationships between sedentary behaviour, physical activity levels and red blood cell distribution width in children and adolescents. *Heal. Promot. Perspect.* **2018**, *8*, 147–154. [CrossRef] [PubMed]

30. Loprinzi, P.D.; Ford, E.S. The Association between Objectively Measured Sedentary Behavior and Red Blood Cell Distribution Width in a National Sample of US Adults. *Am. J. Epidemiol.* **2015**, *181*, 357–359. [CrossRef] [PubMed]

31. Roth, G.A.; Abate, D.; Abate, K.H.; Abay, S.M.; Abbafati, C.; Abbasi, N.; Abbastabar, H.; Abd-Allah, F.; Abdela, J.; Abdelalim, A.; et al. Global, regional, and national age-sex-specific mortality for 282 causes of death in 195 countries and territories, 1980–2017: A systematic analysis for the Global Burden of Disease Study 2017. *Lancet* **2018**, *392*, 1736–1788. [CrossRef]

32. Keyes, K.M.; Westreich, D. UK Biobank, big data, and the consequences of non-representativeness. *Lancet* **2019**, *393*, 1297. [CrossRef]

33. Kichaev, G.; Bhatia, G.; Loh, P.-R.; Gazal, S.; Burch, K.; Freund, M.K.; Schoech, A.; Pasaniuc, B.; Price, A.L. Leveraging Polygenic Functional Enrichment to Improve GWAS Power. *Am. J. Hum. Genet.* **2019**, *104*, 65–75. [CrossRef]

34. Ferrari, R.; Wang, Y.; Vandrovcova, J.; Guelfi, S.; Witeolar, A.; Karch, C.M.; Schork, A.J.; Fan, C.C.; Brewer, J.B.; International FTD-Genomics Consortium (IFGC); et al. Genetic architecture of sporadic frontotemporal dementia and overlap with Alzheimer's and Parkinson's diseases. *J. Neurol. Neurosurg. Psychiatry* **2017**, *88*, 152–164. [CrossRef]

35. Hubel, C.; Gaspar, H.A.; Coleman, J.R.I.; Finucane, H.; Purves, K.L.; Hanscombe, K.B.; Prokopenko, I.; Graff, M.; Ngwa, J.S.; Workalemahu, T.; et al. Genomics of body fat percentage may contribute to sex bias in anorexia nervosa. *Am. J. Med. Genet. B. Neuropsychiatr. Genet.* **2019**, *180*, 428–438. [CrossRef]

36. Akiyama, M.; Okada, Y.; Kanai, M.; Takahashi, A.; Momozawa, Y.; Ikeda, M.; Iwata, N.; Ikegawa, S.; Hirata, M.; Matsuda, K.; et al. Genome-wide association study identifies 112 new loci for body mass index in the Japanese population. *Nat. Genet.* **2017**, *49*, 1458–1467. [CrossRef] [PubMed]

37. Yu, L.; Li, Y.; Du, C.; Zhao, W.; Zhang, H.; Yang, Y.; Sun, A.; Song, X.; Feng, Z. Pattern Recognition Receptor-Mediated Chronic Inflammation in the Development and Progression of Obesity-Related Metabolic Diseases. *Mediators Inflamm.* **2019**, *2019*, 5271295. [CrossRef] [PubMed]

© 2019 by the authors. Licensee MDPI, Basel, Switzerland. This article is an open access article distributed under the terms and conditions of the Creative Commons Attribution (CC BY) license (http://creativecommons.org/licenses/by/4.0/).

 *genes*

Concept Paper

Impact of Acute Aerobic Exercise on Genome-Wide DNA-Methylation in Natural Killer Cells—A Pilot Study

Alexander Schenk [1], Christine Koliamitra [1], Claus Jürgen Bauer [2], Robert Schier [2], Michal R. Schweiger [3,4,5], Wilhelm Bloch [1] and Philipp Zimmer [1,*]

[1] Department of Molecular and Cellular Sport Medicine, Institute for Cardiology and Sports Medicine, German Sport University Cologne, Am Sportpark Müngersdorf 6, 50933 Cologne, Germany; a.schenk@dshs-koeln.de (A.S.); c.koliamitra@dshs-koeln.de (C.K.); w.bloch@dshs-koeln.de (W.B.)
[2] Department of Anesthesiology and Intensive Care Medicine, University Hospital of Cologne, Kerpener Straße 62, 50937 Cologne, Germany; cjbatwork@googlemail.com (C.J.B.); robert.schier@uk-koeln.de (R.S.)
[3] Department of Epigenetics and Tumor Genetics, Medical Faculty, University of Cologne, Kerperner Straße 62, 50937 Cologne, Germany; mschweig@uni-koeln.de
[4] Institute of Human Genetics, University Hospital of Cologne, Kerpener Straße 62, 50937 Cologne, Germany
[5] Center for Molecular Medicine Cologne, Kerperner Straße 62, 50937 Cologne, Germany
* Correspondence: p.zimmer@dshs-koeln.de

Received: 5 April 2019; Accepted: 16 May 2019; Published: 19 May 2019

Abstract: Natural Killer (NK-) cells reveal a keen reaction to acute bouts of exercise, including changes of epigenetic modifications. So far, exercise-induced alterations in NK-cell DNA-methylation were shown for single genes only. Studies analyzing genome-wide DNA-methylation have used conglomerates like peripheral blood mononuclear cells (PBMCs) rather than specific subsets of immune cells. Therefore, the aim of this pilot-study was to generate first insights into the influence of a single bout of exercise on genome-wide DNA-methylation in isolated NK-cells to open the field for such analyses. Five healthy women performed an incremental step test and blood samples were taken before and after exercise. DNA was isolated from magnet bead sorted NK-cells and further analyzed for global DNA-methylation using the Infinium MethylationEPIC BeadChip. DNA-methylation was changed at 33 targets after acute exercise. These targets were annotated to 25 genes. Of the targets, 19 showed decreased and 14 increased methylation. The 25 genes with altered DNA-methylation have different roles in cell regulation and differ in their molecular functions. These data give new insights in the exercise induced regulation of NK-cells. By using isolated NK-cells, exercise induced differences in DNA-methylation could be shown. Whether or not these changes lead to functional adaptions needs to be elucidated.

Keywords: natural killer cell; NK; epigenetics; DNA-methylation; exercise; sport

1. Introduction

Natural killer (NK-) cells are a distinct cell population of the innate immune system that build the first line of defense against virus-infected and neoplastic cells [1]. NK-cells are characterized by their rapid activation through germ line encoded receptors [2] and their ability to discriminate between healthy and diseased cells. NK-cells show a keen reaction to acute bouts of exercise, including changes in circulating NK-cell numbers and NK-cell function [3]. Circulating NK-cell numbers increase during exercise and decrease immediately after cessation of exercise below pre-exercise levels for up to 48 h [4]. Furthermore, the NK-cell pool can be divided into the more regulatory CD56^{bright} subset and the pronounced cytotoxic CD56^{dim} subset. As shown by Bigley et al. [5], exercise predominantly affects the redistribution of CD56^{dim} NK-cells, leading to changes in the distribution of NK-cell subsets. Data for NK-cell function is inconsistent but a tendency regarding higher NK-cell function after intensive exercise is reported [3].

Research of the past decade suggests that several short- and long-term adaptions of various tissues in response to exercise are based on epigenetic modifications [6–9], such as DNA-methylation, histone modification and microRNA expression [10]. In view of immune cells, results from Nakajima et al. [7] and Denham et al. [11] suggested gene-specific changes of DNA-methylation in peripheral blood mononuclear cells (PBMCs) and leukocytes after interval training interventions. Regarding acute exercise, Robson-Ansley et al. [12] found no effects on global DNA-methylation in PBMCs. However, analyzing PBMCs has some major limitations. First, PBMCs represent an agglomerate of several immune cell subsets which have distinct epigenetic patterns. Second, acute exercise and training interventions alter the number and proportions of immune cell subsets. Thereby, detected changes of DNA-methylation may be driven by changes in PBMC composition rather than by changes on the epigenome. Hence, analyzing specific immune cell subsets would be more accurate. Another attempt to overcome these problems was conducted by Horsburgh et al. [13]. PBMCs were cultivated in vitro with plasma collected before and after an intensive exercise bout. They found the nuclear expression of DNA methyltransferase (DNMT) 3B to be decreased after exercise. DNMT3B is one of the DNMTs, accounting for de novo DNA-methylation and therefore changes in DNA-methylation were suggested but unfortunately not tested [13].

Using isolated NK-cells, we [14] have shown that a long-lasting acute bout of endurance exercise (half marathon) did not alter global DNA-methylation. However, this was only tested by global immunocytochemistry. Nevertheless, antibody detected global histone acetylation at the fifth lysine-residue of histone 4 (H4K5) increased after the run and was further accompanied by an elevated gene expression of the activating NK-cell receptor NKG2D. Interestingly, gene expression of NKG2D has previously been described to be regulated by epigenetic modifications [15]. Furthermore, we have recently shown that an acute bout of exercise decreased the promoter DNA-methylation of the activating killer immunoglobulin-like receptor (KIR) KIR2DS4 and increased the corresponding gene expression [16]. In contrast, neither promoter DNA-methylation nor gene expression of the inhibitory *KIR3DL1* gene was affected. In this trial, DNA-methylation was assessed using the high-resolution method targeted deep amplicon bisulfite sequencing (TDBS).

Since immunohistochemically antibody-based assessment of DNA-methylation is relatively vague, and TDBS does not allow the analysis of a wide range of genes, this pilot-study aims to generate first insights into the influence of a single bout of exercise on gene specific DNA-methylation using a microarray approach in isolated NK-cells and open the field for such analyses.

2. Experimental Section

This study was performed in accordance with the declaration of Helsinki and was approved by the ethics committee of the University Hospital of Cologne (IRB#13-274). Five healthy women (age 61.4 ± 8.0 years) were recruited by regional newspaper announcement between December 2014 and September 2015 and provided written informed consent. The term "healthy" was defined as no acute or chronic diseases. Any kind of drug intake was an exclusion criterion. For analyzing the effects of a single bout of exercise on the methylome of NK-cells, venous blood samples were collected before (t0) and 1 min after (t1) an incremental step test on a bicycle ergometer.

2.1. Incremental Step Test

The incremental step test was performed on a bicycle ergometer (Ergoline GmbH, Bitz, Germany) with a spirometry analysis (Cortex Biophysik GmbH, Leipzig, Germany). The incremental step test began with a 1 min rest measurement, followed by a 3 min warm-up at 50 watts power output and an increase of 25 watts every 2 min until exhaustion (Respiratory quotient >1). Heart rate (Promedia Medizintechnik, Siegen, Germany) and self-perceived exhaustion (Borg scale) were assessed in each intensity step.

2.2. Blood Sampling and NK-Cell Isolation

Blood samples were collected before (t0) and 1 min after (t1) the incremental step test. Blood samples were used for isolation of peripheral blood mononuclear cells (PBMCs) with a lymphocyte separation medium (promo Cell, Heidelberg, Germany) based density gradient centrifugation. Isolated PBMCs were used for a magnetic bead associated negative separation of NK-cells (EasySep™ Human NK Cell Enrichment Kit; STEMCELL Technologies Germany GmbH, Cologne, Germany) according to the manufacturer's protocol.

2.3. DNA Isolation

DNA was isolated using a column-based isolation (RNA/DNA/Protein Purification Plus Kit; Norgen Biotek Corp., Thorold, ON, Canada) according to the manufacturer's protocol.

2.4. Genome-Wide DNA-Methylation

Genome-wide DNA-methylation was assessed with the Infinium® MethylationEPIC BeadChip (Illumina, San Diego, CA, USA) according to the manufacturer's protocol. Methylation data was calculated with Genome Studio 2011.1 containing the methylation module (Illumina, San Diego, CA, USA). Methylation was displayed in β-values. The data discussed have been deposited in NCBI's Gene Expression Omnibus and are accessible through GEO Series accession number GSE129376 (https://www.ncbi.nlm.nih.gov/geo/query/acc.cgi?acc=GSE129376).

2.5. Gene Ontology

Functional annotation of genes showing changes in DNA-methylation was performed using the DAVID Bioinformatics Resources [17].

2.6. Flow Cytometry

To control for changes of the NK-cell subsets distribution, isolated PBMCs were used for flow cytometry analysis. PBMCs were stained with anti-CD3 APC-C7 and anti-CD56 PE (BD Bioscience, Heidelberg, Germany). Flow cytometry was performed on a BD FACS Array (BD Bioscience, Heidelberg, Germany) and NK-cells were gated as CD3− and CD56+. Furthermore, CD56^{bright} and CD56^{dim} NK-cells were gated and presented as percentage of the NK-cell pool.

2.7. Statistics

Genome Studio was used to perform a differential methylation analysis, to detect differences between before (t0) and after (t1) acute exercise. The illumina custom model, including consideration of the false discovery rate (FDR) by p-value correction according to Benjamini-Hochberg [18], was used to calculate the differential methylation. FDR corrected DiffScores were computed, with a DiffScore $\geq$ |13| $\approx p \leq 0.05$ as described by Tremblay et al. [19]. In brief, the detection p-value indicated whether the sample signal was distinguishable from the negative control. Targets indicating a detection $p > 0.05$ in more than 10% of the samples were excluded from analyses. Flow cytometry data was compared using a Wilcoxon signed-rank test. The level of significance was set at $p \leq 0.05$.

3. Results

3.1. Participants Characteristics

To analyze the influence of a single bout of acute exercise, five healthy women were recruited with an age of 61.4 ± 8.0 years. The incremental step test revealed a peak oxygen uptake (VO$_{2peak}$) of 32.4 ± 3.8 mL/min/kg with a maximal power output of 165.0 ± 13.7 Watt. Detailed characteristics of the participants are shown in Table 1.

Table 1. Participant characteristics.

Parameter	Mean	Standard Deviation
Age [years]	61.4	8.0
Weight [kg]	60.2	4.0
Height [cm]	166.2	4.1
BMI [kg/m^2]	20.0	1.5
Waist circumference [cm]	84.2	5.5
Maximal power output [Watt]	165.0	13.7
VO$_{2peak}$ [mL/min/kg]	32.4	3.8

3.2. Genome-Wide DNA-Methylation

Analysis with the Infinium MethylationEPIC BeadChip revealed 864,614 targets of a total of 865,918 (99.8%) that could be detected with a detection $p \leq 0.05$. Comparing DNA-methylation before (t0) and after (t1) acute exercise revealed 33 targets being changed with a DiffScore $\geq |13|$ (0.004%). Within the 33 targets affected by acute exercise, 25 were annotated to a gene. As shown in Table 2, in 19 of the 33 targets (57.6%) DNA-methylation was decreased after exercise, whereas 14 targets (42.4%) show increased DNA-methylation. A total of 10 targets (30.3%) were found in a gene body and 8 targets (24.2%) had no gene annotation. Table 2 presents a complete list of targets with a changed methylation, including gene information and β-values.

3.3. Gene Ontology

DAVID Bioinformatics Resources [17] was used for functional annotation of the 25 genes (Table 3). It is noteworthy that *FASLG* is the gene with the most functional annotation. It belongs to the plasma membrane and is involved in regulation of transcription.

3.4. Distribution of NK-Cell Subsets CD56^{bright} and CD56^{dim}

Flow cytometry was used to control blood samples for changes in NK-cell subset distribution. The percent of CD56^{bright} NK-cells within the NK-cell pool, decreased from 14.91% ± 4.64% to 10.38% ± 3.20% comparing before and after acute exercise. Accordingly, the percent of CD56^{dim} NK-cell increased from 84.63% ± 4.78% to 89.12% ± 3.57%. Nevertheless, both changes did not reach statistical significance ($p = 0.080$, $p = 0.080$).

Table 2. Detailed presentation of targets affected by acute exercise. Genomic location of targets is given by the chromosome and location, as well as its orientation to the gene (refgene group). If a gene is annotated to the target cytosine guanine dinucleotide (CpG), the refgene name is presented. The difference in DNA-methylation for each target is shown as $\Delta\beta$-value and the corresponding DiffScore given. The DiffScore is a transformation of the p-value and indicates differences between both time points. A DiffScore of $\geq |13|$ represents a $p \leq 0.05$. Presented is the false discovery rate corrected DiffScore.

Target ID	Chromosome	Location	Refgene Group	Refgene Name	$\Delta\beta$-Value	DiffScore
cg03347334	18	55829444	Body	NEDD4L	−0.16	−60.15
cg05476733	11	128477400			−0.13	−53.12
cg23944105	11	30602030	1stExon	MPPED2	−0.13	−42.75
cg18139362	3	48344301	TSS1500	NME6	−0.14	−41.93
cg21899777	22	46771084	Body	CELSR1	−0.10	−37.09
cg19360943	12	6762431	Body	ING4	−0.11	−36.96
cg22942704	1	20813574	TSS1500	CAMK2N1	−0.15	−35.73
cg13565400	11	73882059	1stExon	C2CD3; PPME1	−0.10	−23.19
cg02295170	6	130718139	5'UTR	TMEM200A	−0.09	−22.73
cg24226193	7	28191663	Body	JAZF1	−0.11	−22.50
cg05119374	6	32399399			−0.09	−20.57
cg27114965	3	57614340	Body	DENND6A	−0.10	−19.06
cg00268500	2	64067540	TSS1500	UGP2	−0.09	−17.98
cg20481542	2	166060635	TSS200	SCN3A	−0.08	−16.45
cg15729230	1	172628514	1stExon	FASLG	−0.10	−15.97
cg03997458	10	125207543			−0.08	−15.61
cg01379853	19	6239836	Body	MLLT1	−0.10	−13.46
cg21895314	21	44593708			−0.10	−13.28
cg20339715	8	27757965	Body	SCARA5	−0.11	−13.20
cg00835758	14	35550189	TSS200	LCC101927178; FAM177A1	0.13	15.61
cg21646955	6	35108921	Body	TCP11	0.08	16.18
cg25540806	4	90815778	TSS1500	MMRN1	0.11	16.61
cg22758714	4	190942739			0.12	17.75
cg07675898	11	41681482			0.08	19.76
cg06716138	8	124857543			0.12	20.77
cg11066566	3	17783373	TSS1500	TBC1D5	0.10	22.87
cg14678442	17	54672540	1stExon	NOG	0.10	23.97
cg17395184	15	42750462	TSS1500	ZFP106	0.13	29.99
cg02777649	1	31474920	Body	PUM1	0.13	31.01
cg20387404	5	34008215	5'UTR	AMACR	0.09	37.09
cg03687640	2	6647183			0.18	51.10
cg02270786	1	45474858	Body	HECTD3	0.11	61.64
cg06095510	17	56764569	5'UTR	TEX14	0.17	314.97

Table 3. Functional Annotation. The functional annotation was performed using the DAVID software. Category gives the database used by the DAVID software. Term gives the functional annotation.

Category	Term	Genes
UP_SEQ_FEATURE	Mutagenesis site	*FASLG, NEDD4L, PUM1, SCN3A, UGP2*
UP_KEYWORDS	Ubl conjugation	*FASLG, NEDD4L, SCN3A, ZNF106*
GOTERM_CC_DIRECT	GO:0005886~plasma membrane	*CELSR1, FASLG, NEDD4L, SCN3A*
UP_KEYWORDS	Metal-binding	*ING4, JAZF1, MPPED2, NME6, UGP2, ZNF106*
UP_KEYWORDS	Zinc-finger	*ING4, JAZF1, ZNF106*
UP_KEYWORDS	Zinc	*ING4, JAZF1, ZNF106*
UP_KEYWORDS	Nucleus	*FASLG, ING4, JAZF1, MLLT1, ZNF106*
UP_SEQ_FEATURE	Glycosylation site:N-linked (GlcNAc …)	*CELSR1, FASLG, MMRN1, NOG, SCARA5, SCN3A, TMEM200A*
UP_KEYWORDS	Glycoprotein	*CELSR1, FASLG, MMRN1, NOG, SCARA5, SCN3A, TMEM200A*
UP_SEQ_FEATURE	Disulfide bond	*CELSR1, MMRN1, NOG, FASLG, SCARA5*
UP_KEYWORDS	Cell membrane	*CAMK2N1, CELSR1, FASLG, SCARA5, SCN3A*
GOTERM_CC_DIRECT	GO:0005887~integral component of plasma membrane	*CELSR1, FASLG, SCARA5*
UP_KEYWORDS	Disulfide bond	*CELSR1, FASLG, MMRN1, NOG, SCARA5*
GOTERM_CC_DIRECT	GO:0005576~extracellular region	*FASLG, MMRN1, NOG*
UP_SEQ_FEATURE	Topological domain:Extracellular	*CELSR1, FASLG, SCARA5, TMEM200A*
UP_KEYWORDS	Secreted	*FASLG, MMRN1, NOG*
UP_SEQ_FEATURE	Transmembrane region	*CELSR1, FASLG, SCARA5, SCN3A, TCP11, TMEM200A*
UP_SEQ_FEATURE	Signal peptide	*AMACR, CELSR1, MMRN1, NOG*
UP_SEQ_FEATURE	Topological domain:Cytoplasmic	*CELSR1, FASLG, SCARA5, TMEM200A*
UP_KEYWORDS	Transmembrane helix	*CELSR1, FASLG, SCARA5, SCN3A, TCP11, TMEM200A*
UP_KEYWORDS	Membrane	*CAMK2N1, CELSR1, FASLG, SCARA5, SCN3A, TBC1D5, TCP11, TMEM200A*
UP_KEYWORDS	Transmembrane	*CELSR1, FASLG, SCARA5, SCN3A, TCP11, TMEM200A*
GOTERM_CC_DIRECT	GO:0005886~plasma membrane	*CELSR1, FASLG, NEDD4L, SCN3A*
GOTERM_CC_DIRECT	GO:0016021~integral component of membrane	*CELSR1, FASLG, SCN3A, TCP11, TMEM200A*
UP_KEYWORDS	Signal	*CELSR1, MMRN1, NOG*
GOTERM_CC_DIRECT	GO:0005634~nucleus	*FASLG, HECTD3, ING4, JAZF1, MLLT1, NEDD4L, UGP2*
UP_KEYWORDS	Transcription regulation	*FASLG, JAZF1, MLLT1*
UP_KEYWORDS	Transcription	*FASLG, JAZF1, MLLT1*
UP_KEYWORDS	Nucleus	*FASLG, ING4, JAZF1, MLLT1, ZNF106*

4. Discussion

4.1. Genome-Wide DNA-Methylation

Investigating the effect of exercise interventions on DNA-methylation in specific immune cells is a novel research field and literature in this context is limited to studies analyzing PBMCs [7,11–13] or whole blood samples [20]. To our knowledge, this is the first study investigating the influence of an acute bout of endurance exercise on genome-wide DNA-methylation on a single cytosine guanine dinucleotide (CpG) level in isolated NK-cells. The results reveal 33 targets being differentially methylated after exercise, with 25 genes corresponding to these targets. Overall, 14 targets show increased DNA-methylation, whereas 19 targets show a decreased methylation.

Our results reveal a fast adaption of NK-cells to an acute bout of endurance exercise by changes of DNA-methylation at 33 targets. So far, DNA-methylation was thought to be a more stable epigenetic modification when compared to histone modifications or miRNAs. Nevertheless, our results indicate a more dynamic function of DNA-methylation in response to external stimuli, e.g., exercise, and this hypothesis is supported by recent research. It was shown that a single bout of exercise changes promoter DNA-methylation in muscle [21] and adipose tissue [22]. In view of blood cells, Robson-Ansley et al. [12] showed keener changes in DNA-methylation immediately after a single bout of exercise compared to a 24 h post-exercise assessment. The authors described a correlation of changes in DNA-methylation of the influenced genes with interleukin 6 (IL-6) levels, indicating changes of cytokine levels as a mediator of changes in DNA-methylation. Horsburgh et al. [13] found a decrease of nuclear concentrations of DNA methyltransferases (DNMTs) of PBMCs after in vitro cultivation with exercise preconditioned plasma. Furthermore, in a recent study we showed that acute exercise resulted in promoter demethylation of the activating NK-cell receptor KIR2DS4 [16]. In the same study also the inhibiting KIR3DL1 was investigated, showing no changes after acute exercise. Interestingly, the presented microarray attempt did not show changes at the *KIR2DS4*-gene. Comparing the *KIR2DS4* and *KIR3DL1* genes in both attempts, we found 6 CpGs (4 in the *KIR2DS4* gene and 2 in the *KIR3DL1* gene), which were analyzed in both attempts (TDBS and EPIC). All 6 commonly analyzed CpGs remained unaltered when comparing both methods and the ones that were changed in the previous study [16] were unfortunately not analyzed within the microarray. It is emphasized that a combination of methods is needed to get detailed information about changes in DNA-methylation.

Analysis of functional annotation reveals *FASLG* with the most annotations. *FASLG* codes for a ligand of the FAS-receptor and induces apoptosis upon binding, one mechanism of NK-cell cytotoxicity. Our results reveal a slightly reduced DNA-methylation of *FASLG* in one CpG within the first exon after acute exercise. As shown by Mooren et al. [23], gene expression of FASLG in leukocytes is not changed immediately after exercise, but it is increased 3 h later. Whether or not there is an association between changes in DNA-methylation and gene expression remains unclear and should be investigated in following studies.

4.2. Distribution of NK-Cell Subsets

The NK-cell subsets CD56^{bright} and CD56^{dim} reveal different functional properties [24] and may also differ in their epigenetic imprinting. It is well described that the NK-cell numbers within the blood stream rapidly increase during exercise and decrease below pre-exercise levels (reviewed by [24]). Moreover, exercise redeploys preferentially CD56^{dim} NK-cells [5], possibly leading to a change in subset proportions and therefore changes in the epigenetic pattern of the entire NK-cell pool. In contrast, we found no statistically significant changes in the NK-cell populations before and directly after exercise. This could be due to the short duration of the exercise bout. While the exercise bout in our study lasted about 15 min, Campbell and Turner [25] describe exercise bouts of 45–60 min duration being characterized by changes in blood lymphocyte numbers. Therefore, it is hypothesized that the short duration of this intensive exercise bout explains the absence of explicit changes in NK-cell subset distribution. Fortunately, this data also supports the finding that changes in DNA-methylation are not driven by changes in NK-cell subsets. Even if the short intensive bout of exercise was not sufficient to induce changes in NK-cell subset distribution, it was sufficient to induce changes in DNA-methylation. It is suggested that a longer bout would induce more pronounced changes, so future studies should include analysis of a single cell population and use longer bouts of exercise.

4.3. Limitations

The results of the presented study should be considered within the context of its limitations. Chip based microarrays for DNA-methylation analyses are a state-of-the-art method and allow the quantification of DNA-methylation at a vast number of single CpGs in parallel. Especially, the Illumina MethylationEPIC BeadChip is described as a robust and reliable platform [26]. Nevertheless, an a priori determined selection of CpGs may also discriminate and possible changes remain undetected. The use of whole genome bisulfate sequencing may provide detailed information of DNA-methylation based gene regulation by measuring all CpGs and all genes. A major limitation is the small sample size of five participants. Nevertheless, it is the first study investigating genome-wide DNA-methylation with a microarray design in isolated NK-cells in the context of physical exercise. In contrast to previous studies investigating cell conglomerates like PBMCs, this attempt produced more precise results for the regulation of a specific cell type. Since NK-cells are further divided into subgroups, analyses of sorted NK-cell subpopulations (CD56^{dim} and CD56^{bright}) provided even more detailed information of differential regulation of the subgroups. However, the amount of DNA isolated from NK-cell subpopulations would be a bottle neck for those analyses. Moreover, the timing of blood sampling after exercise affected the composition of circulating leukocytes and was standardized precisely. By using venipuncture, each blood sampling could be subject to a minor delay and therefore may have influenced the results. Since a change in NK-cell subset proportions was absent, exercise bouts of longer durations should be used to increase the influence of exercise on the NK-cell pool. Furthermore, more measurement time points could give information about the kinetics of exercise induced changes of DNA-methylation and should be considered in following studies. Gene expression of affected genes is not measured. Consecutive studies are warranted, analyzing whether changes in DNA-methylation correspond to changes in gene expression of the affected genes.

5. Conclusions

In conclusion, this is the first study showing effects of a single bout of endurance exercise on genome-wide DNA-methylation in isolated NK-cells. Acute exercise affects the functional genome of NK-cells by changes in DNA-methylation. In order to investigate such changes, it is proposed to use specific cell types rather than conglomerates like PBMCs. More research is needed to combine changes in the functional genome of NK-cells with changes in NK-cell function.

Author Contributions: A.S., P.Z., R.S. and W.B. designed the research. A.S. and P.Z. prepared the analysis and wrote this manuscript. C.K. and C.J.B. performed the exercise intervention and blood sampling. R.S. was responsible for recruitment of participants. M.R.S. performed the methylation analyses. All authors gave editorial support and approved the final version of the manuscript.

Funding: This research received no external funding.

Conflicts of Interest: The authors declare no conflict of interest.

References

1. Cerwenka, A.; Lanier, L.L. Natural killer cells, viruses and cancer. *Nat. Rev. Immunol.* **2001**, *1*, 41–49. [CrossRef] [PubMed]
2. Gleeson, M.; Bosch, J. The human immune system: Innate immunity. The role of natural killer cells in innate immunity. In *Exercise Immunology*; Gleeson, M., Bishop, N., Walsh, N., Eds.; Routledge: Abingdon, UK, 2013; pp. 34–35, ISBN 978-0-415-50726-4.

3. Zimmer, P.; Schenk, A.; Kieven, M.; Holthaus, M.; Lehmann, J.; Lövenich, L.; Bloch, W. Exercise induced alterations in NK-cell cytotoxicity—Methodological issues and future perspectives. *Exerc. Immunol. Rev.* **2017**, *23*, 66–81. [PubMed]

4. Walsh, N.P.; Gleeson, M.; Shephard, R.J.; Gleeson, M.; Woods, J.A.; Bishop, N.C.; Fleshner, M.; Green, C.; Pedersen, B.K.; Hoffman-Goetz, L.; et al. Position statement. Part one: Immune function and exercise. *Exerc. Immunol. Rev.* **2011**, *17*, 6–63. [PubMed]

5. Bigley, A.B.; Rezvani, K.; Chew, C.; Sekine, T.; Pistillo, M.; Crucian, B.; Bollard, C.M.; Simpson, R.J. Acute exercise preferentially redeploys NK-cells with a highly-differentiated phenotype and augments cytotoxicity against lymphoma and multiple myeloma target cells. *Brain Behav. Immun.* **2014**, *39*, 160–171. [CrossRef] [PubMed]

6. McGee, S.L.; Fairlie, E.; Garnham, A.P.; Hargreaves, M. Exercise-induced histone modifications in human skeletal muscle. *J. Physiol.* **2009**, *587*, 5951–5958. [CrossRef] [PubMed]

7. Nakajima, K.; Takeoka, M.; Mori, M.; Hashimoto, S.; Sakurai, A.; Nose, H.; Higuchi, K.; Itano, N.; Shiohara, M.; Oh, T.; et al. Exercise effects on methylation of ASC gene. *Int. J. Sports Med.* **2010**, *31*, 671–675. [CrossRef] [PubMed]

8. Gomez-Pinilla, F.; Zhuang, Y.; Feng, J.; Ying, Z.; Fan, G. Exercise impacts brain-derived neurotrophic factor plasticity by engaging mechanisms of epigenetic regulation. *Eur. J. Neurosci.* **2011**, *33*, 383–390. [CrossRef]

9. Rönn, T.; Volkov, P.; Davegårdh, C.; Dayeh, T.; Hall, E.; Olsson, A.H.; Nilsson, E.; Tornberg, A.; Dekker Nitert, M.; Eriksson, K.-F.; et al. A six months exercise intervention influences the genome-wide DNA-methylation pattern in human adipose tissue. *PLoS Genet.* **2013**, *9*, e1003572. [CrossRef]

10. Holliday, R. Epigenetics: A historical overview. *Epigenetics* **2006**, *1*, 76–80. [CrossRef]

11. Denham, J.; O'Brien, B.J.; Marques, F.Z.; Charchar, F.J. Changes in the leukocyte methylome and its effect on cardiovascular-related genes after exercise. *J. Appl. Physiol.* **2015**, *118*, 475–488. [CrossRef] [PubMed]

12. Robson-Ansley, P.J.; Saini, A.; Toms, C.; Ansley, L.; Walshe, I.H.; Nimmo, M.A.; Curtin, J.A. Dynamic changes in DNA-methylation status in peripheral blood Mononuclear cells following an acute bout of exercise: Potential impact of exercise-induced elevations in interleukin-6 concentration. *J. Biol. Regul. Homeost. Agents* **2014**, *28*, 407–417.

13. Horsburgh, S.; Todryk, S.; Toms, C.; Moran, C.N.; Ansley, L. Exercise-conditioned plasma attenuates nuclear concentrations of DNA methyltransferase 3B in human peripheral blood mononuclear cells. *Physiol. Rep.* **2015**, *3*. [CrossRef] [PubMed]

14. Zimmer, P.; Bloch, W.; Schenk, A.; Zopf, E.M.; Hildebrandt, U.; Streckmann, F.; Beulertz, J.; Koliamitra, C.; Schollmayer, F.; Baumann, F. Exercise-induced Natural Killer Cell Activation is Driven by Epigenetic Modifications. *Int. J. Sports Med.* **2015**, *36*, 510–515. [CrossRef]

15. Fernández-Sánchez, A.; Baragaño Raneros, A.; Carvajal Palao, R.; Sanz, A.B.; Ortiz, A.; Ortega, F.; Suárez-Álvarez, B.; López-Larrea, C. DNA demethylation and histone H3K9 acetylation determine the active transcription of the NKG2D gene in human CD8+ T and NK cells. *Epigenetics* **2013**, *8*, 66–78. [CrossRef] [PubMed]

16. Schenk, A.; Pulverer, W.; Koliamitra, C.; Bauer, C.J.; Ilic, S.; Heer, R.; Schier, R.; Schick, V.; Böttiger, B.W.; Gerhäuser, C.; et al. Acute Exercise Increases the Expression of KIR2DS4 by Promoter Demethylation in NK Cells. *Int. J. Sports Med.* **2018**, *40*, 62–70. [CrossRef] [PubMed]

17. Huang, D.W.; Sherman, B.T.; Lempicki, R.A. Systematic and integrative analysis of large gene lists using DAVID bioinformatics resources. *Nat. Protoc.* **2009**, *4*, 44–57. [CrossRef] [PubMed]

18. Benjamini, Y.; Hochberg, Y. Controlling the False Discovery Rate: A Practical and Powerful Approach to Multiple Testing. *J. R. Stat. Soc. Ser. B Methodol.* **1995**, *57*, 289–300. [CrossRef]

19. Tremblay, B.L.; Guénard, F.; Rudkowska, I.; Lemieux, S.; Couture, P.; Vohl, M.-C. Epigenetic changes in blood leukocytes following an omega-3 fatty acid supplementation. *Clin. Epigenetics* **2017**, *9*, 43. [CrossRef]

20. King-Himmelreich, T.S.; Schramm, S.; Wolters, M.C.; Schmetzer, J.; Möser, C.V.; Knothe, C.; Resch, E.; Peil, J.; Geisslinger, G.; Niederberger, E. The impact of endurance exercise on global and AMPK gene-specific DNA methylation. *Biochem. Biophys. Res. Commun.* **2016**, *474*, 284–290. [CrossRef]

21. Barrès, R.; Yan, J.; Egan, B.; Treebak, J.T.; Rasmussen, M.; Fritz, T.; Caidahl, K.; Krook, A.; O'Gorman, D.J.; Zierath, J.R. Acute exercise remodels promoter methylation in human skeletal muscle. *Cell Metab.* **2012**, *15*, 405–411. [CrossRef]

22. Fabre, O.; Ingerslev, L.R.; Garde, C.; Donkin, I.; Simar, D.; Barrès, R. Exercise training alters the genomic response to acute exercise in human adipose tissue. *Epigenomics* **2018**, *10*, 1033–1050. [CrossRef] [PubMed]
23. Mooren, F.C.; Lechtermann, A.; Völker, K. Exercise-induced apoptosis of lymphocytes depends on training status. *Med. Sci. Sports Exerc.* **2004**, *36*, 1476–1483. [CrossRef]
24. Timmons, B.W.; Cieslak, T. Human natural killer cell subsets and acute exercise: A brief review. *Exerc. Immunol. Rev.* **2008**, *14*, 8–23.
25. Campbell, J.P.; Turner, J.E. Debunking the Myth of Exercise-Induced Immune Suppression: Redefining the Impact of Exercise on Immunological Health Across the Lifespan. *Front. Immunol.* **2018**, *9*, 648. [CrossRef]
26. Pidsley, R.; Zotenko, E.; Peters, T.J.; Lawrence, M.G.; Risbridger, G.P.; Molloy, P.; van Djik, S.; Muhlhausler, B.; Stirzaker, C.; Clark, S.J. Critical evaluation of the Illumina MethylationEPIC BeadChip microarray for whole-genome DNA-methylation profiling. *Genome Biol.* **2016**, *17*, 208. [CrossRef] [PubMed]

© 2019 by the authors. Licensee MDPI, Basel, Switzerland. This article is an open access article distributed under the terms and conditions of the Creative Commons Attribution (CC BY) license (http://creativecommons.org/licenses/by/4.0/).

Review

Ketogenic Diet and Microbiota: Friends or Enemies?

Antonio Paoli [1,2,*], Laura Mancin [3], Antonino Bianco [4], Ewan Thomas [4], João Felipe Mota [5] and Fabio Piccini [3]

1   Department of Biomedical Sciences, University of Padova, 35131 Padova, Italy
2   Research Center for High Performance Sport, UCAM, Catholic University of Murcia, 30107 Murcia, Spain
3   Italian Microbiome Project, 35100 Padova, Italy
4   Sport and Exercise Sciences Research Unit, University of Palermo, 90133 Palermo, Italy
5   Clinical and Sports Nutrition Research Laboratory (LABINCE), Federal University of Goiás,
    74690-900 Goiânia, Goiás, Brazil
*   Correspondence: antonio.paoli@unipd.it; Tel.: +39-049-827-5318

Received: 2 June 2019; Accepted: 10 July 2019; Published: 15 July 2019

Abstract: Over the last years, a growing body of evidence suggests that gut microbial communities play a fundamental role in many aspects of human health and diseases. The gut microbiota is a very dynamic entity influenced by environment and nutritional behaviors. Considering the influence of such a microbial community on human health and its multiple mechanisms of action as the production of bioactive compounds, pathogens protection, energy homeostasis, nutrients metabolism and regulation of immunity, establishing the influences of different nutritional approach is of pivotal importance. The very low carbohydrate ketogenic diet is a very popular dietary approach used for different aims: from weight loss to neurological diseases. The aim of this review is to dissect the complex interactions between ketogenic diet and gut microbiota and how this large network may influence human health.

Keywords: gut microbiota; gut microbiome; intestinal microbiome; ketogenic diet; ketogenic diet and fat

---

1. Introduction

1.1. The Human Gut Microbiota and the Microbiome

The human gut microbiota, that means the types of organisms that are present in an environmental habitat, consisting of trillions of microbial cells and thousands of bacterial species [1]. It encompasses ~10^{-13} microorganisms belonging to the three domains of life Bacteria, Archaea and Eukarya and it is involved in several and different functions [2,3]. Microbiome is the collection of the genes and their functions and, due to the new genetic and bioinformatics technologies, the study of the gut microbiome has been radically transformed. The use of the newest platform next generation sequencing (NGS) enables the sequencing of a thousand to million DNA molecules of bacteria in one sequence run (metagenomics) [4] and through this microbial sequencing has been finally possible the understanding of how different microorganisms are present in different tracts of human body [5]. These new omics-technologies allow scientists to discover the role of bacterial genes in human health [6].

Several studies suggest that a mammalian host establishes their core microbiota at birth [7]; the colonization of the gastrointestinal tract by microorganisms, begins within a few hours of birth and concludes around three to four years of age. The nature of the colonic microbiota is driven by several factors such as breast feeding, geographical location, genetics, age and gender [8].

The impact of food (macronutrients) on gut microbiota composition is growing up in interest, especially with respect of specifically dietary fibers. It has been shown that dietary patterns composed by non-refined foods and a high intake of "microbiota accessible carbohydrate" (MACs), are capable to support the growth of specialist microbes producing short chain fatty acids (SCFAs): the prominent energy source for human colonocytes and the signaling key molecules between the gut microbiota and the host [9]. Controversially, the typical pattern of Western diet, high fat-high sugar and low fibers, reduces the production of SCFAs shifting the gastrointestinal microbiota metabolism to the production of detrimental metabolites, favoring the expansion of bacteria associated with chronic inflammation [10].

The composition of the microbiome is influenced by many factors [11] and the stability of the microbiome, reached between two to five years of age, is overlooked by Bacteroidetes, the largest phylum of gram-negative bacteria associated with both beneficial and detrimental effects on health [12,13]. However, the Firmicutes to Bacteroidetes ratio is regarded to be significant for the gut health, the ratio is clearly linked with increasing body mass index (BMI) [14] and the levels of these two dominant bacterial species are known to shift dramatically with aging, especially *Bifidobacterium* and *Lactobacillus* genera [15].

1.2. Bioactive Products

The microorganisms living in our gut influence the host through the production of bioactive metabolites, which are able to regulate many biological pathways involved in immunity and energy production. The bacterial population of the large intestine digests carbohydrates, proteins and lipids left undigested by the small intestine. Indigested substances, named "microbiota accessible carbohydrates" (MACs), are represented by the walls of plant cell, cellulose, hemicelluloses and pectin and resistant starch; these polymers undergo microbial degradation and subsequent fermentation [3]. It is really fascinating that the genome of gut bacteria, different from the human genome, encoded several highly specified enzymes able to digest and ferment complex biomacromolecules by hydrolyzing the glycosidic bonds [16,17].

More important, microorganisms have the ability to produce a great amount of B_{12} and K vitamins, essential for human health, especially for the daily vitamin K intake that is most frequently insufficient [18,19].

The prominent end-products of fermentation in the colon are short chain fatty acids (SCFAs) such as butyrate ($C_4H_7O_2$-) produced especially by Firmicutes, propionate ($C_3H_5O_2$-) by Bacteroidetes and acetate ($C_2H_4O_2$) by anaerobes; they represent the greatest source of energy for intestinal absorptive cells. [20,21].

SCFAs contribute to the regulation of the systemic immune function, to the direct appropriate immune response to pathogen and they influence the resolution of inflammation [22].

Moreover, specific bacteria have their own ability to produce many neuroendocrine hormones and neuroactive compounds involved in key aspect of neurotransmission, thus, microbial endocrinology interconnects the science of microbiology with neurobiology. As a matter of fact, γ amino butyric acid (GABA), the major inhibitory neurotransmitter of mammalian central nervous system [23], has been demonstrated to be produced by strains of *Lactobacilli* and *Bifidobacteria*, more specifically by *Lactobacillus brevis*, *Bifidobacterium dentium*, *Bifidobacterium adolescentis* and *Bifidobacterium infantis* [24,25]. *Lactobacillus rhamnosus* has been demonstrated for its therapeutical potential in modulating the expression of central GABA receptors, mediating depression and anxiety-like behaviors [26].

Furthermore, another important mediator of the gut-brain axis is serotonin (5-hydroxytryptamine 5-HT) that is produced by the enterochromaffin cells of the gastrointestinal tract. It is a metabolite of the amino acid tryptophan and plays a pivotal role in the regulation of several functions such as the mood.

The 95% of serotonin is stored in enterochromaffin cells and enteric neurons, while only the 5% is found in the central nervous system. Kim and colleagues found that germ-free mice have a two-fold decrease of the serotonin blood's level as compared with commonly mice [27].

However, the gut peripheral serotonin is unable to overstep the blood brain barrier; this serotonin acts on lumen, mucosa, circulating platelets and it is grandly implicated in the gut peristalsis and intestinal anti-inflammation [28,29]. Jun Namking and colleagues suggested that the regulation of the peripheral serotonin might be an adequate tool for the treatment of obesity by the increasing of insulin sensitivity [30].

1.3. Interindividual Variability of Microbiota

The variability among people and the adaptability of gut microbiota to substantial changes have permitted the manipulation of various external factors, restoring both the biological functions and richness of microbiota [31]. The fact that human microbial community is strictly influenced by diet, and, a good ecological community is connected with a better health, offers a range of opportunity for improving human's health by changing the microbiota composition through different patterns of diet [32–34].

The availability of a huge variety and combination of nutrients promotes the selective enrichment of microorganisms, but both the quality and quantity of the macronutrients have an effect on the structure and function of the microbiome [35].

It has been demonstrated the high fat–high sugar Western diet negatively impacts gut health [36] and a high fat diet is closely related to inflammation [37], however, several studies [38–40] suggested the necessity to consider the structure and the function of different fatty acids. De Wit and collaborators [41] showed that specific type of fatty acids affect the gut microbiota in different way and, more recently, it has been said that monounsaturated fatty acid's (MUFA's) and polyunsaturated fatty acid's (PUFA's) omega 3 may be the control key of low-grade systemic inflammation, gut inflammation and as well as obesity [39].

For these reasons, specialized and restricted dietary regimens adopted as a treatment of some diseases, such as low FODMAP for the irritable bowel syndrome and ketogenic diet for refractory epilepsy, should be investigated for their influence on human microbiota [40,42]. These patterns, by reducing or excluding certain type of foods, may affect positively or negatively the microbiota composition and its related influence on host physiology [43–45]. That is the case of very low carbohydrate ketogenic diet (VLCKD), a nutritional approach growing up in interest not only for neurological disorders but also for being a "lose-it-quick-plan" [45,46]. VLCKD, by the drastic reduction of the carbohydrate intake, showed an impairment both on the diversity and richness of gut microbiota [47].

1.4. Very Low Carbohydrate Ketogenic Diet (VLCKD)

The very low carbohydrate ketogenic diet (VLCKD) is a dietary protocol that has been used since the 1920 as a treatment for refractory epilepsy [48] and it is currently getting popularity as a potential therapy for obesity and related metabolic disorders [49]. Due to the typical pattern of VLCKD, a hot topic in research and an evolving area of study has been the effect of ketogenic diet on the gut microbiome [50–53].

Ketogenic diet permits a very low carbohydrate consumption (around 5% to 10% of total caloric intake or below 50 g per day), as a mean to enhance ketone production [54].

Originally, VLCKD had been used as a treatment for epileptic patients that failed to respond to anticonvulsant medication [55]. Currently it has become popular for its benefits extended to neurodegenerative diseases, metabolic diseases and obesity [45]. Recently, VLCKD has been demonstrated to be a powerful tool for some neurodegenerative disease such as autism spectrum disorder (ASD), Alzheimer's disease [46], glucose transporter 1 deficiency syndrome [56] and auto immune multiple sclerosis (AIMS) [57]. Given the fact that VLCKD is a highly restricted dietary pattern, nowadays, there has been the necessity of formulating new features of the VLCKD, such as the popular modified Atkins diet (MAD) and low glycemic index diet (LGIT) [58,59].

These new patterns have been demonstrated as a successful tool able to reduce seizure symptoms, as well as they reveal a similar outcome, with lower side effects, while compared to the traditional regimes of VLCKD [60–62]. LGIT, different from the modified Atkins regime, involves avoiding high glycemic carbohydrates to stabilize blood glucose since it has been shown that stable glucose levels are associated with seizure control [63]. Due to the MAD and LGIT people may choose in a more flexible way the meal they want to consume, they do not have to calculate the specific ketogenic ratio but they may focus on ensuring sufficient and appropriate fats, both in quantity and quality.

1.5. Physiology of Ketosis

The very low carbohydrate ketogenic diet (VLCKD) share several pathways that have been found during fasting state [64]. After several days of drastically reduction of carbohydrate intake, at least <20 g/d or 5% of total daily energy intake, the glucose in the body results insufficient for both fat oxidation (oxaloacetate in tricarboxylic acid cycle TCA) and energy required for the central nervous system forcing the organism to use fats as a primary fuel source [65].

However, fat free acids do not provide energy for the brain because they are not capable to overstep the blood brain barrier: This energy is provided by ketone bodies.

Ketone bodies, 3 hydroxybutyrate (3HB), acetate and acetoacetate (AcAc) are produced in the liver through the process of ketogenesis. Ketogenesis takes place especially in the mitochondria of liver cells where fatty acids reach the mitochondria via carnitine palmitoyltransferase and then breaks down into their metabolites, generating acetyl CoA. The enzyme thiolase (or acetyl coenzyme A acetyltransferase) converts two molecules of acetyl-CoA into acetoacetyl-CoA. Acetoacetyl-CoA is then converted to HMG-CoA due to the enzyme HMG-CoA synthase. Lastly, HMG-CoA lyase converts HMG-CoA to acetoacetate, which can be decarboxylated in acetone or, via β-hydroxybutyrate dehydrogenase, transformed in β-hydroxybutyrate.

The less abundant ketone body is acetone while 3HB plays a main role in the human body under low carbon hydrate diet [66].

The global view of how VLCKD may influence the gut's health is shown in Figure 1.

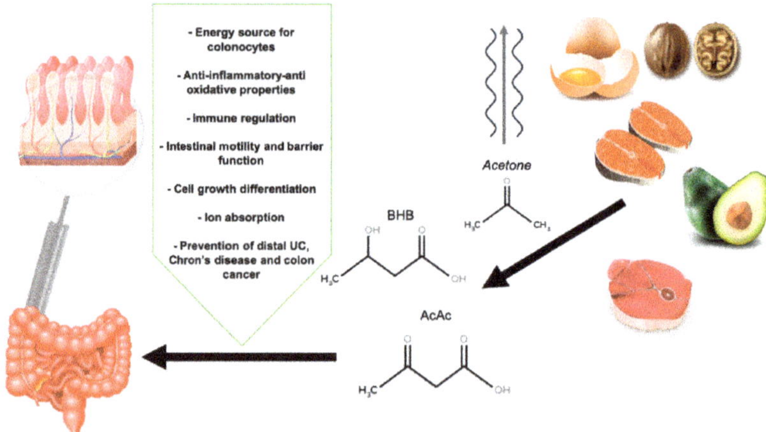

Figure 1. The influence of a very low carbohydrate ketogenic diet and ketone bodies in gut health. BHB: β-hydroxybutyrate, AcAc: Acetoacetate.

2. Methods

We performed a systematic review from February to March 2019; we used the electronic databases Pubmed, (MEDLINE) and Google scholar. We adopted the MeSH term through the function "MeSH Database" within Pubmed. The terms combined with Boolean operators AND, OR, NOT have been "gut microbiota", "gut microbiome", "intestinal microbiome", "ketogenic diet", with "ketogenic", "fat". Eligibility criteria included full-text articles, written in English, available online from 2015 to 2019; specific studies in which authors investigated the effect of the ketogenic diet on gut microbiota and declared no conflict of interest. We decided to include both in vivo and in vitro studies, ranging from randomized controlled trials to case-control and, to emphasize the effects of diet in "fixed" conditions, we included as well animal studies.

3. Results

How VLCKD Affects the Gut Microbiome

As the ketogenic diet seems to gain consensus [63], little is still known about its impact on the gut microbiota.

Only few experimental studies sought the relationship between VLCKD and gut microbiome [47,50,52,53,67–70] investigating how VLCKD impacts composition and characteristics of intestinal microorganisms. The effects of VLCKD on gut microbiome have been explored in mice and humans with mixed results. Our systematic review included nine studies and the major findings have been schematically reported (Table 1).

Table 1. Main findings of the effects of Ketogenic diet (KD) on gut microbiome.

	Subjects	Subjects Characteristics	Duration	Type of KD	Measured KBs (Y/N)	KBs' Level	Genome Analysis Technique	Main Findings of Bacteria Changes
Tagliabue et al. (2017) [50]	6 patients (3 females 3 males) pre-post	Glucose Transporter 1 Deficiency Syndrome	3 months	First 1:1 ratio with gradual increase of 2:1, 3:1 and or 4:1 KD ratio	Ketonuria	Not mentioned	DNA extraction RT-qPCR analysis	INCREASE *Desulfovibrio* spp.
Swidsinki et al. (2017) [52]	25 MS patients and 14 controls	Auto Immune Multiple Sclerosis	6 months	>50 g carbohydrate, >160 g fat, <100 g protein	Ketonemia and ketonuria	β-hydroxybutyric acid ≥ 500 μmol/L; acetoacetate ≥ 500 μmol/L	FISH with ribosomial RNA derived probes	DECREASE β-diversity, DECREASE substantial bacteria groups after two weeks, after six months completely recover the concentration to baseline
Newell et al. (2017) [67]	25 juvenile male C57BL/6 (B6) and 21 BTBR mice	Autism Spectrum Disorder	10–14 days	75% kcal fat	Ketonemia	β-hydroxybutyric acid 5.1 ± 0.8 mmol/L	DNA extraction RT-qPCR analysis	DECREASE in total bacterial content both in cecal and fecal analysis, DECREASE *A. muciniphila* both in cecal and fecal matter, INCREASE Enterobacteriaceae in fecal matter
Burke et al. (2019) [47]	10 LCHF, 10 PCHO, 9 HCHO pre-post	Elite race walkers	3 weeks	78% fat, 2.2 g/kg BM/day protein, <50 g carbohydrate	Ketonemia	β-hydroxybutyric acid ≥ 1.0 mmol/L	16S rRNA-gene amplicon sequencing	INCREASE in *Bacteroides* and *Dorea* spp. DECREASE in *Faecalibacterium* spp.
Lindefeldt et al. (2019) [70]	12 children (parents as controls) pre-post	Therapy-resistant epilepsy	3 months	4:1 in 7 children, 3.5:1 in 2, and 3:1 in 3 KD ratio	Ketonemia	β-hydroxybutyric acid 0.3 ± 0.2 mmol/L	Shotgun metagenomic DNA sequencing	DECREASE in abundance of bifidobacterium, *E. rectale*, *E. dialister*, INCREASE in *E. coli*, changes in 29 SEED subsystem: reduction of seven pathways of carbohydrate metabolism
Olson et al. (2018) [53]	Juvenile SPF wild-type Swiss Webster mice, GF wild type SW mice, SPF C3HeB/FeJ KCNA1 KO mice	6 Hz induced seizure model of refractory epilepsy	3 weeks	6:1 KD ratio	Ketonemia (liver, colon, intestine) and normalized to SPF (specific-pathogen free)	β-hydroxybutyric acid (different levels accepted)	16S rRNA-gene amplicon sequencing	DECREASE in α diversity, INCREASE *A. muciniphila*, Parabacteroides, Suttarella and Erysipelotrichaceae
Zhang et al. (2018) [69]	20 patients (14 males 6 females) pre-post	Refractory epilepsy	6 months	4:1 KD ratio (plant fat 70%, 1 g/kg BM/day from animal source	Ketonemia	β-hydroxybutyric acid 2.85 ± 0.246 and 3.01 ± 0.238 mmol/L (effective and ineffective group)	16S rRNA-gene amplicon sequencing	DECREASE in α diversity, Firmicutes, Actinobacteria, INCREASE in Bacteroidetes
Ma et al. (2017) [51]	C57BL/6 male mice	Healthy mice	4 months	75% fat (saturated, monounsaturated, polyunsaturated), 8.6% protein, 3.2% carbohydrates	Ketonemia	β-hydroxybutyric acid around 1.5 mmol/L	16S rRNA-gene amplicon sequencing	DECREASE in diversity, INCREASE *A. muciniphila*, *Lactobacillus*, DECREASE *Desulfovibrio*, *Turicinabacter*
Xie et al. (2017) [68]	14 patients and 30 healthy infants	Refractory epilepsy	1 week	lipid-to-non-lipid ratio of 4:1 (40% medium chain, 60% long chain), 60–80 kcal/kg per day, 1–1.5 g/kg protein	Not mentioned	Not mentioned	16S rRNA-gene amplicon sequencing	DECREASE Proteobacteria (*Cronobacter*), INCREASE Bacteroidetes (*Bacteroides*, *Prevotella*), *Bifidobacterium*

KD: Ketogenic diet; RT-qPCR: Real-time quantitative polymerase chain reaction; MS: Multiple Sclerosis; FISH: Fluorescent in situ hybridization; rRNA: ribosomial ribonucleic acid; SPF: specific-pathogen-free; SW: Swiss Webster.

Recently, [53] it has been explored the role of VLCKD on gut microbiota related to the anti-seizure effect on mice. In this study, they found that mice, within four days of being on a diet, had significant changes in gut bacterial taxonomy. Two species of bacteria, *Akkermansia* and *Parabacteriodes* were significantly increased in mice that were fed ketogenic diets and gnotobiotic colonization with these microorganisms revealed an anti-seizure effect in germ-free mice or treated with antibiotics.

The increase of these two bacteria species in the gut led to a decreased production of γ-glutamyl transpeptidase by the gut microbiome, the enzyme catalyzes the transfer of γ-glutamyl functional groups from molecules such as glutathione to an acceptor that may be an amino acid forming glutamate [71].

Moreover, they observed a decrease in subset of ketogenic γ-glutatamylated (GG) amino acids (i.e., γ-glutamyl-leucine) both in the gut and blood. GG amino acids are supposed to have transport properties across the blood–brain barrier, different from non-γ-glutamylated forms [72]: This property is involved in glutamate and GABA biosynthesis [73].

This fact, in turn, had the effect of increasing the ratio of GABA to glutamate in the brain of mice. The researchers suggested that VLCKD-microbiota-related limitation in GG amino acids plays a pivotal role on anti-seizure effect, confirmed by the previous studies showing GGT activity to modify the electrical activity of seizure [53].

The ketogenic diet, composed by short fatty acids SFAs, monounsaturated fatty acids MUFAs and polyunsaturated fatty acids PUFAs, has been provided for 16 weeks by Ma and colleagues [51] and it reveals that mice had a variety of neurovascular improvement strictly linked to a lower risk of developing Alzheimer's disease. These beneficial effects may be connected with the changing on gut microbiota composition, more specifically with the growing quota of beneficial bacteria, Akkermansia Muciniphila and Lactobacillus, which have the ability of generating short chain fatty acids SCFAs. Interestingly, they found a reduction in pro inflammatory microbes such as Desulfovibrio and Turicibacter. The VLCKD however, decreased the overall microbial α diversity due to the low carbohydrate (complex carbohydrate) content of diet, which is fundamental for the microorganism in order to breakdown them and producing energy [52].

A 2016 study [67] investigated whether or not a VLCKD could ensure benefits in the gut microbiome in murine model of autism. The authors administrated a VLCKD for several days (10–14) observing changes in gut microbiome; they concluded that the VLCKD had an "anti-microbial" effect by decreasing the overall richness of microorganisms both in cecal and fecal matter, and as well as improved the ratio of Firmicutes to Bacteroides species. A lowered firmicutes: bacteroides ratio is common in ASD and the VLCKD, by improving this ratio, was able to enhance ASD behavioral symptoms. Lastly, different from the above-mentioned studies, the VLCKD decreased the number of *A. muciniphila* bacteria species, resulting in similar levels to those found in the control groups.

It has been described the connection between microbiome, VLCKD and the potential role playing in multiple sclerosis (MS) [52]. A common attribute of the AIMS is the damage and affliction of "colonic bio-fermentative function". The fermentative process which allow the production of beneficial byproducts such as SFCA, is impaired, thus, the dysbiotic colonic bacteria ferment foods into dangerous compounds affecting the organism. The VLCKD completely restored the microbial biofermentative mass and normalizing the concentration of the colonic microbiome. The authors [52] showed a biphasic effect of VLCKD: first there has been a dramatic decrease in richness and bacterial diversity, but, after 12 weeks, bacterial concentration began to recover back to baseline and, after 23 24 weeks, it showed a significant increase in bacterial concentration above baseline.

A study in children by Xie and colleagues [68], investigated the connection between microbiome and refractory epilepsy in 14 epileptic and 30 healthy infants. Patients with epilepsy demonstrated an imbalance of gut microbiota before starting the VLCKD. The authors found a higher amount of pathogenic proteobacteria (Escherichia, Salmonella and Vibrio), which significantly decreased after VLCKD treatment and an increase of Bactrioidetes both in healthy subjects and in patients Bacteroides spp. are strictly connected with the digestion and metabolism of high-fat nutrients and the regulation of the secretion of 6–17 interleukins in dendritic cells, which is connected with the seizure effects on epileptic patients [74]. Researchers suggest that VLCKD can reduce these symptoms by developing changes on microbiota diversity.

Zhang et al. sought the differences in the microbiota of pediatric patients fed a ketogenic diet [69]. They investigated the difference between responders (seizure frequency was reduced or stopped) and non-responders (no effect on seizure). They found increased amount of *Bacteroides* and decreased amounts in *Firmicutes* and *Actinobacteria*, in responders. On the other hand, *Clostridia*, *Ruminococcus* and *Lachnospiraceae* (Firmicutes phylum) increased in non-responders. These data demonstrated that ketogenic diet alters the gut microbiome of pediatric patients, suggesting that the gut microbiome should be taken into account as a biomarker of efficacy of anti-seizure treatment. As regard patients affected by Glucose Transporter 1 Deficiency Syndrome [50], it has been showed a significant increase in *Desulfovibrio* spp. in six patients, after 3 months of intervention. *Desulfovibrio* spp is a genus belonging to a heterogeneous group of sulfate-reducing, motile, anaerobic bacteria related to the inflammatory status of the gut layer mucosa [75]. Authors suggested that in case of dysbiosis, it might be a good option the use of an extra-supplementation with pre or probiotics to maintain the *"ecological balance"* of gut microbiota [50].

Recently, a study in epileptic children found a reduction of *Bifidobacteria*, as well as *E. rectale* and *Dialister*, which are correlated with health promoting benefits such as the prevention of colorectal cancer, IBS and necrotizing entercolitis [76]. Researcher identified a relative abundance of *Actinobacteria* and *Escherichia coli* that may be due to the VLCKD restricted on carbohydrate. It should be stressed that through the analysis of the 29SEED subsystem, scientists revealed a depletion of those pathways responsible of the degradation of carbohydrates [70].

4. Discussion

4.1. Friend or Enemies?

All the papers that have been chosen for depicting the crossing mechanisms, revealed supposed connections between gut microbiome, ketogenic diets and systemic effects. Some findings are demonstrated through "omics" analyses, some are only assumed. As it can be seen, there are several and controversy findings suggesting the necessity of a deeper understanding. The picture (Figure 2) aims to highlight the supposed major effects of ketogenic diet on different tissues and gut microbiota, along with how tissues may be influenced by gut microbiota diversity.

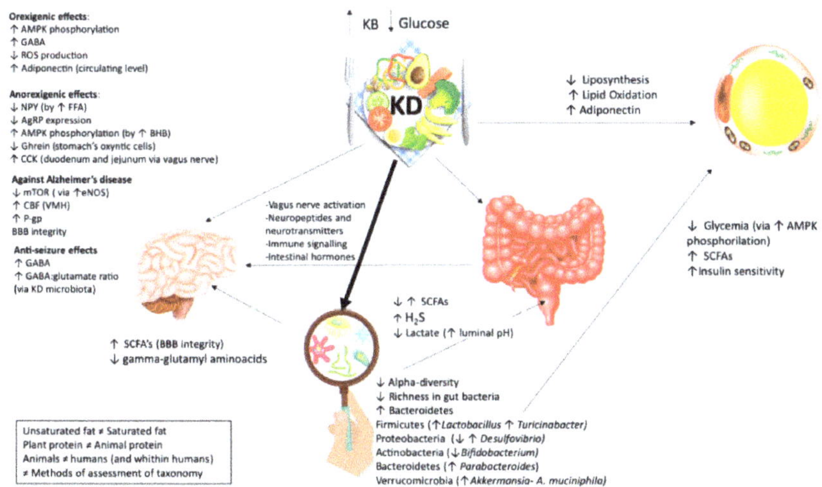

Figure 2. Effects of ketogenic diet on different tissues and the microbiome. KD has a contradictory role on hunger but the net effect is anorexigenic. KD Exerts orexigenic effects: The increase of brain GABA (γ-aminobutyric acid) through BHB (β-hydroxybutyric acid); the increase of AMP (adenosine monophosphate -activated protein) phosphorylation via BHB; the increase of circulating level of adiponectin; the decreases of ROS (reactive oxygen species). KD Exerts anorexigenic effect: the increase of circulating post meal FFA (free fatty acids); a maintained meal's response of CCK (cholecystokinin); a decrease of circulating ghrelin; a decrease of AMP phosphorylation; a decrease of AgRP (agouti-related protein) expression. KD has positive effects on Alzheimer's disease through: an increase levels of CBF (cerebral blood flow) in VMH (ventromedial hypothalamus); a decrease expression of mTOR (mammalian target of rapamycin) by the increase of the level of eNOS (endothelial nitric oxide synthase) protein expression; an increased expression of P-gp (P-glycoprotein), which transport Aβ (amyloid-β) plaques; an improvement of BBB's (blood–brain-barrier) integrity. KD has beneficial effects on epileptic seizure by the modulation of hippocampal GABA/glutamate ratio. It exerts anti-seizure effects through: An increase level of GABA, an increase content of GABA: glutamate ratio. KD plays a main role on fat loss. It exerts positive effects on adipose tissue through: a decrease of liposynthesis, an increase of lipid oxidation and an increase in adiponectin. KD has a contradictory role on microbiome. KD generally exerts its effect through: a decrease in α diversity (the diversity in a single ecosystem/sample) and a decrease in richness (number of different species in a habitat/sample). KD influences the gut health through metabolites produced by different microbes: an increase/decrease in SCFA (short chain fatty acids), an increase in H2S (hydrogen sulfide) and a decrease in lactate. KD to microbiome to the brain: KD may influence the CNS (central nervous system) not only directly but also indirectly. The KD effects on the brain are supposed to be mediated by microbiota through an increase of SCFAs and a decrease of γ-glutamyl amino acid. *A. muciniphila* and *Lactobacillus* are known as SCFAs producers. SCFAs are transported by monocarboxylase transporters expressed at BBB. *Desulfovibrio* has the ability to produce hydrogen sulfide and, as a consequence, impair intestinal mucosal barrier. A reduction in *Desulfovibrio* and an enhancement in *A. muciniphila* and *Lactobacillus* may facilitate BBB and neurovascular amelioration. KD to microbiome to the adipose tissue: KD may indirectly influence the adipose tissue by the microbiota through a decrease in glycemia via adenosine monophosphate-activated protein kinase (AMPK) phosphorylation, an increase in insulin sensitivity and an increase in SCFAs. The great amount of *A. muciniphila* and *Lactobacillus* spp. led to the reduction of body weight and glycemia. It has been demonstrated that patient with type 2 diabetes, treated with metformin, revealed higher level of *A. muciniphila*, may be to the ability of mettormin on decreasing body weight by the activation of AMPK pathways (amp-activated protein kinase). *A. muciniphila* is related with the enhancement of insulin sensitivity and *Lactobacillus* may be playing the same effects through SFCAs production. Several studies showed that *Lactobacillus* is strictly connected with body weight loss.

4.2. Factors Affecting Microbiota during a VLCKD: What Should We Consider?

4.2.1. Fats

The optimal composition of a VLCKD considers both high saturated and mono-polyunsaturated fats [54], whilst the Western diet is rich in saturated-trans fats and poor in mono-polyunsaturated fats [77].

A recent systematic review concluded that diets high in saturated fatty acids led to negative effects on the gut microbiota [78]. The authors observed that diets rich in highly monounsaturated fats affected negatively the gut microbiota decreasing bacteria richness, while diets rich in polyunsaturated fatty acids (with opposite effects when comparing omega 3 vs. omega 6 fats) did not change richness and diversity. However, to notice that only a few studies have been conducted with NGS methods or shotgun sequencing, these new technologies deliver accurate data by avoiding experimental pitfalls and biases created by the "old fashioned" sequencing methods [79]. Recently, a randomized controlled trial study [80] has revealed that a diet with a high content in fat increased *Bacteroides* while reducing the number of butyrate producers (*Faecalibacterium* and *Blautia* bacteria) compared with a middle-lower-fat group. The differences in fecal SCFA could be explained by the high content of carbohydrates in the middle to low-fat diets, made up of resistant starches that have been broken down and fermented. It has to be stressed that the source of fat came from soybean oil, which is highly rich in omega 6 polyunsaturated fatty acids [81]; a higher omega-6: omega-3 long-chain PUFA ratio is associated with many health risks and chronic state of inflammation [82–84]. Another RCT study [85] showed that a supplementation with omega 3 PUFA did not disclose any taxonomic changes both in α and β diversity (at family and genus levels) including short-chain fatty acid producers.

According to these results, different studies demonstrated that each type of fatty acid may induce different effects: The saturated fats (palm oil) induce higher liver triglyceride content in mice, as opposed to polyunsaturated fats (olive oil) [41]. Moreover, genetically modified mice, characterized by the ability of producing omega 3 (PUFAs) and fed with high fat and high sugar diet, showed a higher microbial diversity and a normal gut layer function in the distal intestine, different from non-modified-mice fed with the same macronutrients [86].

The source of fats (omega 6: Omega3, PUFAa and MUFAs) and their own quality should be highly considered when performing a very low carbohydrate dietary plan and as well as when giving general nutritional advices.

4.2.2. Sweeteners

An area of controversy in the ketogenic diet is the consumption of artificial sweeteners replacing natural sugars. Several evidences demonstrated that artificial sweeteners have a negative impact on both host and gut health. Nettleton at al. found that low calorie sweeteners, such as acesulfame potassium (Ace-K) and sucralose, disrupted the structure and function of gut microbiota and gut mucosa [87]. More recently Qiao-Ping Wang investigated, through the use of NGS, the effects of non-nutritive sweeteners (NNSs) on the gut microbioma of mice at the organism level; the study reveals that artificial sweeteners has bacteriostatic effects and as well as change the composition of microbiota [88]. These findings, according to the fact that the routine consumption of NNSs may increase the risk of cardiometabolic diseases [89], suggested that these chemical substitutes may be detrimental for human health and should be avoided [90]. However, recently, the use of stevia (also called *Stevia rebaudiana*) has been widely adopted as a non-nutrient but natural sweeteners. The use of Stevia lowered insulin and glucose level in 19 healthy lean and 12 obese individuals and left them satisfied and full after eating, despite the lower calorie intake [91]. Accordingly, Sharma and colleagues [92] showed a reduction of cholesterol level, triglyceride, low-density lipoprotein (LDL) and an enhancement of high-density lipoprotein (HDL) on 20 hypercholesterolemic women consuming stevia extracts. In a 2008 review, authors suggest that there are not enough information concerning the effect of stevia on gut microbioma [93], whilst others reported a possible link between nonnutritive sweeteners, including stevia, and the disruption of beneficial intestinal flora [94].

Given the fact that there is no explicit data available on gut microbiome, but, The Food and Drug Administration (FDA) considered it as "generally safe" [95], stevia might slightly be used in place of artificial and chemical sweeteners, within coffee, tea or in a unsweetened yogurt. However, further investigation need to be done considering the effect of low calorie sweeteners on gut and human health.

4.2.3. Pre and Probiotics

A proper suggestion for maintaining a healthy gut microbiota during the ketogenic diet may be the use of pre and probiotics: Increasing evidences [96,97] demonstrate their positive benefits. The major source of prebiotics is represented by fructo-oligosaccharides, inulin, lactulose galacto-oligosaccharides and trans-galacto-oligosaccharides [98]. Fermentation of prebiotics by gut microbiota produces SCFAs, which positively modulate the composition of microbiota (by increasing intestinal bifidobacteria and lactic acid bacteria), providing an energy source for colonocytes [99]. Differently, probiotics are living bacteria (especially from the genera *Bifidobacterium* and *Lactobacillus)* and yeasts that, when administrated in an adequate amount, show positive effect on human health; they are usually added to yogurts or found in "specialty food" [100–102]. It has been reported [103,104] that foods enriched with these microorganisms are able to recovery and improve gut microbiota, reaching the state of eubiosis. Cultured-milk products (kefir, Greek yogurt), traditional buttermilk, water kefir, fermented cheese, sauerkraut, kimchi, miso, kombucha and pickles contain several and different "friendly bacteria" such as *Lactobacillus acidophilus, Lactobacillus delbrueckii* subsp. *bulgarius, Lactobacillus reuteri, Saccharomyces boulardii* and *Bifidobacterium bifidum* [105–108].

However, despite the growing interest on fermented foods, there is a lack of epidemiological studies [104] and the majority have focused only on yogurt and cultured dairy foods [109,110]. The paucity arises from the difficulty of understanding if health benefits come from the fermentation operated by microbes or other bioactive compounds. As regard the usefulness of fermented foods during a VLCKD, they represent an excellent and palatable source of dietary fiber and essential micronutrients [111], which should be moderately provided during a VLCKD.

In our opinion since foods that have undergone deep fermentation seem to improve the gut microbiome diversity and gut health index [112] adding small portions of fermented foods to the diet may be a useful prebiotic/probiotic supplementation as well as an effective aid to digestion. A caveat should be done: It is mandatory to verify that fermented foods and beverages are able to not modify in a significant manner glycaemia and insulinaemia while maintaining a sufficient ketonemia.

It has been recently shown that parmesan (an Italian hard and dry cheese), contains "friendly bacteria" acting as probiotics and able to colonize and live in the gut of those individuals who daily consume it [113]. Thus, the moderate consumption of a high-fat fermented food is well recommended for human gut and human health.

4.2.4. Proteins

Several considerations have to be done to the different impact of different protein on gut microbiome.

The source and type of protein must be considered, especially in the field of sports, in which the intake of protein within VLCKD is fundamental to maintain lean body mass [114].

Several studies investigated how and how much different kind of protein (plant versus animal) modify microbiome [115–117], showing that, even though high protein diet generally impair gut health (decrease abundance and change composition) [118], several and disparate effects appear on the gut microbiota [119].

Plant-derived protein, such as mung bean protein (as a part of high fat diet), increased Bacteroidetes while decreasing *Firmicutes* as well as pea protein increased strains of Bifidobacterium and lactobacillus [115].

These studies demonstrated that plant-derived protein get better benefits on gut microbiome along with positive effects on the host metabolism.

To note that we did consider that no studies investigated how protein have been processed, such as thermal treatment, and the effect of the processing treatment on microbiome composition.

During a period of VLCKD, we recommend the use of a source of plant protein (veg protein) since these are more beneficial in terms of health gut microbiota.

5. Conclusions, Perspective and Future Research

In the recent years, the interest regarding the benefits of ketogenic diets is growing up and expanding well beyond the seizure control. Ketogenic diet, as well as the more flexible and less restrictive regimens MAD, LGIT is commonly adopted for weight loss in both obese patients and athlete populations. Bacteria taxa, richness and diversity are strictly influenced by ketogenic diet. A few human and animal studies have shown different results demonstrating positive effects on reshaping bacterial architecture and gut biological functions, while others reporting negative effects as a lowered diversity and an increased amount of pro-inflammatory bacteria.

Nevertheless, short period studies and with specific disease conditions have been carried out [50,52,67,68], limiting generalization to the overall population. Additionally, the microbiota of many environments may be highly variable and its plasticity could be dependent on past and specific dietary patterns [120]. In agreement with these considerations, Healey and colleagues concluded that because of the high variability among people of microbiome composition, it is actually difficult to identify how microbiota may change the diversity in relation to a specific dietary pattern [121]. According to different authors [50,70], there is the necessity to find better strategies to maximize the benefit of VLCKD. It may be useful implementing VLCKD with specific pre and probiotics, which has been found to be drastically reduced during VLCKD [50]. Additionally, promising evidence comes from randomized control trials suggesting that quality dietary fats highly affects the gut microbiota composition. Diets with a high fat content and good quality of polyunsaturated fats and plant-derived protein are able to maintain normal gut function [80,86]. In parallel, the abolition of artificial sweeteners [90] should be recommended to avoid negative effects on general health caused by alteration of gut microbiota. It has been suggested that a supplementation with prebiotics, such as inulin, lactulose, fruttooligosaccharides (FOS) and galactooligosaccharides (GOS) that increases Bifidobacteria, may prevent undesired changes in the gut microbiota [122].

Nonetheless, it is essential to point out that the modified microbiota composition, changed by VLCKD, plays a pivotal role on the itself activity of VLCKD [53,67,68]; the changes have been demonstrated to be necessary in order to provide positive effects such as the anti-seizure effect and amelioration of neurovascular function [53,69,70].

Although there are still many questions limiting the practical research on microbiome, several new developments carried on advancement in this field. Integration of omics science with the newest metagenomic methods of microbiota assessment (next generation sequencing, shotgun sequencing 16S rRNA) shall be helpful to define healthy versus unhealthy microbial operational taxonomic units (OTUs). For this purpose, the Italian Microbiome Project (http://progettomicrobiomaitaliano.org) focuses his research on the advantages and disadvantages that may arise from the genes of bacterial origin, by combining bioinformatic tools with algorithms to better link microbiota data to human health outcomes. Recently, it has been developed a machine e-learning algorithm that is able to predict a specific post-prandial glycemic response by analyzing microbiome profiling [123,124].

The observations that a ketogenic diet can modulate and reshape gut microbiota represents a potential and promising future therapeutic approach. VLCKD has been demonstrated to be a powerful tool and needs to be further refined and well formulated considering its impact on gut health. In conclusion, further research with long-term clinical trials has to be performed in order to establish safer and healthier specific dietary interventions for patients.

Take Home Message:

Practical recommendations to preserve gut health during a VLCKD:

- Introduce the use of whey and plant proteins (i.e., pea protein);
- Reduce the intake of animal protein;
- Implement fermented food and beverages (yoghurt, water and milk kefir, kimchi, fermented vegetables);
- Introduce properly prebiotics and specific probiotics (if needed);
- Reduce omega 3 to omega 6 fatty acids ratio (increase omega 3 while decreasing omega 6);
- Introduce an accurate quantity and quality of unsaturated fatty acids;
- Avoid artificial sweeteners (stevia?) and processed foods;
- Test your microbiome if needed (analysis of 16S rRNA to identify biodiversity and richness).

It is mandatory to verify that fermented foods and beverages and proteins should not modify (in a significant manner) glycaemia and insulinaemia while maintaining a sufficient ketonemia.

We need to remember as well as that the modified microbiota composition induced by VLCKD, plays a pivotal role on the itself activity of diet.

Author Contributions: All authors contributed equally to the review.

Funding: This research received no external funding.

Conflicts of Interest: The authors declare no conflict of interest.

References

1. Thursby, E.; Juge, N. Introduction to the human gut microbiota. *Biochem. J.* **2017**, *474*, 1823–1836. [CrossRef] [PubMed]
2. Gill, S.R.; Pop, M.; Deboy, R.T.; Eckburg, P.B.; Turnbaugh, P.J.; Samuel, B.S.; Gordon, J.I.; Relman, D.A.; Fraser-Liggett, C.M.; Nelson, K.E. Metagenomic analysis of the human distal gut microbiome. *Science* **2006**, *312*, 1355–1359. [CrossRef] [PubMed]
3. Backhed, F.; Ley, R.E.; Sonnenburg, J.L.; Peterson, D.A.; Gordon, J.I. Host-bacterial mutualism in the human intestine. *Science* **2005**, *307*, 1915–1920. [CrossRef] [PubMed]
4. Weinstock, G.M. Genomic approaches to studying the human microbiota. *Nature* **2012**, *489*, 250–256. [CrossRef] [PubMed]
5. Koedooder, R.; Mackens, S.; Budding, A.; Fares, D.; Blockeel, C.; Laven, J.; Schoenmakers, S. Identification and evaluation of the microbiome in the female and male reproductive tracts. *Hum. Reprod. Update* **2019**, *25*, 298–325. [CrossRef]
6. Oulas, A.; Pavloudi, C.; Polymenakou, P.; Pavlopoulos, G.A.; Papanikolaou, N.; Kotoulas, G.; Arvanitidis, C.; Iliopoulos, I. Metagenomics: Tools and insights for analyzing next-generation sequencing data derived from biodiversity studies. *Bioinform. Biol. Insights* **2015**, *9*, 75–88. [CrossRef] [PubMed]
7. Group, N.H.W.; Peterson, J.; Garges, S.; Giovanni, M.; McInnes, P.; Wang, L.; Schloss, J.A.; Bonazzi, V.; McEwen, J.E.; Wetterstrand, K.A.; et al. The nih human microbiome project. *Genome Res.* **2009**, *19*, 2317–2323. [CrossRef]
8. Taneja, V. Arthritis susceptibility and the gut microbiome. *Febs Lett.* **2014**, *588*, 4244–4249. [CrossRef]
9. Morrison, D.J.; Preston, T. Formation of short chain fatty acids by the gut microbiota and their impact on human metabolism. *Gut Microbes* **2016**, *7*, 189–200. [CrossRef]
10. Annalisa, N.; Alessio, T.; Claudette, T.D.; Erald, V.; Antonino, d.L.; Nicola, D.D. Gut microbioma population: An indicator really sensible to any change in age, diet, metabolic syndrome, and life-style. *Mediat. Inflamm.* **2014**, *2014*, 901308. [CrossRef]
11. Rodríguez, J.M.; Murphy, K.; Stanton, C.; Ross, R.P.; Kober, O.I.; Juge, N.; Avershina, E.; Rudi, K.; Narbad, A.; Jenmalm, M.C.; et al. The composition of the gut microbiota throughout life, with an emphasis on early life. *Microb. Ecol. Health Dis.* **2015**, *26*, 26050. [CrossRef] [PubMed]

12. Gibiino, G.; Lopetuso, L.R.; Scaldaferri, F.; Rizzatti, G.; Binda, C.; Gasbarrini, A. Exploring bacteroidetes: Metabolic key points and immunological tricks of our gut commensals. *Dig. Liver Dis.* **2018**, *50*, 635–639. [CrossRef] [PubMed]

13. Fouhy, F.; Watkins, C.; Hill, C.J.; O'Shea, C.A.; Nagle, B.; Dempsey, E.M.; O'Toole, P.W.; Ross, R.P.; Ryan, C.A.; Stanton, C. Perinatal factors affect the gut microbiota up to four years after birth. *Nat. Commun.* **2019**, *10*, 1517. [CrossRef] [PubMed]

14. Koliada, A.; Syzenko, G.; Moseiko, V.; Budovska, L.; Puchkov, K.; Perederiy, V.; Gavalko, Y.; Dorofeyev, A.; Romanenko, M.; Tkach, S.; et al. Association between body mass index and firmicutes/bacteroidetes ratio in an adult ukrainian population. *BMC Microbiol.* **2017**, *17*, 120. [CrossRef] [PubMed]

15. Nagpal, R.; Mainali, R.; Ahmadi, S.; Wang, S.; Singh, R.; Kavanagh, K.; Kitzman, D.W.; Kushugulova, A.; Marotta, F.; Yadav, H. Gut microbiome and aging: Physiological and mechanistic insights. *Nutr Healthy Aging* **2018**, *4*, 267–285. [CrossRef] [PubMed]

16. El Kaoutari, A.; Armougom, F.; Gordon, J.I.; Raoult, D.; Henrissat, B. The abundance and variety of carbohydrate-active enzymes in the human gut microbiota. *Nat. Rev. Microbiol.* **2013**, *11*, 497–504. [CrossRef] [PubMed]

17. Sonnenburg, E.D.; Sonnenburg, J.L. Starving our microbial self: The deleterious consequences of a diet deficient in microbiota-accessible carbohydrates. *Cell Metab.* **2014**, *20*, 779–786. [CrossRef] [PubMed]

18. LeBlanc, J.G.; Milani, C.; de Giori, G.S.; Sesma, F.; van Sinderen, D.; Ventura, M. Bacteria as vitamin suppliers to their host: A gut microbiota perspective. *Curr. Opin. Biotechnol.* **2013**, *24*, 160–168. [CrossRef]

19. Vrieze, A.; Holleman, F.; Zoetendal, E.G.; de Vos, W.M.; Hoekstra, J.B.; Nieuwdorp, M. The environment within: How gut microbiota may influence metabolism and body composition. *Diabetologia* **2010**, *53*, 606–613. [CrossRef]

20. Topping, D.L.; Clifton, P.M. Short-chain fatty acids and human colonic function: Roles of resistant starch and nonstarch polysaccharides. *Physiol. Rev.* **2001**, *81*, 1031–1064. [CrossRef]

21. Louis, P.; Flint, H.J. Formation of propionate and butyrate by the human colonic microbiota. *Env. Microbiol.* **2017**, *19*, 29–41. [CrossRef] [PubMed]

22. Lundin, A.; Bok, C.M.; Aronsson, L.; Bjorkholm, B.; Gustafsson, J.A.; Pott, S.; Arulampalam, V.; Hibberd, M.; Rafter, J.; Pettersson, S. Gut flora, toll-like receptors and nuclear receptors: A tripartite communication that tunes innate immunity in large intestine. *Cell. Microbiol.* **2008**, *10*, 1093–1103. [CrossRef] [PubMed]

23. Nemeroff, C.B. The role of gaba in the pathophysiology and treatment of anxiety disorders. *Psychopharmacol. Bull.* **2003**, *37*, 133–146. [PubMed]

24. Cryan, J.F.; Kaupmann, K. Don't worry 'b' happy!: A role for gaba(b) receptors in anxiety and depression. *Trends Pharm. Sci.* **2005**, *26*, 36–43. [CrossRef] [PubMed]

25. Barrett, E.; Ross, R.P.; O'Toole, P.W.; Fitzgerald, G.F.; Stanton, C. γ-aminobutyric acid production by culturable bacteria from the human intestine. *J. Appl. Microbiol.* **2012**, *113*, 411–417. [CrossRef] [PubMed]

26. Bravo, J.A.; Forsythe, P.; Chew, M.V.; Escaravage, E.; Savignac, H.M.; Dinan, T.G.; Bienenstock, J.; Cryan, J.F. Ingestion of lactobacillus strain regulates emotional behavior and central gaba receptor expression in a mouse via the vagus nerve. *Proc. Natl. Acad. Sci. USA* **2011**, *108*, 16050–16055. [CrossRef]

27. Kim, D.Y.; Camilleri, M. Serotonin: A mediator of the brain-gut connection. *Am. J. Gastroenterol.* **2000**, *95*, 2698–2709.

28. Yano, J.M.; Yu, K.; Donaldson, G.P.; Shastri, G.G.; Ann, P.; Ma, L.; Nagler, C.R.; Ismagilov, R.F.; Mazmanian, S.K.; Hsiao, E.Y. Indigenous bacteria from the gut microbiota regulate host serotonin biosynthesis. *Cell* **2015**, *161*, 264–276. [CrossRef]

29. Agus, A.; Planchais, J.; Sokol, H. Gut microbiota regulation of tryptophan metabolism in health and disease. *Cell Host Microbe* **2018**, *23*, 716–724. [CrossRef]

30. Namkung, J.; Kim, H.; Park, S. Peripheral serotonin: A new player in systemic energy homeostasis. *Mol. Cells* **2015**, *38*, 1023–1028.

31. Gentile, C.L.; Weir, T.L. The gut microbiota at the intersection of diet and human health. *Science* **2018**, *362*, 776–780. [CrossRef] [PubMed]

32. David, L.A.; Maurice, C.F.; Carmody, R.N.; Gootenberg, D.B.; Button, J.E.; Wolfe, B.E.; Ling, A.V.; Devlin, A.S.; Varma, Y.; Fischbach, M.A.; et al. Diet rapidly and reproducibly alters the human gut microbiome. *Nature* **2014**, *505*, 559–563. [CrossRef] [PubMed]

33. Klimenko, N.S.; Tyakht, A.V.; Popenko, A.S.; Vasiliev, A.S.; Altukhov, I.A.; Ischenko, D.S.; Shashkova, T.I.; Efimova, D.A.; Nikogosov, D.A.; Osipenko, D.A.; et al. Microbiome responses to an uncontrolled short-term diet intervention in the frame of the citizen science project. *Nutrients* **2018**, *10*, 567. [CrossRef] [PubMed]

34. De Filippo, C.; Cavalieri, D.; Di Paola, M.; Ramazzotti, M.; Poullet, J.B.; Massart, S.; Collini, S.; Pieraccini, G.; Lionetti, P. Impact of diet in shaping gut microbiota revealed by a comparative study in children from europe and rural africa. *Proc. Natl. Acad. Sci. USA* **2010**, *107*, 14691–14696. [CrossRef] [PubMed]

35. Rowland, I.; Gibson, G.; Heinken, A.; Scott, K.; Swann, J.; Thiele, I.; Tuohy, K. Gut microbiota functions: Metabolism of nutrients and other food components. *Eur. J. Nutr.* **2018**, *57*, 1–24. [CrossRef] [PubMed]

36. Zinocker, M.K.; Lindseth, I.A. The western diet-microbiome-host interaction and its role in metabolic disease. *Nutrients* **2018**, *10*, 365. [CrossRef]

37. Duan, Y.; Zeng, L.; Zheng, C.; Song, B.; Li, F.; Kong, X.; Xu, K. Inflammatory links between high fat diets and diseases. *Front. Immunol.* **2018**, *9*, 2649. [CrossRef]

38. Mills, S.; Stanton, C.; Lane, J.A.; Smith, G.J.; Ross, R.P. Precision nutrition and the microbiome, part I: Current state of the science. *Nutrients* **2019**, *11*, 923. [CrossRef]

39. Candido, F.G.; Valente, F.X.; Grzeskowiak, L.M.; Moreira, A.P.B.; Rocha, D.; Alfenas, R.C.G. Impact of dietary fat on gut microbiota and low-grade systemic inflammation: Mechanisms and clinical implications on obesity. *Int. J. Food Sci. Nutr.* **2018**, *69*, 125–143. [CrossRef]

40. Reddel, S.; Putignani, L.; Del Chierico, F. The impact of low-fodmaps, gluten-free, and ketogenic diets on gut microbiota modulation in pathological conditions. *Nutrients* **2019**, *11*, 373. [CrossRef]

41. De Wit, N.; Derrien, M.; Bosch-Vermeulen, H.; Oosterink, E.; Keshtkar, S.; Duval, C.; de Vogel-van den Bosch, J.; Kleerebezem, M.; Muller, M.; van der Meer, R. Saturated fat stimulates obesity and hepatic steatosis and affects gut microbiota composition by an enhanced overflow of dietary fat to the distal intestine. *Am. J. Physiol. Gastrointest. Liver Physiol.* **2012**, *303*, G589–G599. [CrossRef] [PubMed]

42. Varju, P.; Farkas, N.; Hegyi, P.; Garami, A.; Szabo, I.; Illes, A.; Solymar, M.; Vincze, A.; Balasko, M.; Par, G.; et al. Low fermentable oligosaccharides, disaccharides, monosaccharides and polyols (FODMAP) diet improves symptoms in adults suffering from irritable bowel syndrome (IBS) compared to standard IBS diet: A meta-analysis of clinical studies. *PLoS ONE* **2017**, *12*, e0182942. [CrossRef] [PubMed]

43. Wong, W.M. Restriction of fodmap in the management of bloating in irritable bowel syndrome. *Singap. Med. J.* **2016**, *57*, 476–484. [CrossRef] [PubMed]

44. Bascunan, K.A.; Vespa, M.C.; Araya, M. Celiac disease: Understanding the gluten-free diet. *Eur. J. Nutr.* **2017**, *56*, 449–459. [CrossRef] [PubMed]

45. Paoli, A.; Rubini, A.; Volek, J.S.; Grimaldi, K.A. Beyond weight loss: A review of the therapeutic uses of very-low-carbohydrate (ketogenic) diets. *Eur. J. Clin. Nutr.* **2013**, *67*, 789–796. [CrossRef] [PubMed]

46. Stafstrom, C.E.; Rho, J.M. The ketogenic diet as a treatment paradigm for diverse neurological disorders. *Front. Pharm.* **2012**, *3*, 59. [CrossRef] [PubMed]

47. Murtaza, N.; Burke, L.M.; Vlahovich, N.; Charlesson, B.; O'Neill, H.; Ross, M.L.; Campbell, K.L.; Krause, L.; Morrison, M. The effects of dietary pattern during intensified training on stool microbiota of elite race walkers. *Nutrients* **2019**, *11*, 261. [CrossRef] [PubMed]

48. Cooder, H.R. Epilepsy in children: With particular reference to the ketogenic diet. *Cal. West. Med.* **1933**, *39*, 169–173.

49. Perez-Guisado, J. Ketogenic diets: Additional benefits to the weight loss and unfounded secondary effects. *Arch. Lat. Nutr.* **2008**, *58*, 323–329.

50. Tagliabue, A.; Ferraris, C.; Uggeri, F.; Trentani, C.; Bertoli, S.; de Giorgis, V.; Veggiotti, P.; Elli, M. Short-term impact of a classical ketogenic diet on gut microbiota in GLUT1 deficiency syndrome: A 3-month prospective observational study. *Clin. Nutr. Espen* **2017**, *17*, 33–37. [CrossRef]

51. Ma, D.; Wang, A.C.; Parikh, I.; Green, S.J.; Hoffman, J.D.; Chlipala, G.; Murphy, M.P.; Sokola, B.S.; Bauer, B.; Hartz, A.M.S.; et al. Ketogenic diet enhances neurovascular function with altered gut microbiome in young healthy mice. *Sci. Rep.* **2018**, *8*, 6670. [CrossRef] [PubMed]

52. Swidsinski, A.; Dorffel, Y.; Loening-Baucke, V.; Gille, C.; Goktas, O.; Reisshauer, A.; Neuhaus, J.; Weylandt, K.H.; Guschin, A.; Bock, M. Reduced mass and diversity of the colonic microbiome in patients with multiple sclerosis and their improvement with ketogenic diet. *Front. Microbiol.* **2017**, *8*, 1141. [CrossRef] [PubMed]

53. Olson, C.A.; Vuong, H.E.; Yano, J.M.; Liang, Q.Y.; Nusbaum, D.J.; Hsiao, E.Y. The gut microbiota mediates the anti-seizure effects of the ketogenic diet. *Cell* **2018**, *174*, 497. [CrossRef] [PubMed]

54. Veech, R.L. The therapeutic implications of ketone bodies: The effects of ketone bodies in pathological conditions: Ketosis, ketogenic diet, redox states, insulin resistance, and mitochondrial metabolism. *Prostaglandins Leukot. Essent. Fat. Acids* **2004**, *70*, 309–319. [CrossRef] [PubMed]

55. Kwan, P.; Brodie, M.J. Early identification of refractory epilepsy. *N. Engl. J. Med.* **2000**, *342*, 314–319. [CrossRef] [PubMed]

56. Klepper, J. GLUT1 deficiency syndrome in clinical practice. *Epilepsy Res.* **2012**, *100*, 272–277. [CrossRef] [PubMed]

57. Choi, I.Y.; Piccio, L.; Childress, P.; Bollman, B.; Ghosh, A.; Brandhorst, S.; Suarez, J.; Michalsen, A.; Cross, A.H.; Morgan, T.E.; et al. A diet mimicking fasting promotes regeneration and reduces autoimmunity and multiple sclerosis symptoms. *Cell Rep.* **2016**, *15*, 2136–2146. [CrossRef] [PubMed]

58. Liu, Y.M.; Wang, H.S. Medium-chain triglyceride ketogenic diet, an effective treatment for drug-resistant epilepsy and a comparison with other ketogenic diets. *Biomed. J.* **2013**, *36*, 9–15. [CrossRef]

59. Miranda, M.J.; Turner, Z.; Magrath, G. Alternative diets to the classical ketogenic diet—Can we be more liberal? *Epilepsy Res.* **2012**, *100*, 278–285. [CrossRef]

60. Sharma, S.; Sankhyan, N.; Gulati, S.; Agarwala, A. Use of the modified atkins diet for treatment of refractory childhood epilepsy: A randomized controlled trial. *Epilepsia* **2013**, *54*, 481–486. [CrossRef]

61. Kossoff, E.H.; Cervenka, M.C.; Henry, B.J.; Haney, C.A.; Turner, Z. A decade of the modified Atkins diet (2003–2013): Results, insights, and future directions. *Epilepsy Behav.* **2013**, *29*, 437–442. [CrossRef] [PubMed]

62. Muzykewicz, D.A.; Lyczkowski, D.A.; Memon, N.; Conant, K.D.; Pfeifer, H.H.; Thiele, E.A. Efficacy, safety, and tolerability of the low glycemic index treatment in pediatric epilepsy. *Epilepsia* **2009**, *50*, 1118–1126. [CrossRef] [PubMed]

63. Paoli, A.; Canato, M.; Toniolo, L.; Bargossi, A.M.; Neri, M.; Mediati, M.; Alesso, D.; Sanna, G.; Grimaldi, K.A.; Fazzari, A.L.; et al. The ketogenic diet: An underappreciated therapeutic option? *La Clinica Terapeutica* **2011**, *162*, e145–e153. [PubMed]

64. Paoli, A.; Tinsley, G.; Bianco, A.; Moro, T. The influence of meal frequency and timing on health in humans: The role of fasting. *Nutrients* **2019**, *11*, 719. [CrossRef] [PubMed]

65. Fukao, T.; Lopaschuk, G.D.; Mitchell, G.A. Pathways and control of ketone body metabolism: On the fringe of lipid biochemistry. *Prostaglandins Leukot. Essent. Fat. Acids* **2004**, *70*, 243–251. [CrossRef] [PubMed]

66. Dhillon, K.K.; Gupta, S. *Biochemistry, Ketogenesis*; Statpearls: Treasure Island, FL, USA, 2019.

67. Newell, C.; Bomhof, M.R.; Reimer, R.A.; Hittel, D.S.; Rho, J.M.; Shearer, J. Ketogenic diet modifies the gut microbiota in a murine model of autism spectrum disorder. *Mol. Autism* **2016**, *7*, 37. [CrossRef] [PubMed]

68. Xie, G.; Zhou, Q.; Qiu, C.Z.; Dai, W.K.; Wang, H.P.; Li, Y.H.; Liao, J.X.; Lu, X.G.; Lin, S.F.; Ye, J.H.; et al. Ketogenic diet poses a significant effect on imbalanced gut microbiota in infants with refractory epilepsy. *World J. Gastroenterol.* **2017**, *23*, 6164–6171. [CrossRef] [PubMed]

69. Zhang, Y.; Zhou, S.; Zhou, Y.; Yu, L.; Zhang, L.; Wang, Y. Altered gut microbiome composition in children with refractory epilepsy after ketogenic diet. *Epilepsy Res.* **2018**, *145*, 163–168. [CrossRef] [PubMed]

70. Lindefeldt, M.; Eng, A.; Darban, H.; Bjerkner, A.; Zetterstrom, C.K.; Allander, T.; Andersson, B.; Borenstein, E.; Dahlin, M.; Prast-Nielsen, S. The ketogenic diet influences taxonomic and functional composition of the gut microbiota in children with severe epilepsy. *npj Biofilms Microbiomes* **2019**, *5*, 5. [CrossRef] [PubMed]

71. Whitfield, J.B. γ glutamyl transferase. *Crit. Rev. Clin. Lab. Sci.* **2001**, *38*, 263–355. [CrossRef] [PubMed]

72. Pica, A.; Chi, M.C.; Chen, Y.Y.; d'Ischia, M.; Lin, L.L.; Merlino, A. The maturation mechanism of γ-glutamyl transpeptidases: Insights from the crystal structure of a precursor mimic of the enzyme from bacillus licheniformis and from site-directed mutagenesis studies. *Biochim. Biophys. Acta* **2016**, *1864*, 195–203. [CrossRef] [PubMed]

73. Hertz, L. The glutamate-glutamine (gaba) cycle: Importance of late postnatal development and potential reciprocal interactions between biosynthesis and degradation. *Front. Endocrinol. (Lausanne)* **2013**, *4*, 59. [CrossRef]

74. Arumugam, M.; Raes, J.; Pelletier, E.; Le Paslier, D.; Yamada, T.; Mende, D.R.; Fernandes, G.R.; Tap, J.; Bruls, T.; Batto, J.M.; et al. Enterotypes of the human gut microbiome. *Nature* **2011**, *473*, 174–180. [CrossRef] [PubMed]

75. Devkota, S.; Wang, Y.; Musch, M.W.; Leone, V.; Fehlner-Peach, H.; Nadimpalli, A.; Antonopoulos, D.A.; Jabri, B.; Chang, E.B. Dietary-fat-induced taurocholic acid promotes pathobiont expansion and colitis in IL-10$^{-/-}$ mice. *Nature* **2012**, *487*, 104–108. [CrossRef]

76. O'Callaghan, A.; van Sinderen, D. Bifidobacteria and their role as members of the human gut microbiota. *Front. Microbiol.* **2016**, *7*, 925. [CrossRef] [PubMed]

77. Graham, L.C.; Harder, J.M.; Soto, I.; de Vries, W.N.; John, S.W.; Howell, G.R. Chronic consumption of a western diet induces robust glial activation in aging mice and in a mouse model of alzheimer's disease. *Sci. Rep.* **2016**, *6*, 21568. [CrossRef]

78. Wolters, M.; Ahrens, J.; Romani-Perez, M.; Watkins, C.; Sanz, Y.; Benitez-Paez, A.; Stanton, C.; Gunther, K. Dietary fat, the gut microbiota, and metabolic health—A systematic review conducted within the mynewgut project. *Clin. Nutr.* **2018**. [CrossRef]

79. Cao, Y.; Fanning, S.; Proos, S.; Jordan, K.; Srikumar, S. A review on the applications of next generation sequencing technologies as applied to food-related microbiome studies. *Front. Microbiol.* **2017**, *8*, 1829. [CrossRef]

80. Wan, Y.; Wang, F.; Yuan, J.; Li, J.; Jiang, D.; Zhang, J.; Li, H.; Wang, R.; Tang, J.; Huang, T.; et al. Effects of dietary fat on gut microbiota and faecal metabolites, and their relationship with cardiometabolic risk factors: A 6-month randomised controlled-feeding trial. *Gut* **2019**, *68*, 1417–1429. [CrossRef]

81. Isaac, D.M.; Alzaben, A.S.; Mazurak, V.C.; Yap, J.; Wizzard, P.R.; Nation, P.N.; Zhao, Y.Y.; Curtis, J.M.; Sergi, C.; Wales, P.W.; et al. Mixed lipid, fish oil and soybean oil parenteral lipids impact cholestasis, hepatic phytosterol and lipid composition. *J. Pediatr. Gastroenterol. Nutr.* **2019**, *68*, 861–867. [CrossRef]

82. Jang, H.; Park, K. Omega-3 and omega-6 polyunsaturated fatty acids and metabolic syndrome: A systematic review and meta-analysis. *Clin. Nutr.* **2019**. [CrossRef]

83. Noori, N.; Dukkipati, R.; Kovesdy, C.P.; Sim, J.J.; Feroze, U.; Murali, S.B.; Bross, R.; Benner, D.; Kopple, J.D.; Kalantar-Zadeh, K. Dietary omega-3 fatty acid, ratio of omega-6 to omega-3 intake, inflammation, and survival in long-term hemodialysis patients. *Am. J. Kidney Dis.* **2011**, *58*, 248–256. [CrossRef]

84. Lee, H.J.; Han, Y.M.; An, J.M.; Kang, E.A.; Park, Y.J.; Cha, J.Y.; Hahm, K.B. Role of omega-3 polyunsaturated fatty acids in preventing gastrointestinal cancers: Current status and future perspectives. *Expert Rev. Anticancer* **2018**, *18*, 1189–1203. [CrossRef]

85. Watson, H.; Mitra, S.; Croden, F.C.; Taylor, M.; Wood, H.M.; Perry, S.L.; Spencer, J.A.; Quirke, P.; Toogood, G.J.; Lawton, C.L.; et al. A randomised trial of the effect of omega-3 polyunsaturated fatty acid supplements on the human intestinal microbiota. *Gut* **2018**, *67*, 1974–1983. [CrossRef]

86. Bidu, C.; Escoula, Q.; Bellenger, S.; Spor, A.; Galan, M.; Geissler, A.; Bouchot, A.; Dardevet, D.; Morio, B.; Cani, P.D.; et al. The transplantation of ω3 PUFA-altered gut microbiota of FAT-1 mice to wild-type littermates prevents obesity and associated metabolic disorders. *Diabetes* **2018**, *67*, 1512–1523. [CrossRef]

87. Nettleton, J.E.; Reimer, R.A.; Shearer, J. Reshaping the gut microbiota: Impact of low calorie sweeteners and the link to insulin resistance? *Physiol. Behav.* **2016**, *164*, 488–493. [CrossRef]

88. Wang, Q.P.; Browman, D.; Herzog, H.; Neely, G.G. Non-nutritive sweeteners possess a bacteriostatic effect and alter gut microbiota in mice. *PLoS ONE* **2018**, *13*, e0199080. [CrossRef]

89. Azad, M.B.; Abou-Setta, A.M.; Chauhan, B.F.; Rabbani, R.; Lys, J.; Copstein, L.; Mann, A.; Jeyaraman, M.M.; Reid, A.E.; Fiander, M.; et al. Nonnutritive sweeteners and cardiometabolic health: A systematic review and meta-analysis of randomized controlled trials and prospective cohort studies. *CMAJ* **2017**, *189*, E929–E939. [CrossRef]

90. Suez, J.; Korem, T.; Zeevi, D.; Zilberman-Schapira, G.; Thaiss, C.A.; Maza, O.; Israeli, D.; Zmora, N.; Gilad, S.; Weinberger, A.; et al. Artificial sweeteners induce glucose intolerance by altering the gut microbiota. *Nature* **2014**, *514*, 181–186. [CrossRef]

91. Anton, S.D.; Martin, C.K.; Han, H.; Coulon, S.; Cefalu, W.T.; Geiselman, P.; Williamson, D.A. Effects of stevia, aspartame, and sucrose on food intake, satiety, and postprandial glucose and insulin levels. *Appetite* **2010**, *55*, 37–43. [CrossRef]

92. Sharma, N.; Mogra, R.; Upadhyay, B. Effect of stevia extract intervention on lipid profile. *Stud. Ethno-Med.* **2009**, *3*, 137–140. [CrossRef]

93. Samuel, P.; Ayoob, K.T.; Magnuson, B.A.; Wolwer-Rieck, U.; Jeppesen, P.B.; Rogers, P.J.; Rowland, I.; Mathews, R. Stevia leaf to stevia sweetener: Exploring its science, benefits, and future potential. *J. Nutr.* **2018**, *148*, 1186S–1205S. [CrossRef]

94. Pepino, M.Y. Metabolic effects of non-nutritive sweeteners. *Physiol. Behav.* **2015**, *152*, 450–455. [CrossRef]

95. Perrier, J.D.; Mihalov, J.J.; Carlson, S.J. Fda regulatory approach to steviol glycosides. *Food Chem. Toxicol.* **2018**, *122*, 132–142. [CrossRef]

96. Finamore, A.; Roselli, M.; Donini, L.; Brasili, D.E.; Rami, R.; Carnevali, P.; Mistura, L.; Pinto, A.; Giusti, A.; Mengheri, E. Supplementation with Bifidobacterium longum Bar33 and lactobacillus helveticus bar13 mixture improves immunity in elderly humans (over 75 years) and aged mice. *Nutrition* **2019**, *63–64*, 184–192. [CrossRef]

97. Bagheri, S.; Heydari, A.; Alinaghipour, A.; Salami, M. Effect of probiotic supplementation on seizure activity and cognitive performance in PTZ-induced chemical kindling. *Epilepsy Behav.* **2019**, *95*, 43–50. [CrossRef]

98. Davani-Davari, D.; Negahdaripour, M.; Karimzadeh, I.; Seifan, M.; Mohkam, M.; Masoumi, S.J.; Berenjian, A.; Ghasemi, Y. Prebiotics: Definition, types, sources, mechanisms, and clinical applications. *Foods* **2019**, *8*, 92. [CrossRef]

99. Flint, H.J.; Duncan, S.H.; Scott, K.P.; Louis, P. Interactions and competition within the microbial community of the human colon: Links between diet and health. *Env. Microbiol.* **2007**, *9*, 1101–1111. [CrossRef]

100. Quigley, E.M.M. Prebiotics and probiotics in digestive health. *Clin. Gastroenterol. Hepatol.* **2019**, *17*, 333–344. [CrossRef]

101. Nath, A.; Haktanirlar, G.; Varga, A.; Molnar, M.A.; Albert, K.; Galambos, I.; Koris, A.; Vatai, G. Biological activities of lactose-derived prebiotics and symbiotic with probiotics on gastrointestinal system. *Medicina (Kaunas)* **2018**, *54*, 18. [CrossRef]

102. Valdes, A.M.; Walter, J.; Segal, E.; Spector, T.D. Role of the gut microbiota in nutrition and health. *BMJ* **2018**, *361*, k2179. [CrossRef]

103. Marco, M.L.; Heeney, D.; Binda, S.; Cifelli, C.J.; Cotter, P.D.; Foligne, B.; Ganzle, M.; Kort, R.; Pasin, G.; Pihlanto, A.; et al. Health benefits of fermented foods: Microbiota and beyond. *Curr. Opin. Biotechnol.* **2017**, *44*, 94–102. [CrossRef]

104. Borresen, E.C.; Henderson, A.J.; Kumar, A.; Weir, T.L.; Ryan, E.P. Fermented foods: Patented approaches and formulations for nutritional supplementation and health promotion. *Recent Pat. Food Nutr. Agric.* **2012**, *4*, 134–140. [CrossRef]

105. Dong, Y.; Xu, M.; Chen, L.; Bhochhibhoya, A. Probiotic foods and supplements interventions for metabolic syndromes: A systematic review and meta-analysis of recent clinical trials. *Ann. Nutr. Metab.* **2019**, *74*, 224–241. [CrossRef]

106. Sairanen, U.; Piirainen, L.; Grasten, S.; Tompuri, T.; Matto, J.; Saarela, M.; Korpela, R. The effect of probiotic fermented milk and inulin on the functions and microecology of the intestine. *J. Dairy Res.* **2007**, *74*, 367–373. [CrossRef]

107. Yoon, H.; Park, Y.S.; Lee, D.H.; Seo, J.G.; Shin, C.M.; Kim, N. Effect of administering a multi-species probiotic mixture on the changes in fecal microbiota and symptoms of irritable bowel syndrome: A randomized, double-blind, placebo-controlled trial. *J. Clin. Biochem. Nutr.* **2015**, *57*, 129–134. [CrossRef]

108. Yang, S.J.; Lee, J.E.; Lim, S.M.; Kim, Y.J.; Lee, N.K.; Paik, H.D. Antioxidant and immune-enhancing effects of probiotic *Lactobacillus plantarum* 200655 isolated from kimchi. *Food Sci. Biotechnol.* **2019**, *28*, 491–499. [CrossRef]

109. Shiby, V.K.; Mishra, H.N. Fermented milks and milk products as functional foods—A review. *Crit. Rev. Food Sci. Nutr.* **2013**, *53*, 482–496. [CrossRef]

110. Matsumoto, K.; Takada, T.; Shimizu, K.; Moriyama, K.; Kawakami, K.; Hirano, K.; Kajimoto, O.; Nomoto, K. Effects of a probiotic fermented milk beverage containing *Lactobacillus casei* strain shirota on defecation frequency, intestinal microbiota, and the intestinal environment of healthy individuals with soft stools. *J. Biosci. Bioeng.* **2010**, *110*, 547–552. [CrossRef]

111. Bell, V.; Ferrao, J.; Pimentel, L.; Pintado, M.; Fernandes, T. One health, fermented foods, and gut microbiota. *Foods* **2018**, *7*, 195. [CrossRef]

112. Singh, R.K.; Chang, H.W.; Yan, D.; Lee, K.M.; Ucmak, D.; Wong, K.; Abrouk, M.; Farahnik, B.; Nakamura, M.; Zhu, T.H.; et al. Influence of diet on the gut microbiome and implications for human health. *J. Transl. Med.* **2017**, *15*, 73. [CrossRef]

113. Milani, C.; Duranti, S.; Napoli, S.; Alessandri, G.; Mancabelli, L.; Anzalone, R.; Longhi, G.; Viappiani, A.; Mangifesta, M.; Lugli, G.A.; et al. Colonization of the human gut by bovine bacteria present in parmesan cheese. *Nat. Commun* **2019**, *10*, 1286. [CrossRef]

114. Paoli, A.; Bianco, A.; Grimaldi, K.A. The ketogenic diet and sport: A possible marriage? *Exerc. Sport Sci. Rev.* **2015**, *43*, 153–162. [CrossRef]
115. Nakatani, A.; Li, X.; Miyamoto, J.; Igarashi, M.; Watanabe, H.; Sutou, A.; Watanabe, K.; Motoyama, T.; Tachibana, N.; Kohno, M.; et al. Dietary mung bean protein reduces high-fat diet-induced weight gain by modulating host bile acid metabolism in a gut microbiota-dependent manner. *Biochem. Biophys. Res. Commun.* **2018**, *501*, 955–961. [CrossRef]
116. Meddah, A.T.; Yazourh, A.; Desmet, I.; Risbourg, B.; Verstraete, W.; Romond, M.B. The regulatory effects of whey retentate from bifidobacteria fermented milk on the microbiota of the simulator of the human intestinal microbial ecosystem (SHIME). *J. Appl. Microbiol.* **2001**, *91*, 1110–1117. [CrossRef]
117. Swiatecka, D.; Narbad, A.; Ridgway, K.P.; Kostyra, H. The study on the impact of glycated pea proteins on human intestinal bacteria. *Int. J. Food Microbiol.* **2011**, *145*, 267–272.
118. Mu, C.; Yang, Y.; Luo, Z.; Guan, L.; Zhu, W. The colonic microbiome and epithelial transcriptome are altered in rats fed a high-protein diet compared with a normal-protein diet. *J. Nutr.* **2016**, *146*, 474–483. [CrossRef]
119. Zhu, Y.; Lin, X.; Zhao, F.; Shi, X.; Li, H.; Li, Y.; Zhu, W.; Xu, X.; Li, C.; Zhou, G. Erratum: Meat, dairy and plant proteins alter bacterial composition of rat gut bacteria. *Sci. Rep.* **2015**, *5*, 16546. [CrossRef]
120. Griffin, N.W.; Ahern, P.P.; Cheng, J.; Heath, A.C.; Ilkayeva, O.; Newgard, C.B.; Fontana, L.; Gordon, J.I. Prior dietary practices and connections to a human gut microbial metacommunity alter responses to diet interventions. *Cell Host Microbe* **2017**, *21*, 84–96. [CrossRef]
121. Healey, G.R.; Murphy, R.; Brough, L.; Butts, C.A.; Coad, J. Interindividual variability in gut microbiota and host response to dietary interventions. *Nutr. Rev.* **2017**, *75*, 1059–1080. [CrossRef]
122. Turroni, F.; Milani, C.; Duranti, S.; Mancabelli, L.; Mangifesta, M.; Viappiani, A.; Lugli, G.A.; Ferrario, C.; Gioiosa, L.; Ferrarini, A.; et al. Deciphering bifidobacterial-mediated metabolic interactions and their impact on gut microbiota by a multi-omics approach. *ISME J.* **2016**, *10*, 1656–1668. [CrossRef]
123. Integrative HMP (iHMP) Research Network Consortium. The integrative human microbiome project: Dynamic analysis of microbiome host omics profiles during periods of human health and disease. *Cell Host Microbe* **2014**, *16*, 276–289. [CrossRef]
124. Zeevi, D.; Korem, T.; Zmora, N.; Israeli, D.; Rothschild, D.; Weinberger, A.; Ben-Yacov, O.; Lador, D.; Avnit-Sagi, T.; Lotan-Pompan, M.; et al. Personalized nutrition by prediction of glycemic responses. *Cell* **2015**, *163*, 1079–1094. [CrossRef]

© 2019 by the authors. Licensee MDPI, Basel, Switzerland. This article is an open access article distributed under the terms and conditions of the Creative Commons Attribution (CC BY) license (http://creativecommons.org/licenses/by/4.0/).

 *genes*

Review

Physical Activity Might Reduce the Adverse Impacts of the FTO Gene Variant rs3751812 on the Body Mass Index of Adults in Taiwan

Yi-Ching Liaw [1,2], Yung-Po Liaw [3,4,*] and Tsuo-Hung Lan [2,5,*]

1   School of Nutrition and Health Sciences, Taipei Medical University, Taipei 11031, Taiwan;
    b506100085@tmu.edu.tw
2   Institute of Clinical Medicine, National Yang-Ming University, Taipei 11221, Taiwan
3   Department of Public Health and Institute of Public Health, Chung Shan Medical University, No. 110, Sec. 1
    Jianguo N. Road, Taichung 40201, Taiwan
4   Department of Family and Community Medicine, Chung Shan Medical University Hospital,
    Taichung 40201, Taiwan
5   Department of Psychiatry, Taichung Veterans General Hospital, 1650 Taiwan Boulevard Sect. 4,
    Xitun District, Taichung 407, Taiwan
*   Correspondence: Liawyp@csmu.edu.tw (Y.-P.L.); tosafish@gmail.com (T.-H.L.);
    Tel: +8864-2473-0022 (ext. 11838) (Y.-P.L.); +8864-2359-2525 (ext. 3460) (T.-H.L.); Fax: +8864-2324-8179 (Y.-P.L.)

Received: 11 April 2019; Accepted: 6 May 2019; Published: 9 May 2019

Abstract: The fat mass and obesity-associated (*FTO*) gene is a significant genetic contributor to polygenic obesity. We investigated whether physical activity (PA) modulates the effect of *FTO* rs3751812 on body mass index (BMI) among Taiwanese adults. Analytic samples included 10,853 Taiwan biobank participants. Association of the single-nucleotide polymorphism (SNP) with BMI was assessed using linear regression models. Physical activity was defined as any kind of exercise lasting 30 min each session, at least three times a week. Participants with heterozygous (TG) and homozygous (TT) genotypes had higher BMI compared to those with wild-type (GG) genotypes. The β value was $0.381(p < 0.0001)$ for TG individuals and 0.684 ($p = 0.0204$) for TT individuals. There was a significant dose-response effect among carriers of different risk alleles (p trend <0.0001). Active individuals had lower BMI than their inactive counterparts ($\beta = -0.389$, $p < 0.0001$). Among the active individuals, significant associations were found only with the TG genotype ($\beta = 0.360$, $p = 0.0032$). Inactive individuals with TG and TT genotypes had increased levels of BMI compared to those with GG genotypes: Their β values were 0.381 ($p = 0.0021$) and 0.950 ($p = 0.0188$), respectively. There was an interaction between the three genotypes, physical inactivity, and BMI (p trend = 0.0002). Our data indicated that increased BMI owing to genetic susceptibility by *FTO* rs3751812 may be reduced by physical activity.

Keywords: body mass index; Taiwan biobank; obesity; physical exercise

1. Introduction

Obesity is a global public health issue associated with an unhealthy lifestyle and genetic factors. According to the World Health Organization (WHO), more than 1.9 billion adults are either overweight or obese [1]. Based on criteria established by the Department of Health in Taiwan, a body mass index (BMI) of 24–26.9 kg/m^2 indicates overweight while BMI $\geq$ 27 kg/m^2 indicates obesity. Results from the 1993–1996 and 2005–2008 Nutrition and Health Surveys showed an increased prevalence of obesity and overweight among Taiwanese individuals (that is, 33.4–51% in men and 33.5–35.9% in women) [2]. About 17.1% of children and adolescents in the United States were overweight, while 32.2% of adults were obese in 2003–2004 [3]. The etiology of obesity is complex and multifactorial, involving genetic background, hormones, and lifestyle and environmental issues [4].

The fat mass and obesity-associated (*FTO*) gene variants rs1121980, rs17817449, rs8050136, rs9935401, rs3751812, rs9939609, rs9930506, and rs9922708 were previously associated with obesity [5,6]. Several other variants have been associated with obesity risk in different population and age groups [7–11]. The role played by *FTO* and other obesity-related genes in the etiology of obesity has been reported [12]. The mechanisms responsible for the effect of the *FTO* gene on obesity remain unknown [13]. Multiple single-nucleotide polymorphisms (SNPs) in the first intron of the *FTO* gene have been associated with BMI [14]. Because the intron lacks protein-altering variants and the biological pathways involved are not well-known, several studies have provided a candidate mechanism of obesity and *FTO* [15–19].

Physical activity (PA) prevents obesity in many ways. Several studies have associated the *FTO* gene with PA. In one of its variants, rs1477196, the C allele has been associated with a 1.22 increase in BMI among Old Order Amish (OOA) individuals who were engaged in low levels of physical activity [20]. However, only a 0.27 increase was noted in people who were engaged in high levels of physical activity. In the same study, the difference in BMI across rs1861868 genotypes was large in less physically active individuals but was small (although not significant) in the more physically active individuals. In another study, A-allele homozygotes who were physically inactive had a 1.95 increase in BMI compared with T-allele homozygotes [21].

The European Prospective Investigation into Cancer and Nutrition-Norfolk Study also confirmed that physical activity attenuated the effect of rs1121980 on BMI. However, in the physically active group, the risk allele increased BMI by 0.25 per allele. The increase was significantly more pronounced in inactive individuals (0.44 per risk allele) [22]. However, other studies found no interaction between BMI and physical activity [23–25].

Based on previous studies, *FTO* gene effects on obesity may be modified by physical activity [26,27]. In Taiwan, obesity (defined by a BMI at or greater than 27 kg/m^2) is a serious health issue. How physical activity modulates the BMI-increasing influence of *FTO* variants in Asian populations is not well-understood. Therefore, the purpose of this study was to examine the role PA plays in modifying the effect of *FTO* variants on BMI. We hypothesized that increased BMI due to genetic susceptibility by an *FTO* variant (rs3751812) may be attenuated by physical activity.

2. Materials and Methods

2.1. Participants and Measurements

Study data were obtained from Taiwan biobank, a national large-scale data source with genetic and demographic data from Taiwanese individuals between the ages of 30 and 70. Initially, data were collected from 10,853 participants. After excluding those with incomplete information (n = 21), the final enrollment included 10,832 participants. Anthropometric measures included body weight and height (measured in accordance with the standard procedures), as well as BMI (weight/height^2). Other variables included age, sex, physical activity/inactivity, total cholesterol, smoking (defined as never/former and current smoking), alcohol consumption (defined as weekly drinking of at least 150 cc of alcohol continuously for 6 months), vegetarian diet (defined as never/former and current vegetarian), coffee intake (that is, more than three times per week), and tea consumption (more than one time per day). Physical activity was defined as any kind of exercise lasting 30 min each session, at least three times a week. All study participants provided written informed consent, according to protocols approved by the institutional review board. All methods were carried out in accordance with relevant guidelines and regulations. The Institutional Review Board of Chung Shan Medical University approved this study (CS2-16114, 18 October 2016).

2.2. Genotyping

Eight SNPs in the *FTO* gene that have been consistently associated with obesity in European populations were selected. The source of SNPs was the human genome database (accessed at https://www.ncbi.nlm.nih.gov/genome/guide/human/), which contains findings from international research programs, such as the HapMap and 1000 Genomes Projects. Included in our analysis were the following variants: rs1121980, rs17817449, rs8050136, rs9935401, rs3751812, rs9939609, rs9930506, and rs9922708. SNPs were excluded if their minor allele frequencies (MAFs) were less than 0.01, or call rates less than 98%. Also excluded were SNPs that were not in Hardy–Weinberg equilibrium. In the final model, one tagging SNP (rs3751812) was selected for genotyping.

2.3. Statistical Analysis

All analyses were conducted using SAS 9.3 statistical software (SAS Institute, Cary, NC, USA). The X^2 test was used to compare the differences between the three genotypes. Data were expressed as X ± standard error (S.E.) and %. Multicollinearity was measured using the variance inflation factor (VIF). Values that exceeded 10 suggested multicollinearity. Hardy–Weinberg equilibrium (HWE) was tested for each SNP using a 1 degree of freedom χ^2-test. LD and its correlation coefficients (D values) were calculated using the Haploview software. Linear regression models were used to test the association between the tag-SNP and BMI. Adjustments were made for potential confounding variables (age, sex, physical activity, alcohol drinking, smoking, total cholesterol, tea consumption, coffee consumption, and vegetarian diets). $p < 0.05$ was considered statistically significant.

3. Results

Demographic characteristics of participants are shown in Table 1. Among individuals with the *FTO* rs3751812 variant, 168 (1.55%) were homozygous (TT), 2343 (21.63%) were heterozygous (TG), and 8321 (76.82%) were wild-type (GG). The mean BMI was 25.21 ± 0.26 kg/m^2 for TT carriers, 24.58 ± 0.08 kg/m^2 for TG carriers, and 24.27 ± 0.04 kg/m^2 for GG carriers. There were more women with GG and TG genotypes than men (that is 51.91% versus 48.06% for GG and 52.80% versus 47.20% for TG genotype) except for those with the TT genotype (that is 39.88% versus 60.12%). Fasting blood glucose levels differed significantly across genotypes ($p = 0.0398$). However, values were within normal ranges. Other variables did not differ significantly across genotypes. Table 2 shows the demographic and lifestyle variables of study participants based on physical exercise. Among physically active individuals, 66 (1.49%) were homozygous (TT), 975 (22.07%) were heterozygous (TG), and 3376 (76.43%) were wild-type (GG). Likewise, among their inactive counterparts, 102 (1.59%) were homozygous (TT), 1368 (21.33%) were heterozygous (TG), and 4944 (77.08%) were wild-type (GG). There were no significant differences between physically active and inactive participants based on genotype distributions ($p = 0.6154$). Fasting glucose and total cholesterol differed among physically active and inactive individuals though values were within normal ranges. Table 3 shows the association between rs3751812 and BMI. After adjusting for potential confounders, individuals with TG and TT genotypes exhibited higher BMI compared to those with GG genotypes. Their β values were β = 0.381 ($p < 0.0001$) and 0.684 ($p = 0.0204$), respectively. BMI significantly increased with an increase in the number of risk alleles (p trend <0.0001). In addition, active individuals had a lower BMI than their inactive counterparts (β = −0.389, $p < 0.0001$). Table 4 shows the association of analyzed variables with BMI in groups with different genotypes. Decreased BMI was associated with GG (β = −0.368, $p < 0.0001$) and TG (β = −0.414, $p = 0.0175$) carriers who were physically active compared to their physically inactive counterparts. However, the decreased BMI in TT carriers was not significant (β = −1.059, $p = 0.1099$). This may have been due to the small sample size.

Table 5 shows the association between rs3751812 and BMI based on exercise status. Among physically active individuals, the TG genotype was significantly associated with increased BMI (β = 0.360, $p = 0.0032$). However, the effect of the TT genotype on BMI was not significant (β = 0.245,

$p = 0.5606$). Among physically inactive individuals, both TG ($\beta = 0.381$, $p = 0.0021$) and TT ($\beta = 0.95036$, $p = 0.0188$) were significantly associated with increased BMI (p trend = 0.0002). However, no dose-related trend was observed among physically active individuals.

Table 1. Characteristics of study participants according to rs3751812 genotypes.

Parameters	Total	GG (n = 8321)	TG (n = 2343)	TT (n = 168)	*p*-Value
Age (years)	48.68 ± 0.11	48.64 ± 0.12	48.78 ± 0.23	49.13 ± 0.90	0.7433
BMI (kg/m^2)	24.35 ± 0.03	24.27 ± 0.04	24.58 ± 0.08	25.21 ± 0.26	<0.0001
Fasting blood glucose(mg/dl)	96.49 ± 0.20	96.36 ± 0.23	96.69 ± 0.43	100.43 ± 2.18	0.0398
Total cholesterol (mg/dl)	193.76 ± 0.34	193.93 ± 0.39	193.10 ± 0.74	194.58 ± 2.43	0.5763
Sex (n, %)					
Male	5219 (48.09)	4002 (48.06)	1106 (47.20)	101 (60.12)	0.0053
Female	5634 (51.91)	4319 (51.90)	1237 (52.80)	67 (39.88)	
Alcohol intake					
Never/Former	9992 (92.08)	7641 (91.84)	2180 (93.04)	151 (89.88)	0.0930
Current	860 (7.92)	679 (8.16)	163 (6.96)	17 (10.12)	
Smoking					
No	7451 (68.70)	5743 (69.50)	1583 (67.65)	111 (66.07)	0.3299
Yes	3395 (31.30)	2574 (30.95)	757 (32.35)	57 (33.93)	
Physical activity					
No	6426 (59.21)	4944 (59.42)	1368 (58.39)	102 (60.71)	0.6154
Yes	4426 (40.79)	3376 (40.58)	975 (41.61)	66 (39.29)	
Tea consumption					
No 5655 (63.06)		4353 (63.26)	1213 (62.82)	78 (56.52)	0.2591
Yes	3312 (36.94)	2528 (36.74)	718 (37.18)	60 (43.48)	
Coffee consumption					
No	6086 (67.87)	4668 (67.84)	1320 (68.36)	84 (60.87)	0.1910
Yes	2881 (32.13)	2213 (32.16)	611 (31.64)	54 (39.13)	
Vegetarian Diet					
Never/Former	8551 (95.36)	6570 (95.48)	1836 (95.08)	129 (93.48)	0.4335
Current	416 (4.64)	311 (4.52)	95 (4.92)	9 (6.52)	

All variables are presented as mean ± standard error (S.E.) (continuous variables) or numbers (%). BMI: body mass index, GG: wild-type, TG: heterozygous; TT: homozygous.

Table 2. Demographic and lifestyle variables of study participants under stratification based on physical activity.

Parameters	Physically Active (n = 4426)	Physically Inactive (n = 6426)	*p*-Value
rs3751812 (n, %)			
GG	3376 (76.43)	4944 (77.08)	0.6154
TG	975 (22.07)	1368 (21.33)	
TT	66 (1.49)	102 (1.59)	
Age (years)	53.36 ± 0.15	45.45 ± 0.13	<0.0001
BMI (kg/m^2)	24.24 ± 0.05	24.42 ± 0.05	0.0083
Fasting glucose (mg/dl)	97.93 ± 0.32	95.50 ± 0.26	<0.0001
Total cholesterol (mg/dl)	195.68 ± 0.53	192.44 ± 0.45	<0.0001
Sex (n, %)			
Male	2158 (48.76)	3061 (47.63)	0.2500
Female	2268 (51.24)	3365 (52.37)	
Alcohol drinking			
Never/Former	4080 (92.18)	5911 (92.00)	0.7294
Current	346 (7.82)	514 (8.00)	
Smoking			
No	3144 (71.05)	4306 (67.07)	<0.0001
Yes	1281 (28.95)	2114 (32.93)	
Tea consumption			
No	2340 (62.37)	3315 (63.57)	0.2455
Yes	1412 (37.63)	1900 (36.43)	
Coffee consumption			
No	2581 (68.79)	3505 (67.21)	0.1140
Yes	1171 (31.21)	1710 (32.79)	
Vegetarian diet			
Never/Former	3598 (95.90)	4953 (94.98)	0.0111
Current	154 (4.10)	262 (5.02)	

Table 3. Linear regression analysis showing the association between rs3751812 and BMI.

	β	*p*-Value
rs3751812		
GG		-
TG	0.381	<0.0001
TT	0.684	0.0204
	p for trend < 0.0001	
Physical activity	−0.389	<0.0001
Sex	1.384	<0.0001
Age	0.020	<0.0001
Total cholesterol	0.008	<0.0001
Alcohol intake	0.092	0.5267
Smoking	0.501	<0.0001
Tea consumption	0.492	<0.0001
Coffee consumption	0.108	0.1723
Vegetarian diet	−0.343	0.0493

β: beta coefficient.

Table 4. Association of analyzed variables with BMI in groups with different genotypes.

	GG		TG		TT	
	β	*p*-Value	β	*p*-Value	β	*p*-Value
Physical activity	−0.368	<0.0001	−0.414	0.0175	−1.059	0.1099
Sex	1.461	<0.0001	1.196	<0.0001	0.850	0.1984
Age	0.019	<0.0001	0.026	0.0013	0.014	0.6146
Total cholesterol	0.008	<0.0001	0.008	0.0003	0.004	0.6724
Alcohol intake	0.060	0.7148	−0.008	0.9819	2.065	0.0337
Smoking	0.387	0.0003	0.860	<0.0001	0.885	0.2209
Tea consumption	0.488	<0.0001	0.516	0.0031	0.499	0.3737
Coffee consumption	0.212	0.0179	−0.215	0.2260	−0.641	0.2823
Vegetarian diet	−0.516	0.0098	0.166	0.6626	0.401	0.7291

Table 5. Association between rs3751812 and obesity based on physical activity.

	Physical Activity		Physical Inactivity	
	β	*p*-Value	β-	*p*-Value
rs3751812				
GG	-	-	-	-
TG	0.360	0.0032	0.381	0.0021
TT	0.245	0.5606	0.950	0.0188
				p trend = 0.0002
Sex	1.150	<0.0001	1.517	<0.0001
Age	0.006	0.2077	0.030	<0.0001
Total cholesterol	0.003	0.0741	0.011	<0.0001
Alcohol drinking	0.460	0.0241	−0.192	0.3374
Smoking	0.439	0.0012	0.514	<0.0001
Tea consumption	0.565	<0.0001	0.417	0.0001
Coffee consumption	0.270	0.0150	−0.017	0.8746
Vegetarian diet	−0.626	0.0152	−0.224	0.3401

β: beta coefficient.

4. Discussion

In the current study, we found that there was an association between *FTO* SNP (rs3751812) and BMI among Taiwanese adults. Compared with the GG genotype, carriers of the TT genotype had a higher BMI than those with the GT genotype. Similar findings were reported among the Han Chinese, where increased mean values of BMI were seen among GT + TT than GG carriers [28]. In addition, we found that the effect on BMI was in an allele-dose-dependent manner (p trend < 0.0001).

After stratification, we found that physical activity was significantly associated with a decreased BMI. The decrease was significant in carriers of GG and TG genotypes. Carriers of the TT genotype also exhibited decreased BMI even though the effect was not significant. Latinos with TT alleles (considered as carriers of two risk alleles) who were engaged in regular PA exhibited significant reductions in BMI [26].

Efforts have been made to understand the biological mechanisms underlying body weight regulation: SNPs in FTO are believed to be associated with obesity through an effect on RPGRIP1L [29]. Reports from another study indicated that RPGRIP1L may be partly or exclusively responsible for the obesity susceptibility signal at the *FTO* locus [16]. It has also been suggested that the homeobox gene Iroquois homeobox 3 (IRX3) is a functional long-range target of obesity-associated variants within *FTO* [17].

Genetic, environmental, and lifestyle factors affect body mass index. Such variables as tea, coffee, and cholesterol were included in the current study to understand their modulating role on BMI. These variables were selected based on previous associations with obesity. Findings from a previous study showed that the risk of hypercholesterolaemia was modified by BMI in adults aged 25–39 years [30]. Unlike green tea, coffee consumption has been strongly associated with a higher blood cholesterol and BMI [31]. Findings from another study showed that higher coffee drinking attenuated genetic associations with BMI and obesity [32]. Further analysis of our data showed that vegetarian diet was associated with a lower BMI mainly among participants with the GG genotype, as well those that were physically active.

To our knowledge, studies investigating the interactive influence of genes and physical activity have focused on populations other than Asians [26,27,33]. Associations between *FTO* variants with obesity risk (measured by BMI) have not been widely investigated in Taiwan. Therefore, our findings are relevant and may serve as a reference for future studies.

Recent studies have also examined the interactive influence of *FTO* variants and lifestyle factors on obesity risk. Lower levels of BMI have been reported among smokers compared to nonsmokers [34,35]. However, we observed a significant association between smoking and increased BMI (β = 0.501, p < 0.0021). This may be associated with other health risks in individuals engaged in heavy smoking. There was no significant association between alcohol intake and BMI. After our stratified analysis, coffee consumers who were engaged in physical activity exhibited increased BMI. Lower levels were observed among consumers that were inactive, though not significant. In addition, we found that vegetarian diet was associated with a decreased BMI. Contrasting results have been reported. According to a study conducted in North America, the mean BMI was lowest in vegans (23.6 kg/m^2) and was incrementally higher in lacto-ovo vegetarians (25.7 kg/m^2), pesco-vegetarians (26.3 kg/m^2), semi-vegetarians (27.3 kg/m^2), and nonvegetarians (28.8 kg/m^2) [36]. Our data suggested that BMI is associated with physical activity and *FTO* rs3751812 variants in Taiwanese individuals. The strength of the study included the use of a large-scale data source with genetic and demographic information. In addition, we included information on smoking and drinking habits. However, the study is limited in that we did not consider the gene–gene interaction and the effect on BMI.

In conclusion, our study validated the association between an *FTO* variant and BMI in Taiwanese individuals. In addition, individuals with TG and TT genotypes who were physically active had a decreased BMI. These results indicate that physical activity might be necessary to mitigate the deleterious effect of BMI among genetically susceptible Taiwanese individuals.

Author Contributions: Conceptualization, Y.-C.L., Y.-P.L., and T.-H.L.; Formal analysis, Y.-C.L. and Y.-P.L.; Investigation, T.-H.L.; Methodology, Y.-C.L., Y.-P.L., and T.-H.L.; Resources, T.-H.L.; Supervision, T.-H.L.; Writing—original draft, Y.-C.L.; Writing—review & editing, Y.-P.L. and T.-H.L.

Funding: This study was supported by grants from the Ministry of Science and Technology (MOST 105-2627-M-040-002).

Conflicts of Interest: The authors declare no conflict of interest.

References

1. Organization, W.H. Obesity and Overweight: Fact Sheet N 311. 2015. Available online: http://wwwwhoint/mediacentre/factsheets/fs311/en (accessed on 1 March 2018).

2. Pan, W.H.; Wu, H.J.; Yeh, C.J.; Chuang, H.Y.; Yeh, N.H.; Hsieh, Y.T. Diet and Health Trends in Taiwan: Comparison of Two Nutrition and Health Surveys from 1993–1996 and 2005–2008. *Asia Pac. J. Clin. Nutr.* **2011**, *20*, 238–250. [PubMed]

3. Oqden, C.L.; Carroll, M.D.; Curtin, L.R.; McDowell, M.A.; Tabak, C.J.; Fleqal, K.M. Prevalence of Overweight and Obesity in the United States, 1999–2004. *JAMA* **2006**, *295*, 1549–1555. [CrossRef] [PubMed]

4. Chan, R.S.; Woo, J. Prevention of Overweight and Obesity: How Effective is the Current Public Health Approach. *Int. J. Environ. Res. Public Health* **2010**, *7*, 765–783. [CrossRef] [PubMed]

5. Saldaña-Alvarez, Y.; Salas-Martínez, M.G.; García-Ortiz, H.; Luckie-Duque, A.; García-Cárdenas, G.; Vicenteño-Ayala, H.; Cordova, E.J.; Esparza-Aguilar, M.; Contreras-Cubas, C.; Carnevale, A.; et al. Gender-Dependent Association of *FTO* Polymorphisms with Body Mass Index in Mexicans. *PLoS ONE* **2016**, *11*, e0145984. [CrossRef] [PubMed]

6. Consortium, W.T.C.C. Genome-Wide Association Study of 14,000 Cases of Seven Common Diseases and 3000 Shared Controls. *Nature* **2007**, *447*, 661–678.

7. Berulava, T.; Horsthemke, B. The Obesity-Associated SNPs in Intron 1 of the *FTO* Gene Affect Primary Transcript Levels. *Eur. J. Hum. Genet.* **2010**, *18*, 1054–1056. [CrossRef]

8. Hertel, J.K.; Johansson, S.; Raeder, H.; Midthjell, K.; Lyssenko, V.; Groop, L.; Molven, A.; Njølstad, P.R. Genetic Analysis of Recently Identified Type 2 Diabetes Loci in 1,638 Unselected Patients with Type 2 Diabetes and 1,858 Control Participants from a Norwegian Population-Based Cohort (the HUNT Study). *Diabetologia* **2008**, *51*, 971–977. [CrossRef]

9. Hotta, K.; Nakata, Y.; Matsuo, T.; Kamohara, S.; Kotani, K.; Komatsu, R.; Itoh, N.; Mineo, I.; Wada, J.; Masuzaki, H.; et al. Variations in the *FTO* Gene are Associated with Severe Obesity in the Japanese. *J. Hum. Genet.* **2008**, *53*, 546–553. [CrossRef]

10. Villalobos-Comparán, M.; Flores-Dorantes, M.T.; Villarreal-Molina, M.T.; Rodríguez-Cruz, M.; García-Ulloa, A.C.; Robles, L.; Huertas-Vázquez, A.; Saucedo-Villarreal, N.; López-Alarcón, M.; Sánchez-Muñoz, F.; et al. The *FTO* Gene is Associated with Adulthood Obesity in the Mexican Population. *Obesity* **2008**, *16*, 2296–2301. [CrossRef]

11. Liu, G.; Zhu, H.; Lagou, V.; Gutin, B.; Stallmann-Jorgensen, I.S.; Treiber, F.A.; Dong, Y.; Snieder, H. *FTO* Variant Rs9939609 is Associated with Body Mass Index and Waist Circumference, but not with Energy Intake or Physical Activity in European-And African-American Youth. *BMC Med. Genet.* **2010**, *11*, 57. [CrossRef]

12. Speliotes, E.K.; Willer, C.J.; Berndt, S.I.; Monda, K.L.; Thorleifsson, G.; Jackson, A.U.; Lango Allen, H.; Lindgren, C.M.; Luan, J.; Mägi, R.; et al. Association Analyses of 249,796 Individuals Reveal 18 New Loci Associated with Body Mass Index. *Nat. Genet.* **2010**, *42*, 937–948. [CrossRef] [PubMed]

13. Zhao, X.; Yang, Y.; Sun, B.-F.; Zhao, Y.-L.; Yang, Y.-G. FTO and Obesity: Mechanisms of Association. *Curr. Diabetes Rep.* **2014**, *14*, 486. [CrossRef]

14. Yeo, G.S.; O'Rahilly, S. Uncovering the Biology of FTO. *Mol. Metab.* **2012**, *1*, 32–36. [CrossRef]

15. Fischer, J.; Koch, L.; Emmerling, C.; Vierkotten, J.; Peters, T.; Brüning, J.C.; Rüther, U. Inactivation of the *FTO* Gene Protects from Obesity. *Nature* **2009**, *458*, 894–898. [CrossRef]

16. Stratigopoulos, G.; Carli, J.F.M.; O'Day, D.R.; Wang, L.; Leduc, C.A.; Lanzano, P.; Chung, W.K.; Rosenbaum, M.; Egli, D.; Doherty, D.A.; et al. Hypomorphism for *RPGRIP1L*, a Ciliary Gene Vicinal to the *FTO* Locus, Causes Increased Adiposity in Mice. *Cell Metab.* **2014**, *19*, 767–779. [CrossRef]

17. Smemo, S.; Tena, J.J.; Kim, K.-H.; Gamazon, E.R.; Sakabe, N.J.; Gómez-Marín, C.; Aneas, I.; Credidio, F.L.; Sobreira, D.R.; Wasserman, N.F.; et al. Obesity-Associated Variants within *FTO* form Long-Range Functional Connections with IRX3. *Nature* **2014**, *507*, 371–375. [CrossRef]

18. Ragvina, A.; Moroc, E.; Fredmand, D.; Navratilova, P.; Drivenes, Ø.; Engström, P.G.; Alonso, M.E.; de la Calle Mustienes, E.; Gómez Skarmeta, J.L.; Tavares, M.J.; et al. Long-Range Gene Regulation Links Genomic Type 2 Diabetes and Obesity Risk Regions to *HHEX*, *SOX4*, and *IRX3*. *Proc. Natl. Acad. Sci. USA* **2011**, *108*, 775–780. [CrossRef]

19. Jowett, J.B.; Curran, J.E.; Johnson, M.P.; Carless, M.A.; Göring, H.H.H.; Dyer, T.D.; Cole, S.A.; Comuzzie, A.G.; MacCluer, J.W.; Moses, E.K.; et al. Genetic Variation at the *FTO* Locus Influences *RBL2* Gene Expression. *Diabetes* **2010**, *59*, 726–732. [CrossRef]

20. Rampersaud, E.; Mitchell, B.D.; Pollin, T.I.; Fu, M.; Shen, H.; O'Connell, J.R.; Ducharme, J.L.; Hines, S.; Sack, P.; Naglieri, R.; et al. Physical Activity and the Association of Common *FTO* Gene Variants with Body Mass Index and Obesity. *Arch. Intern. Med.* **2008**, *168*, 1791–1797. [CrossRef] [PubMed]

21. Andreasen, C.H.; Stender-Petersen, K.L.; Mogensen, M.S.; Torekov, S.S.; Wegner, L.; Andersen, G.; Nielsen, A.L.; Albrechtsen, A.; Borch-Johnsen, K.; Rasmussen, S.S.; et al. Low Physical Activity Accentuates the Effect of the *FTO* Rs9939609 Polymorphism on Body Fat Accumulation. *Diabetes* **2008**, *57*, 95–101. [CrossRef] [PubMed]

22. Vimaleswaran, K.S.; Li, S.; Zhao, J.H.; Luan, J.A.; Bingham, S.; Khaw, K.-T.; Ekelund, U.; Wareham, N.J.; Loos, R.J. Physical Activity Attenuates the Body Mass Index–Increasing Influence of Genetic Variation in the *FTO* Gene. *Am. J. Clin. Nutr.* **2009**, *90*, 425–428. [CrossRef]

23. Tan, J.T.; Dorajoo, R.; Seielstad, M.; Sim, X.L.; Ong, R.T.-H.; Chia, K.S.; Wong, T.Y.; Saw, S.M.; Chew, S.K.; Aung, T.; et al. *FTO* Variants Are Associated with Obesity in the Chinese and Malay Populations in Singapore. *Diabetes* **2008**, *57*, 2851–2857. [CrossRef]

24. Jónsson, Á.; Renström, F.; Lyssenko, V.; Brito, E.C.; Isomaa, B.; Berglund, G.; Nilsson, P.M.; Groop, L.; Franks, P.W. Assessing the Effect of Interaction between an *FTO* Variant (Rs9939609) and Physical Activity on Obesity in 15,925 Swedish and 2,511 Finnish adults. *Diabetologia* **2009**, *52*, 1334–1338. [CrossRef] [PubMed]

25. Harbron, J.; Van Der Merwe, L.; Zaahl, M.G.; Kotze, M.J.; Senekal, M. Fat Mass and Obesity-Associated (*FTO*) Gene Polymorphisms Are Associated with Physical Activity, Food Intake, Eating Behaviors, Psychological Health, and Modeled Change in Body Mass Index in Overweight/Obese Caucasian Adults. *Nutrients* **2014**, *6*, 3130–3152. [CrossRef] [PubMed]

26. Kim, J.Y.; DeMenna, J.T.; Puppala, S.; Chittoor, G.; Schneider, J.; Duggirala, R.; Mandarino, L.J.; Shaibi, G.Q.; Coletta, D.K. Physical Activity and *FTO* Genotype by Physical Activity Interactive Influences on Obesity. *BMC Genet.* **2016**, *17*, 1549. [CrossRef] [PubMed]

27. Kilpeläinen, T.O.; Qi, L.; Brage, S.; Sharp, S.J.; Sonestedt, E.; Demerath, E.; Ahmad, T.; Mora, S.; Kaakinen, M.; Sandholt, C.H.; et al. Physical Activity Attenuates the Influence of *FTO* Variants on Obesity Risk: A Meta-Analysis of 218,166 Adults and 19,268 Children. *PLoS Med.* **2011**, *8*, e1001116. [CrossRef]

28. Wu, J.; Xu, J.; Zhang, Z.; Ren, J.; Li, Y.; Wang, J.; Cao, Y.; Rong, F.; Zhao, R.; Huang, X.; et al. Association of *FTO* Polymorphisms with Obesity and Metabolic Parameters in Han Chinese Adolescents. *PLoS ONE* **2014**, *9*, e98984. [CrossRef]

29. Fawcett, K.A.; Barroso, I. The Genetics of Obesity: *FTO* Leads the Way. *Trends Genet.* **2010**, *26*, 266–274. [CrossRef]

30. Gostynski, M.; Gutzwiller, F.; Kuulasmaa, K.; Döring, A.; Ferrario, M.; Grafnetter, D.; Pajak, A. Analysis of the Relationship between Total Cholesterol, Age, Body Mass Index among Males and Females in the WHO MONICA Project. *Int. J. Obes.* **2004**, *28*, 1082–1090. [CrossRef] [PubMed]

31. Walli, R.R.; Almosrati, R.A.; Zaied, A.A.; Shummakhi, F.M.E.; Dredae, E.G.; Shalaka, O.K. The Relationship between Habitual Coffee and Tea Consumption and Type 2 Diabetes Mellitus among Libyan Adults. *Int. J. Pharma Res. Rev.* **2015**, *4*, 34–39.

32. Wang, T.; Huang, T.; Kang, J.H.; Zheng, Y.; Jensen, M.K.; Wiggs, J.L.; Pasquale, L.R.; Fuchs, C.S.; Campos, H.; Rimm, E.B.; et al. Habitual Coffee Consumption and Genetic Predisposition to Obesity: Gene-Diet Interaction Analyses in Three US Prospective Studies. *BMC Med.* **2017**, *15*, 97. [CrossRef]

33. Ahmad, S.; Rukh, G.; Varga, T.V.; Ali, A.; Kurbasic, A.; Shungin, D.; Ericson, U.; Koivula, R.W.; Chu, A.Y.; Rose, L.M.; et al. Gene × Physical Activity Interactions in Obesity: Combined Analysis of 111,421 Individuals of European Ancestry. *PLoS Genet.* **2013**, *9*, e1003607. [CrossRef] [PubMed]

34. Plurphanswat, N.; Rodu, B. The Association of Smoking and Demographic Characteristics on Body Mass Index and Obesity among Adults in the U.S., 1999–2012. *BMC Obes.* **2014**, *1*, 18. [CrossRef] [PubMed]
35. Chiolero, A.; Faeh, D.; Paccaud, F.; Cornuz, J. Consequences of Smoking for Body Weight, Body Fat Distribution, and Insulin Resistance. *Am. J. Clin. Nutr.* **2008**, *87*, 801–809. [CrossRef]
36. Tonstad, S.; Butler, T.; Yan, R.; Fraser, G.E. Type of Vegetarian Diet, Body Weight, and Prevalence of Type 2 Diabetes. *Diabetes Care* **2009**, *32*, 791–796. [CrossRef] [PubMed]

© 2019 by the authors. Licensee MDPI, Basel, Switzerland. This article is an open access article distributed under the terms and conditions of the Creative Commons Attribution (CC BY) license (http://creativecommons.org/licenses/by/4.0/).

Article

Association between Aerobic Exercise and High-Density Lipoprotein Cholesterol Levels across Various Ranges of Body Mass Index and Waist-Hip Ratio and the Modulating Role of the Hepatic Lipase rs1800588 Variant

Yasser Nassef [1], Oswald Ndi Nfor [2], Kuan-Jung Lee [2], Ming-Chih Chou [1,*] and Yung-Po Llaw [2,3,*]

1   Institute of Medicine, Chung Shan Medical University, Taichung City 40201, Taiwan; yasser@nassef.org
2   Department of Public Health and Institute of Public Health, Chung Shan Medical University, Taichung City 40201, Taiwan; nforoswald2@yahoo.com (O.N.N.); jasminemachi@gmail.com (K.-J.L.)
3   Department of Family and Community Medicine, Chung Shan Medical University Hospital, Taichung City 40201, Taiwan
*   Correspondence: cs1601@csmu.edu.tw (M.-C.C.); Liawyp@csmu.edu.tw (Y.-P.L.); Tel.: +886-4-2473-0022 (ext. 11191) (M.-C.C.); +886-4-2473-0022 (ext. 11838) (Y.-P.L.)

Received: 29 April 2019; Accepted: 3 June 2019; Published: 10 June 2019

Abstract: Changes in concentrations of high-density lipoprotein cholesterol (HDL-C) are modified by several factors. We examined the relationship between aerobic exercise and HDL-C among different categories of body mass index (BMI) and waist-hip ratio (WHR) and the impact of rs1800588 variant in the hepatic lipase (LIPC) gene. We analyzed data from 6184 men and 8353 women aged 30–70 years. Participants were grouped into two WHR categories: Normal ($0 <$ WHR < 0.9 for men and $0 <$ WHR < 0.8 for women) and abnormal (WHR ≥ 0.9 for men and WHR ≥ 0.8 for women). The BMI categories were: Underweight (BMI < 18.5 kg/m^2), normal weight ($18.5 \leq$ BMI < 24 kg/m^2), overweight ($24 \leq$ BMI < 27 kg/m^2), and obese (BMI ≥ 27 kg/m^2). Multivariate linear regression models were used to investigate associations between HDL-C and exercise. Aerobic exercise was significantly associated with higher HDL-C ($\beta = 1.18325$; $p < 0.0001$) when compared with no exercise. HDL-C was significantly lower in persons with abnormal compared to those with normal WHR ($\beta = -3.06689$; $p < 0.0001$). Compared with normal weight, overweight and obese categories were associated with lower HDL-C, with β values of -4.31095 and -6.44230, respectively ($p < 0.0001$). Unlike rs1800588 CT and TT genotypes, associations between aerobic exercise and HDL were not significant among CC carriers no matter their BMI or WHR.

Keywords: high-density lipoprotein; aerobic exercise; body fat; hepatic lipase; Taiwan Biobank

1. Introduction

Prospective cohort studies have consistently demonstrated that high-density lipoprotein-cholesterol (HDL-C) is a strong predictor of cardiovascular diseases in different populations [1–3]. It has antioxidative, anti-inflammatory, antidiabetic, and anti-thrombotic activities [4,5], and plays an essential role in the management of coronary heart disease (CHD) and risk reduction [6]. A higher level of HDL-C is protective against heart disease. On the other hand, lower levels HDL-C (defined as <40 mg/dL in men and <50 mg/dL in women) [7] are associated with higher risks for heart disease.

Environmental and genetic factors contribute to variations in HDL-C levels. Exercise training is one of the strategies suggested to improve HDL-C function via proprotein convertase subtilisin/kexin type 9 (PCSK9) and/or sterol regulatory element binding protein 2 (SREBP2) [1]. Besides modifying HDL subclass distribution, exercise training has also resulted in a decrease in the body mass index of obese women [4]. Aerobic exercise has been recommended for the prevention of coronary heart disease [1,8], which is a serious health issue in Taiwan. This exercise training also reduces stress, has fewer side effects compared to medications, and is easier to carry out [1]. Of note, many studies have investigated the benefits of exercise on HDL-C. In our recently published study, we found that aerobic exercise was associated with a higher level of HDL-C ($\beta = 1.3154$; $p < 0.0001$) among Taiwanese adults [9]. Findings from a previous study have suggested that compared to other lipid fractions, HDL-C levels are more sensitive to aerobic exercise [1].

Several variants have been associated with HDL-C [10]. The hepatic lipase (LIPC) gene located on chromosome 15 (q21–q23) influences the production of the hepatic lipase enzyme that plays a vital role in lipid metabolism [11]. Based on previous data, genetic variation in hepatic lipase activity is an important determinant of plasma HDL-C concentrations [12]. Rs1800588, a common variant in the LIPC gene has been associated with a higher concentration of HDL-C [13]. Prior work involving female carriers of this variant demonstrated that those carrying at least one copy of the minor allele had higher HDL- levels than those that were homozygous for the major allele [14]. In addition, variations in several genes (including the LIPC gene) are reported to influence interindividual variability in the HDL-C response to exercise [14]. However, the impact of physical exercise on the relationship between hepatic lipase activity and HDL-C levels has not been reported in Taiwan.

Obesity indices including body mass index (BMI) and waist-hip ratio (WHR) increase with increasing categories of abnormal serum lipids [15]. BMI is defined as a person's weight in kilograms divided by the square of height in meters while WHR is a dimensionless ratio that is calculated as the waist circumference divided by the hip circumference. Among the anthropometric measures, WHR has shown good correlations with serum lipids especially among elderly women [16]. On the other hand, negative associations have been found between BMI and HDL-C [17]. Based on previous literature, significant differences have been found between pre and post-test values of HDL-C, BMI, and WHR of individuals who were engaged in 8 weeks of aerobic training [18]. Considering that BMI and WHR are independently associated with HDL-C and that aerobic training has improved both anthropometric variables, we investigated the association between aerobic exercise and HDL-C across different categories of BMI and WHR. Furthermore, we tested whether this association is modified by a selected HDL-C raising variant (LIPC rs1800588).

2. Materials and Methods

2.1. Data Source

Phenotypic and genotypic data were collected from participants (aged 30–70 years) that were enrolled in Taiwan Biobank from 2008–2016. Recruitment methods in the Biobank are in accordance with relevant guidelines and regulations. Written informed consents are obtained from all participants prior to data collection. Data collection was through questionnaires as well as physical and biochemical examinations. The Institutional Review Board of Chung Shan Medical University approved this study (project identification code CS2-16114).

2.2. Study Participants

We analyzed data from 6184 men and 8353 women aged 30–70 years recruited in the Taiwan Biobank project from 2012–2016. Age, sex, BMI, WHR, and lifestyle (physical activity, coffee drinking, smoking, alcohol consumption, and vegetarian diet) measures were determined from the database. Participants were grouped into WHR categories as follows: Normal ($0 < $ WHR $ < 0.9$ for men and $0 < $ WHR $ < 0.8$ for women), and abnormal (WHR ≥ 0.9 for men and WHR ≥ 0.8 for women). Likewise, the BMI categories included the following: Normal weight ($18.5 \leq$ BMI < 24 kg/m^2), overweight ($24 \leq$ BMI < 27 kg/m^2), and obesity (BMI ≥ 27 kg/m^2). Information on aerobic exercise was self-reported.

Using questionnaires in the Biobank, participants selected at most 3 types of their habitual aerobic activities, which included jogging, strolling, swimming, yoga, taijiquan, biking, and aerobic dance. The minimum amount of exercise was 30 min per session, at least 3 times per week, for the last 3 months. "No exercise" was defined as participation in exercise for less than 30 min per day and less than two times per week.

2.3. SNP Selection and Genotyping

We selected the LIPC variant (rs1800588) that has been consistently associated with elevated levels of HDL-C through a literature search. Genotyping was performed using TaqMan SNP Genotyping Assays from Applied Biosystems (ABI; Foster City, CA, USA). We included only participants with call rates greater than 90%. Polymorphic variants with minor allele frequency (MAF) <0.05, as well as those whose genotypes deviated from the Hardy-Weinberg equilibrium (HWE) were excluded.

2.4. Statistical Analysis

Analyses were performed using the SAS 9.4 software (SAS Institute, Cary, NC, USA). Differences in HDL-C among the body fat indicators (BMI and WHR) were compared using the t-test. The association between HDL-C and exercise was determined using multivariate linear regression models. Data were presented as mean ± standard error (SE) for continuous variables. Values of $p < 0.05$ were considered statistically significant.

3. Results

Average levels of HDL-C in study participants were determined among different categories of WHR and BMI as shown in Tables 1 and 2. Among participants who were engaged in aerobic exercise, HDL level was 56.43 ± 0.35 mg/dL in those with normal WHR and 53.89 ± 0.23 mg/dL in those with abnormal WHR ($p < 0.0001$). Mean HDL levels differed significantly among the different categories of BMI ($p < 0.0001$). Individuals who had aerobic exercise had higher HDL-C than those who did not exercise. That is 66.83 ± 1.55 mg/dL vs. 65.32 ± 0.80 mg/dL for underweight; 58.60 ± 0.28 mg/dL vs. 57.61 ± 0.19 mg/dL for normal weight; 52.02 ± 0.32 mg/dL vs. 49.51 ± 0.21 mg/dL for overweight; and 47.30 ± 0.35 mg/dL vs. 46.49 ± 0.23 mg/dL for the obese category. The overall effect of aerobic exercise on HDL-C is shown in Table 3. Aerobic exercise was significantly associated with higher HDL-C ($\beta = 1.18325$; $p < 0.0001$) when compared with no exercise. HDL-C was significantly lower in persons with abnormal compared to those with normal WHR ($\beta = -3.06689$; $p < 0.0001$). Compared with normal weight individuals, overweight and obese groups were also associated with lower HDL-C, with β values of -4.31095 and -6.44230, respectively ($p < 0.0001$). Table 4 is the association of HDL-C based on WHR. There was an interaction between WHR and aerobic exercise on HDL-C ($p = 0.0421$). After the stratification, aerobic exercise was associated with a higher HDL-C especially in those with normal WHR ($\beta = 1.69668$, $p < 0.0001$ vs. 0.97921, $p < 0.0001$). Table 5 is an association of HDL-C with aerobic exercise based on BMI. After stratification by BMI, significant associations of aerobic exercise and HDL-C were found only for normal ($\beta = 1.03261$, $p = 0.0019$) and overweight ($\beta = 2.01758$, $p < 0.0001$) categories. Rs1800588 CT and TT carriers who had aerobic exercise were associated with a higher HDL-C compared to their inactive counterparts. That is, significant increases in HDL were noticed only among aerobically active CT carriers with normal weight ($\beta = 1.99961$, $p = 0.0027$), overweight ($\beta = 1.59362$, $p = 1.1371$), and abnormal WHR ($\beta = 1.48073$, $p = 0.0063$), as well as in TT carriers with both normal and abnormal WHR ($\beta = 4.04073$, $p = 0.0094$ and $\beta = 2.19244$, $p = 0.0445$), and those in the overweight category (5.54693, $p = 0.0003$) (Table 6).

Table 1. Mean HDL-C levels of participants categorized by waist-hip ratio (WHR).

| | Normal WHR | | Abnormal WHR | | *p*-Value |
| | (n = 5190) | | (n = 9347) | | |
	n	Mean ± SE	n	Mean ± SE	
Exercise					
No exercise	3565	55.01 ± 0.23	6148	52.13 ± 0.16	<0.0001
Aerobic	1625	56.43 ± 0.35	3199	53.89 ± 0.23	<0.0001
BMI					
Underweight	276	66.49 ± 0.87	132	63.98 ± 1.23	0.0993
Normal	3159	58.28 ± 0.24	3799	57.67 ± 0.21	0.0587
Overweight	1332	49.38 ± 0.30	2939	50.84 ± 0.22	<0.0001
Obese	423	46.32 ± 0.48	2477	46.80 ± 0.21	0.3770
Body fat rate					
Normal	3941	56.50 ± 0.22	3600	55.00 ± 0.23	<0.0001
Abnormal	1249	52.15 ± 0.36	5747	51.31 ± 0.16	0.0330
Sex					
Women	2096	62.74 ± 0.30	6257	56.46 ± 0.16	<0.0001
Men	3094	50.52 ± 0.21	3090	45.19 ± 0.18	<0.0001
Age, year					
30–40	2066	55.56 ± 0.30	2046	52.06 ± 0.28	<0.0001
40–50	1495	55.26 ± 0.36	2592	52.86 ± 0.26	<0.0001
51–60	1141	55.70 ± 0.42	3000	53.11 ± 0.24	<0.0001
61–70	488	55.07 ± 0.63	1709	52.68 ± 0.31	0.0004
Smoking					
Never	3980	57.06 ± 0.22	7489	54.29 ± 0.15	<0.0001
Former	603	51.30 ± 0.48	946	47.50 ± 0.37	<0.0001
Current	307	49.09 ± 0.50	912	45.37 ± 0.37	<0.0001
Drinking					
Never	4720	55.72 ± 0.20	8464	53.15 ± 0.14	<0.0001
Former	107	49.41 ± 1.05	261	44.30 ± 0.65	<0.0001
Current	363	53.85 ± 0.67	622	50.64 ± 0.54	0.0003
Coffee drinking					
No	3380	55.01 ± 0.23	6322	52.03 ± 0.16	<0.0001
Yes	1810	56.29 ± 0.33	3025	54.20 ± 0.24	<0.0001
Vegetarian diet					
Non	4729	55.69 ± 0.20	8418	52.93 ± 0.14	<0.0001
Former	227	55.08 ± 0.96	458	52.53 ± 0.63	0.0234
Current	234	51.01 ± 0.74	471	49.38 ± 0.56	0.0841

SE = standard error, BMI = body mass index, HDL-C = high-density lipoprotein cholesterol.

Table 2. Mean HDL-C levels of participants categorized by BMI.

	Underweight (n = 408)		Normal Weight (n = 6958)		Overweight (n = 4271)		Obese (n = 2900)		*p*-Value
	n	Mean ± SE	n	Mean ± SE	n	Mean ± SE	n	Mean ± SE	
Exercise									
No exercise	312	65.32 ± 0.80	4563	57.61 ± 0.19	2791	49.51 ± 0.21	2047	46.49 ± 0.23	<0.0001
Aerobic	96	66.83 ± 1.55	2395	58.60 ± 0.28	1480	52.02 ± 0.32	853	47.30 ± 0.35	<0.0001
Waist-hip ratio									
Normal	276	66.49 ± 0.87	3159	58.28 ± 0.24	1332	49.38 ± 0.30	423	46.32 ± 0.48	<0.0001
Abnormal	132	63.98 ± 1.23	3799	57.67 ± 0.21	2939	50.84 ± 0.22	2477	46.80 ± 0.21	<0.0001
Body fat rate									
Normal	408	65.68 ± 0.71	5191	58.20 ± 0.19	1686	47.64 ± 0.26	256	44.66 ± 0.63	<0.0001
Abnormal	0	-	1767	57.20 ± 0.30	2585	52.17 ± 0.23	2644	46.93 ± 0.20	<0.0001
Sex									
Women	333	66.55 ± 0.79	4766	60.72 ± 0.19	1963	54.66 ± 0.27	1291	51.06 ± 0.30	<0.0001
Men	75	61.83 ± 1.52	2192	51.92 ± 0.25	2308	46.75 ± 0.21	1609	43.25 ± 0.21	<0.0001
Age									
30–40	183	64.13 ± 1.01	2156	57.83 ± 0.27	941	49.69 ± 0.36	832	45.82 ± 0.35	<0.0001
40–50	99	66.72 ± 1.44	1966	58.60 ± 0.30	1209	49.75 ± 0.33	813	46.35 ± 0.35	<0.0001
51–60	81	67.20 ± 1.66	1846	58.16 ± 0.32	1355	51.19 ± 0.31	859	47.40 ± 0.37	<0.0001
61–70	45	66.96 ± 2.40	990	56.55 ± 0.43	766	50.79 ± 0.45	396	47.96 ± 0.51	<0.0001
Smoking									
Never	375	66.05 ± 0.74	5900	58.97 ± 0.17	3169	51.54 ± 0.21	2025	48.20 ± 0.23	<0.0001
Former	12	59.00 ± 4.60	489	53.24 ± 0.56	604	48.66 ± 0.45	444	44.45 + 0.44	<0.0001
Current	21	62.86 ± 2.97	569	51.38 ± 0.53	498	45.10 ± 0.48	431	42.13 ± 0.42	<0.0001
Drinking									
Never	392	65.64 ± 0.73	6512	58.10 ± 0.16	3778	50.56 ± 0.19	2502	47.04 ± 0.21	<0.0001
Former	1	62.00	100	51.37 ± 1.16	127	45.49 ± 0.91	140	41.94 ± 0.75	<0.0001
Current	15	67.07 ± 3.73	346	56.97 ± 0.76	366	50.26 ± 0.66	258	46.26 ± 0.60	<0.0001
Coffee drinking									
No	301	65.30 ± 0.79	4648	57.21 ± 0.19	2834	49.60 ± 0.21	1919	46.24 ± 0.23	<0.0001
Yes	107	66.74 ± 1.56	2310	59.45 ± 0.28	1437	51.93 ± 0.33	981	47.67 ± 0.34	<0.0001
Vegetarian diet									
No	364	66.12 ± 0.74	6240	58.28 ± 0.17	3900	50.57 ± 0.19	2643	46.91 ± 0.20	<0.0001
Former	9	69.22 ± 5.18	343	58.69 ± 0.76	183	49.02 ± 0.79	150	45.59 ± 0.85	<0.0001
Current	35	60.20 ± 2.54	375	51.76 ± 0.60	188	47.82 ± 0.75	107	43.79 ± 0.97	<0.0001

Table 3. Overall effect of aerobic exercise, WHR, and BMI on HDL-C levels.

	β-Coefficient	*p*-Value
Exercise (ref: No exercise)		
Aerobic	1.18325	<0.0001
Waist-hip ratio (ref: Normal)		
Abnormal	−3.06689	<0.0001
BMI (ref: Normal)		
Underweight	5.64809	<0.0001
Overweight	−4.31095	<0.0001
Obese	−6.4423	<0.0001
Body fat rate (ref: Normal)		
Abnormal	−2.1956	<0.0001
Sex (ref: Women)		
Men	−9.87778	<0.0001
Age (ref: 30–40)		
40–50	0.75841	0.003
51–60	1.18484	<0.0001
61–70	0.87361	0.0068
Smoking (ref: Never)		
Former	−0.16071	0.6368
Current	−2.92965	<0.0001
Drinking (ref: Never)		
Former	−1.35588	0.0292
Current	4.0583	<0.0001
Coffee drinking (ref: No)		
Yes	1.29158	<0.0001
Vegetarian diet (ref: No)		
Former	−0.88308	0.0484
Current	−5.69481	<0.0001

Table 4. Multiple linear regression showing the effect of aerobic exercise on HDL-C based on WHR.

	Normal WHR		Abnormal WHR	
	β	*p*-Value	β	*p*-Value
Exercise (ref: No exercise)				
Aerobic	1.69668	<0.0001	0.97921	0.0002
BMI (ref: Normal)				
Underweight	5.59243	<0.0001	5.02426	<0.0001
Overweight	−5.27181	<0.0001	−3.73275	<0.0001
Obese	−7.09311	<0.0001	−6.20423	<0.0001
Body fat rate (ref: Normal)				
Abnormal	−2.42606	<0.0001	−2.10034	<0.0001
Sex (ref: Women)				
Men	−10.28598	<0.0001	−9.68871	<0.0001
Age (ref: 30–40)				
40–50	0.63513	0.1145	0.8325	0.0128
51–60	1.43776	0.0015	1.04859	0.0017
61–70	1.37112	0.0267	0.71197	0.0659
Smoking (ref: Never)				
Former	−0.61703	0.2593	0.10417	0.8122
Current	−3.36699	<0.0001	−2.6385	<0.0001
Drinking (ref: Never)				
Former	−2.05145	0.0784	−1.29173	0.0784
Current	3.60076	<0.0001	4.30474	<0.0001
Coffee drinking (ref: No)				
Yes	0.80019	0.0203	1.59424	<0.0001
Vegetarian diet (ref: No)				
Former	−1.0724	0.178	−0.79071	0.1425
Current	−6.41356	<0.0001	−5.30493	<0.0001
WHR*exercise			*p*-value = 0.0421	

β = β value, ref. = reference.

Table 5. Multiple linear regression showing the effect of aerobic exercise on HDL-C based on BMI.

	Underweight		Normal Weight		Overweight		Obese	
	β	*p*-Value	β	*p*-Value	β	*p*-Value	β	*p*-Value
Exercise (ref: No exercise)								
Aerobic	0.25906	0.8905	1.03261	0.0019	2.01758	<0.0001	0.45652	0.2547
WHR (ref: Normal)								
Abnormal	−4.15959	0.0097	−3.93381	<0.0001	−2.04105	<0.0001	−1.81686	0.0004
Body fat rate (ref: Normal)								
Abnormal	-	-	−3.48363	<0.0001	−0.44724	0.3138	−1.17241	0.0718
Sex (ref: Women)								
Men	−6.10877	0.0038	−10.93384	<0.0001	−9.12807	<0.0001	−7.89031	<0.0001
Age (ref: 30–40)								
40–50	4.01615	0.0298	1.37246	0.0004	−0.10543	0.8231	0.32885	0.4854
51–60	4.58714	0.0229	1.63715	<0.0001	0.60932	0.2041	0.90307	0.0615
61–70	4.24658	0.1094	1.03037	0.0436	0.65788	0.2371	1.17587	0.0531
Smoking (ref: Never)								
Former	−6.12763	0.1880	−1.08682	0.0830	0.75294	0.1476	−0.1813	0.7427
Current	−2.77427	0.4150	−3.40832	<0.0001	−2.86523	<0.0001	−2.51485	<0.0001
Drinking (ref: Never)								
Former	−2.20863	0.8829	−1.67239	0.1854	−1.57299	0.1130	−1.55493	0.0702
Current	5.43379	0.1797	4.53503	<0.0001	4.27903	<0.0001	2.86789	<0.0001
Coffee drinking (ref: No)								
Yes	0.77143	0.6411	1.33778	<0.0001	1.73784	<0.0001	0.74399	0.0472
Vegetarian diet (ref: No)								
Former	2.80844	0.5567	0.11192	0.8694	−2.72699	0.0008	−1.35844	0.0877
Current	−5.40391	0.0338	−7.19076	<0.0001	−3.77391	<0.0001	−4.27177	<0.0001

Table 6. The impact of aerobic exercise on HDL-C stratified by rs1800588 variant and obesity indexes.

	WHR Stratification				BMI Stratification							
rs1800588	Normal WHR		Abnormal WHR		Underweight		Normal		Overweight		Obese	
	β	*p*-Value	β	*p*-Value	B	*p*-Value	β	*p*-Value	β	*p*-Value	β	*p*-Value
CC	1.36476	0.0798	0.78743	0.1622	2.38864	0.5136	0.63897	0.3730	1.22861	0.1093	1.47101	0.1052
CT	1.22236	0.0870	1.48073	0.0063	−1.45958	0.7809	1.99961	0.0027	1.59362	0.0371	0.00127	0.9987
TT	4.04073	0.0094	2.19244	0.0445	15.7138	0.2848	1.18290	0.3920	5.54693	0.0003	2.02256	0.2826
P-interaction	0.5865		0.0792		0.4733		0.9596		0.0281		0.9294	

The *p*-interaction shown is for hepatic lipase (LIPC) Rs1800588 and exercise (Rs1800588*exercise).

4. Discussion

The primary objective of this study was to determine the association between aerobic exercise and HDL-C among different categories of BMI and WHR and also to highlight the modulating role of rs1800588 variant in the hepatic lipase gene. We found that (1) consistent with our previous findings [9], aerobic exercise was better than no exercise for improving HDL-C in Taiwanese adults. (2) Aerobic exercise was associated with a lower HDL-C in persons with abnormal compared to normal WHR (β = −306,689, *p* < 0.0001). In addition, there was an interaction between WHR and aerobic exercise. (3) Compared with aerobically active normal weight individuals, their overweight and obese counterparts were associated with lower HDL-C levels. (4) Unlike the LIPC rs1800588 TT and CT genotype, the effect of CC genotype on HDL was not modified by aerobic exercise no matter the BMI or WHR category. Our study findings highlight the impact of aerobic exercise on HDL-C. This in part is mediated by liver X receptor (LXR) [1] and liver ATP-binding cassette transporters A-1 (ABCA1) [19] as previously reported.

Previously published articles have discussed associations of HDL-C with anthropometric measures [20] and physical exercise [1,9]. Findings from a study of 28,000 men and women suggested that HDL-cholesterol decreased concurrently with increases in BMI [21]. In another study, WHR was found to be a good predictor of the lipid profile ($\beta = 3.51$, $p = 0.005$) [22]. Aerobic exercise has resulted in significant changes in body fat measures like BMI and WHR among young Taiwanese adults who were obese [23]. Despite the numerous findings, the impact of aerobic exercise on HDL levels based on anthropometric measures and genetic factors have not been reported in Taiwan. In the current study, we included rs1800588 variant in the LIPC gene in the model and found that the effect of CC genotype on HDL was not modified by aerobic exercise no matter the BMI or WHR category. However, the magnitude of the association between CT and TT genotypes on HDL-C differed with respect to BMI and WHR categories. For instance, we found that the effect of CT genotype on HDL was significant only among aerobically active normal weight and overweight adults as well as those with abnormal WHR, while the effect of TT genotype on HDL-C was significant only among aerobically active overweight adults and those with both normal and abnormal WHR. The mechanisms explaining these differences in HDL-C response with respect to body fat measures are still to be clearly understood. However, it has been reported that HDL levels of certain individuals do not necessarily increase no matter the exercise regimen [14,24]. Another study including Caucasian women found that the effect of rs1800588 variant on HDL-C was modified by physical activity [14]. However, stratifications were not made based on genotypes.

The T allele of rs1800588 has been associated with higher baseline levels of HDL-C [25]. As stated earlier female carriers of the rs1800588 variant with at least one copy of the minor allele had higher concentrations of HDL-C than those that were homozygous for the major allele [14]. Further analysis of data from those women demonstrated that the per-minor allele increase in HDL-C was greater in active than inactive women. This aligns with other findings which suggested that LIPC polymorphisms might serve as useful indicators of higher HDL-C in women [25,26]. Prior findings from studies investigating the relationship between LIPC rs1800588 and HDL-C differ according to gender and ethnicity [11]. In our study, there was the presence of LIPC rs1800588 CC, CT and TT genotypes. However, in a study by Brinkley and his colleagues, there were no subjects with the rs1800588 TT genotype [25]. This highlights the diverse effect of the variant on HDL-C levels.

Anthropometric, lifestyle, environmental, and genetic factors influence changes in Lipid fractions [1]. Moderate intensity aerobic exercise is related to a higher HDL-C [9]. Variations in HDL-C responses to exercise are influenced by several factors including sex, changes in body composition and genetic effects [27]. Based on our analyzed data, there was a significant LIPC rs1800588*exercise effect on HDL-C particularly in overweight adults, with higher levels in CT and TT compared to CC carriers ($p = 0.0281$ for the interaction). There were no genotype*exercise interactions for HDL-C across other BMI and WHR categories.

In summary, we report evidence that associations between aerobic exercise and HDL-C levels in Taiwanese adults differed not only across different ranges of body mass index and waist-to-hip ratios but also among carriers of the rs1800588 variant located in the hepatic lipase gene. However, unlike CT and TT genotypes, the effects of aerobic exercise on HDL-C levels were not significant among rs1800588 CC carriers no matter their BMI or WHR.

Author Contributions: Conceptualization, Y.N., M.-C.C. and Y.-P.L.; Formal analysis, O.N.N. and K.-J.L.; Investigation, M.C.C.; Methodology, Y.N., O.N.N., K.-J.L. and Y.-P.L.; Resources, M.-C.C. and Y.-P.L.; Supervision, M.-C.C. and Y.-P.L.; Writing—original draft, Y.N.; Writing—review & editing, O.N.N., K.-J.L., M.-C.C. and Y.-P.L.

Funding: This work was supported by the Ministry of Science and Technology (MOST 105-2627-M-040-002, 106-2627-M-040-002, 107-2627-M-040-002).

Conflicts of Interest: The authors declare no conflict of interest. The funders had no role in the design of the study; in the collection, analyses, or interpretation of data; in the writing of the manuscript, or in the decision to publish the results.

References

1. Wang, Y.; Xu, D. Effects of aerobic exercise on lipids and lipoproteins. *Lipids Health Dis.* **2017**, *16*, 132. [CrossRef] [PubMed]
2. Rye, K.-A.; Ong, K.L. HDL function as a predictor of coronary heart disease events: Time to re-assess the HDL hypothesis? *Lancet Diabetes Endocrinol.* **2015**, *3*, 488–489. [CrossRef]
3. Bartlett, J.; Predazzi, I.M.; Williams, S.M.; Bush, W.S.; Kim, Y.; Havas, S.; Toth, P.P.; Fazio, S.; Miller, M. Is isolated low high-density lipoprotein cholesterol a cardiovascular disease risk factor? New insights from the Framingham offspring study. *Circ. Cardiovasc. Qual. Outcomes* **2016**, *9*, 206–212. [CrossRef] [PubMed]
4. Woudberg, N.J.; Mendham, A.E.; Katz, A.A.; Goedecke, J.H.; Lecour, S. Exercise intervention alters HDL subclass distribution and function in obese women. *Lipids Health Dis.* **2018**, *17*, 232. [CrossRef] [PubMed]
5. Rye, K.-A.; Barter, P.J. Cardioprotective functions of HDLs. *J. Lipid Res.* **2014**, *55*, 168–179. [CrossRef] [PubMed]
6. Gadi, R.; Amanullah, A.; Figueredo, V.M. HDL-C: Does it matter? An update on novel HDL-directed pharmaco-therapeutic strategies. *Int. J. Cardiol.* **2013**, *167*, 646–655. [CrossRef]
7. März, W.; Kleber, M.E.; Scharnagl, H.; Speer, T.; Zewinger, S.; Ritsch, A.; Parhofer, K.G.; von Eckardstein, A.; Landmesser, U.; Laufs, U. HDL cholesterol: Reappraisal of its clinical relevance. *Clin. Res. Cardiol.* **2017**, *106*, 663–675. [CrossRef]
8. National Institutes of Health. *Third Report of the National Cholesterol Education Program Expert Panel on Detection, Evaluation, and Treatment of High Blood Cholesterol in Adults (Adult Treatment Panel III)*; National Institutes of Health: Bethesda, MD, USA, 2001; Volume 285, pp. 2486–2497.
9. Nassef, Y.; Lee, K.-J.; Nfor, O.N.; Tantoh, D.M.; Chou, M.-C.; Liaw, Y.-P. The impact of aerobic exercise and badminton on HDL cholesterol levels in adult Taiwanese. *Nutrients* **2019**, *11*, 515. [CrossRef]
10. Qasim, A.; Rader, D.J. Human genetics of variation in high-density lipoprotein cholesterol. *Curr. Atheroscler. Rep.* **2006**, *8*, 198–205. [CrossRef]
11. Goodarzynejad, H.; Boroumand, M.; Behmanesh, M.; Ziaee, S.; Jalali, A.; Pourgholi, L. Association between the hepatic lipase promoter region polymorphism (-514 C/T) and the presence and severity of premature coronary artery disease. *J. Tehran Univ. Heart Cent.* **2017**, *12*, 119.
12. Guerra, R.; Wang, J.; Grundy, S.M.; Cohen, J.C. A hepatic lipase (LIPC) allele associated with high plasma concentrations of high density lipoprotein cholesterol. *Proc. Natl. Acad. Sci. USA* **1997**, *94*, 4532–4537. [CrossRef] [PubMed]
13. Fan, Y.M.; Raitakari, O.; Kähönen, M.; Hutri-Kähönen, N.; Juonala, M.; Marniemi, J.; Viikari, J.; Lehtimäki, T. Hepatic lipase promoter C-480T polymorphism is associated with serum lipids levels, but not subclinical atherosclerosis: The cardiovascular risk in young Finns study. *Clin. Genet.* **2009**, *76*, 46–53. [CrossRef] [PubMed]
14. Ahmad, T.; Chasman, D.I.; Buring, J.E.; Lee, I.-M.; Ridker, P.M.; Everett, B.M. Physical activity modifies the effect of LPL, LIPC, and CETP polymorphisms on HDL-C levels and the risk of myocardial infarction in women of European ancestry. *Circ. Cardiovasc. Genet.* **2011**, *4*, 74–80. [CrossRef] [PubMed]
15. Zhang, K.; Zhao, Q.; Li, Y.; Zhen, Q.; Yu, Y.; Tao, Y.; Cheng, Y.; Liu, Y. Feasibility of anthropometric indices to identify dyslipidemia among adults in Jilin Province: A cross-sectional study. *Lipids Health Dis.* **2018**, *17*, 16. [CrossRef] [PubMed]
16. Rocha, F.L.; de Menezes, T.N.; de Melo, R.L.P.; Pedraza, D.F. Correlation between indicators of abdominal obesity and serum lipids in the elderly. *Rev. Assoc. Médica Bras.* **2013**, *59*, 48–55.
17. Özkaya, İ.; Bavunoglu, I.; Tunçkale, A. Body mass index and waist circumference affect lipid parameters negatively in Turkish women. *Am. J. Public Health Res.* **2014**, *2*, 226–231. [CrossRef]
18. Passos, G.S.; Poyares, D.; Santana, M.G.; D'Aurea, C.V.R.; Youngstedt, S.D.; Tufik, S.; de Mello, M.T. Effects of moderate aerobic exercise training on chronic primary insomnia. *Sleep Med.* **2011**, *12*, 1018–1027. [CrossRef]
19. Ghanbari-Niaki, A.; Khabazian, B.M.; Hossaini-Kakhak, S.A.; Rahbarizadeh, F.; Hedayati, M. Treadmill exercise enhances ABCA1 expression in rat liver. *Biochem. Biophys. Res. Commun.* **2007**, *361*, 841–846. [CrossRef]
20. Lee, B.J.; Kim, J.Y. Identification of the best anthropometric predictors of serum high-and low-density lipoproteins using machine learning. *IEEE J. Biomed. Health Inform.* **2015**, *19*, 1747–1756. [CrossRef]

21. Dansinger, M.; Williams, P.T.; Superko, H.R.; Asztalos, B.F.; Schaefer, E.J. Effects of weight change on HDL-cholesterol and its subfractions in over 28,000 men and women. *J. Clin. Lipidol.* **2019**, *13*, 308–316. [CrossRef]

22. Lemos-Santos, M.G.F.; Valente, J.G.; Gonçalves-Silva, R.M.V.; Sichieri, R. Waist circumference and waist-to-hip ratio as predictors of serum concentration of lipids in Brazilian men. *Nutrition* **2004**, *20*, 857–862. [CrossRef] [PubMed]

23. Chiu, C.-H.; Ko, M.-C.; Wu, L.-S.; Yeh, D.-P.; Kan, N.-W.; Lee, P.-F.; Hsieh, J.-W.; Tseng, C.-Y.; Ho, C.-C. Benefits of different intensity of aerobic exercise in modulating body composition among obese young adults: A pilot randomized controlled trial. *Health Qual. Life Outcomes* **2017**, *15*, 168. [CrossRef] [PubMed]

24. Leon, A.; Gaskill, S.; Rice, T.; Bergeron, J.; Gagnon, J.; Rao, D.C.; Skinner, J.S.; Wilmore, J.H.; Bouchard, C. Variability in the response of HDL cholesterol to exercise training in the HERITAGE family study. *Int. J. Sports Med.* **2002**, *23*, 1–9. [CrossRef] [PubMed]

25. Brinkley, T.E.; Halverstadt, A.; Phares, D.A.; Ferrell, R.E.; Prigeon, R.L.; Hagberg, J.M.; Goldberg, A.P. Hepatic lipase gene-514C> T variant is associated with exercise training-induced changes in VLDL and HDL by lipoprotein lipase. *J. Appl. Physiol.* **2011**, *111*, 1871–1876. [CrossRef] [PubMed]

26. Couture, P.; Otvos, J.D.; Cupples, L.A.; Lahoz, C.; Wilson, P.W.; Schaefer, E.J.; Ordovas, J.M. Association of the C−514T polymorphism in the hepatic lipase gene with variations in lipoprotein subclass profiles: The Framingham offspring study. *Arterioscler. Thromb. Vasc. Biol.* **2000**, *20*, 815–822. [CrossRef]

27. Rice, T.; Després, J.-P.; Pérusse, L.; Hong, Y.; Province, M.A.; Bergeron, J.; Gagnon, J.; Leon, A.S.; Skinner, J.S.; Wilmore, J.H.; et al. Familial aggregation of blood lipid response to exercise training in the health, risk factors, exercise training, and genetics (HERITAGE) family study. *Circulation* **2002**, *105*, 1904–1908. [CrossRef]

© 2019 by the authors. Licensee MDPI, Basel, Switzerland. This article is an open access article distributed under the terms and conditions of the Creative Commons Attribution (CC BY) license (http://creativecommons.org/licenses/by/4.0/).

 *genes*

Article

ACTN3 R577X Genotype and Exercise Phenotypes in Recreational Marathon Runners

Juan Del Coso [1,*], Victor Moreno [2], Jorge Gutiérrez-Hellín [1], Gabriel Baltazar-Martins [1], Carlos Ruíz-Moreno [1], Millán Aguilar-Navarro [1,3], Beatriz Lara [1] and Alejandro Lucía [4,5]

[1] Exercise Physiology Laboratory, Camilo José Cela University, 28692 Madrid, Spain; jhellin@ucjc.edu (J.G.-H.); jgsoares@ucjc.edu (G.B.-M.); cruizm@ucjc.edu (C.R.-M.); millan.aguilar@ufv.es (M.A.-N.); blara@ucjc.edu (B.L.)

[2] Sports Research Centre, Miguel Hernandez University of Elche, 03202 Alicante, Spain; vmoreno@goumh.umh.es

[3] Sport Science Department, Francisco de Vitoria University, 28223 Madrid, Spain

[4] Faculty of Sport Sciences, Research Institute i+12, Universidad Europea de Madrid, 28670 Madrid, Spain; alejandro.lucia@universidadeuropea.es

[5] Centro de Investigación Biomédica en Red de Fragilidad y Envejecimiento Saludable, 28029 Madrid, Spain

* Correspondence: jdelcoso@ucjc.edu; Tel.: +34-91815-3131

Received: 11 April 2019; Accepted: 23 May 2019; Published: 29 May 2019

Abstract: Background: Homozygosity for the X-allele in the *ACTN3* R577X (rs1815739) polymorphism results in the complete absence of α-actinin-3 in sarcomeres of fast-type muscle fibers. In elite athletes, the *ACTN3* XX genotype has been related to inferior performance in speed and power-oriented sports; however, its influence on exercise phenotypes in recreational athletes has received less attention. We sought to determine the influence of *ACTN3* genotypes on common exercise phenotypes in recreational marathon runners. **Methods:** A total of 136 marathoners (116 men and 20 women) were subjected to laboratory testing that included measurements of body composition, isometric muscle force, muscle flexibility, ankle dorsiflexion, and the energy cost of running. *ACTN3* genotyping was performed using TaqMan probes. **Results:** 37 runners (27.2%) had the RR genotype, 67 (49.3%) were RX and 32 (23.5%) were XX. There was a difference in body fat percentage between RR and XX genotype groups (15.7 ± 5.8 vs. 18.8 ± 5.5%; effect size, ES, = 0.5 ± 0.4, p = 0.024), whereas the distance obtained in the sit-and-reach-test was likely lower in the RX than in the XX group (15.3 ± 7.8 vs. 18.4 ± 9.9 cm; ES = 0.4 ± 0.4, p = 0.046). Maximal dorsiflexion during the weight-bearing lunge test was different in the RR and XX groups (54.8 ± 5.8 vs. 57.7 ± 5.1 degree; ES = 0.5 ± 0.5, p = 0.044). Maximal isometric force was higher in the RR than in the XX group (16.7 ± 4.7 vs. 14.7 ± 4.0 N/kg; ES = −0.5 ± 0.3, p = 0.038). There was no difference in the energy cost of running between genotypes (~4.8 J/kg/min for all three groups, ES ~0.2 ± 0.4). **Conclusions:** The *ACTN3* genotype might influence several exercise phenotypes in recreational marathoners. Deficiency in α-actinin-3 might be accompanied by higher body fatness, lower muscle strength and higher muscle flexibility and range of motion. Although there is not yet a scientific rationale for the use of commercial genetic tests to predict sports performance, recreational marathon runners who have performed such types of testing and have the *ACTN3* XX genotype might perhaps benefit from personalized strength training to improve their performance more than their counterparts with other *ACTN3* genotypes.

Keywords: α-actinin; exercise; performance; endurance; genetics; single nucleotide polymorphism

---

1. Introduction

α-Actinin-2 and α-actinin-3 are key structural proteins in the contractile apparatus of the skeletal muscle fiber, as they bind and possibly cross-link the ends of F-actin filaments at the Z-line [1].

Whereas α-actinin-2 is ubiquitously expressed in all muscle fiber types, α-actinin-3 expression is largely restricted to fast-type muscle fibers [2]. Homozygosity for the null X-allele of the R577X polymorphism in the α-actinin-3 gene, *ACTN3*, results in the complete absence of α-actinin-3 in fast-type muscle fibers [3]. Individuals with the *ACTN3* XX genotype compensate for the deficiency of α-actinin-3 through elevated expression of α-actinin-2 in fast-type muscle fibers [4], although several specific muscle phenotypes have been related to α-actinin-3 deficiency [5].

α-Actinin-3 deficiency is believed to affect the muscle's ability to generate rapid, forceful contractions and thus might be detrimental for the production of fast and explosive movements. This notion has been verified in almost 20 case-control studies, as recently reviewed by Houweling et al., (2018), with the frequency of the XX genotype being lower in elite athletes participating in sprint and power-based sports than in the general non-athletic population. By contrast, the RR genotype, which is associated with full expression of α-actinin-3 in fast-type muscle fibers, is highly prevalent among elite athletes in sprint/power disciplines. However, the effect of the *ACTN3* XX genotype on the sports performance of recreational athletes is unexplored. The study of such a relationship might be particularly interesting given that ~20% of the world's population is α-actinin-3 deficient [6], and because genetic testing of this polymorphism has recently become a commercially available diagnostic test [7], which can inform exercise recommendations.

In untrained populations, *ACTN3* XX individuals produce less handgrip strength and less muscle force and power than their RR counterparts [8–10], but this difference is lost when the same genotypes are compared in active/trained individuals [11–13]. Furthermore, whereas muscle fiber composition is not affected by α-actinin-3 deficiency [14]; muscle volume [8], and especially the size of fast-type muscle fibers [15], is lower in XX than in RR counterparts. Finally, a higher response to strength training has been found in RR than in XX individuals [16], coupled with a lower signaling for muscle hypertrophy in XX subjects [14]. Given this information, it might be speculated that α-actinin-3 deficiency derived from *ACTN3* XX homozygosity might also affect force and power production in recreational athletes and affect sports performance.

In addition to affecting exercise performance, *ACTN3* genotypes might also influence exercise-induced muscle damage, particularly after endurance events such as marathon running. Indeed, the X-allele has been associated with higher levels of several markers of muscle damage after exercise in amateur athletes [17–19]. Conversely, a higher muscle flexibility and a superior range of motion has been reported in XX individuals versus their RR referents [20–22], although some authors have failed to replicate this finding [23]. Although more flexible muscles are less susceptible to eccentric exercise-induced damage [24], higher muscle flexibility values do not seem to attenuate marathon-induced muscle damage in XX runners [17–19]. Finally, the *ACTN3* XX genotype has been related to lower body mass and lower fat-free mass [8,23], likely due to a reduction in muscle mass as a result of smaller fast-type fiber size [15,25]. However, the effect of *ACTN3* genotypes on fat mass and body composition in sedentary and clinical populations is unclear [25–27], and is unknown in recreational athletes. The methodological differences in assessing these exercise phenotypes, the relatively small study samples in some investigations, and the wide range of age and fitness levels under investigation make it difficult to ascertain whether the effect of *ACTN3* genotypes is of sufficient magnitude to represent a variable that affect sports performance and training in recreational athletes, as seems to be the case in elite athlete populations.

α-Actinin-3 deficiency has also been related to positive phenotypes that would explain the perpetuation of the *ACTN3* XX genotype through natural selection in human evolution. Particularly, it has been proposed that the high frequency of the X allele in some human populations could be the result of increased metabolic efficiency, possibly enhancing the capability for endurance running [28]. This theory is supported by studies in mouse models, because a shift towards a more efficient aerobic muscle metabolism has been found in *Actn3* knockout (KO) mice [6,29]. This has fueled the notion that the X allele might act as a thrifty allele [30], although this theory has little support in humans [31].

Indeed, recent case-control investigations suggest that it is unlikely that the *ACTN3* XX genotype provides an advantage in competitive endurance running performance [32,33].

The aim of the present study was to determine the influence of *ACTN3* genotypes on common exercise phenotypes in recreational marathon runners. Our main hypothesis was that, compared with their RR counterparts, *ACTN3* XX runners would present with lower values of muscle force, but higher values of running efficiency.

2. Materials and Methods

2.1. Subjects

One hundred thirty-six healthy experienced recreational marathon runners (116 men and 20 women) volunteered to participate in this study. Participants were either recruited by email from a group of runners that had participated in previous investigations or were recruited at the time of race registration. Inclusion criteria were as follows: Age 18–65 years; being free of any history of muscle, cardiac or kidney disorders; participating in the marathon at maximal possible intensity; and having a running experience of at least 3 years, with at least three marathons completed during this period. Exclusion criteria were: taking medications during the 2 weeks prior to competing or having had a musculoskeletal injury in the month prior to the competition. The fulfillment of inclusion/exclusion criteria was verified through an ad hoc questionnaire. Age and main morphological and physical characteristics of the participants in this investigation are shown in Table 1. Before enrollment, each participant was informed about the risks and discomforts associated with the investigation and signed an informed consent document. The study was approved by the Camilo Jose Cela University Ethics Committee (ID ACTN3 approved 18/4/2018) in accordance with the latest version of the Declaration of Helsinki. Participants' rights and confidentiality were protected during the whole experiment, and the genetic information was used only for the purposes included in this investigation.

Table 1. Age, anthropometric characteristics, running experience, and training status of marathoners with different *ACTN3* R577X genotypes. Data are mean ± standard deviation (SD) for each genotype. Degrees of freedom = 2, between-groups; 133, intra-groups; 135, total.

Variable (Units)	RR	RX	XX	p	η^2
n (frequency)	37 (27.2%)	67 (49.3%)	32 (23.5%)	-	-
Men/women (frequency)	31/6 (83.8/16.2%)	58/9 (86.6/13.4%)	27/5 (84.4/15.6%)	0.922	-
Age (years)	41.2 ± 10.2	40.3 ± 8.8	40.7 ± 9.8	0.880	<0.01
Body mass (kg)	70.9 ± 7.1	71.6 ± 10.8	72.8 ± 10.5	0.731	<0.01
Body height (m)	1.73 ± 0.06	1.73 ± 0.08	1.72 ± 0.10	0.723	<0.01
Body mass index (kg/m^2)	23.7 ± 1.7	23.8 ± 1.2	24.5 ± 1.5	0.26.9	0.03
Running experience (years)	9.0 ± 7.5	8.1 ± 7.8	8.3 ± 6.0	0.880	<0.01
Best race time in the marathon (min)	218 ± 27	223 ± 38	219 ± 39	0.881	<0.01
Completed marathons (number)	5 ± 4	5 ± 3	5 ± 3	0.776	<0.01
Average training distance /week (km)	50.7 ± 14.6	52.5 ± 17.0	51.7 ± 16.9	0.889	<0.01
Training sessions /week (number)	4 ± 1	4 ± 1	4 ± 1	0.794	<0.01

2.2. Experimental Design

All participants underwent the same testing under identical experimental conditions. Participants were registered in the 2018 edition of the Rock'n'Roll Madrid Marathon and once they had completed all the testing and finished the marathon, they were included into a common database. Subsequently, participants were divided into three groups, established according to their individual *ACTN3* R577X genotype (RR, RX or XX groups). Because the men and women responded in the same manner when comparing the three genotypes, and the frequency of men/women was similar in all three groups (Table 1), we analyzed all the data without considering sex as a covariable.

2.3. Experimental Protocol

At least 1 week before the marathon, each participant received information about the benefits and risks of the investigation and the standardization procedures. At this time, they filled out the pre-participation ad hoc questionnaire. Participants were instructed to avoid strenuous exercise, caffeine and alcohol for the 24 h before the onset of testing, which was performed the day before the race. On this day, participants signed the informed consent and anthropometric characteristics were registered by an ISAK-certified anthropometrist following international standards [34]. Anthropometric measurements included body mass and height (±50 g scale; Radwag, Radom, Poland), skinfold thickness (±0.1 mm skinfold caliper, Holtain Ltd., Crosswell, UK: triceps, subscapular, iliac crest, abdominal, anterior and posterior thigh and medial calf) and thigh circumference (±0.5 mm fiber glass measuring tape; Holtain Ltd.: above the knee, at the maximum thigh circumference and at the gluteal furrow). Three measurements were obtained on the dominant side of the body and the mean was used for data analysis. Relative adiposity (in %) was calculated from the sum of skinfolds [35]. The mean fat-free volume of the dominant thigh (in mL/kg) was measured according to the protocol described by Jones & Pearson (1969) and normalized by body mass to allow a better comparison among groups [36].

Participants underwent a standardized 10-min warm-up including low-intensity running at 8 km/h on a treadmill. Treadmill velocity was progressively increased until 10 km/h and oxygen uptake (VO_2) and carbon dioxide production (VCO_2) were measured at this velocity for 5 min. Expired gases were collected breath-by-breath with a metabolic cart (Metalyzer 3B, Cortex, Leipzig, Germany), and gas exchange data of the last minute was used as a representative value. Certified calibration gases (16% O_2, 5% CO_2, Cortex) and a 3-L syringe were used to calibrate the gas analyzer and the flowmeter, respectively. Gas measurements were made with the clothes and shoes used during the marathon competition. The energy cost of running (in J/kg/m) was calculated using the non-protein respiratory quotient [37] and was normalized by body mass to allow a better between-subject comparison [38].

After 5 min of recovery, participants performed two maximal countermovement vertical jumps on a force platform (Quattrojump, Kistler, Wintherthur, Switzerland), as previously described [17]. The jumps were separated by a 1-min rest period. The jump with the highest height (in cm) was used for statistical analysis. Then, participants performed a whole-body isometric force test [39]. The isometric muscle strength was measured using a hand-held pull gauge (Isocontrol, Isometrico, Madrid, Spain) set at a frequency of 1000 Hz. For this measurement, participants were asked to stand on a 50 × 50 cm iron base connected to a handle-bar by a non-elastic cable. The isometric gauge was inserted within the cable, and the height of the cable was individually set to provide a 135° knee flexion while the back and the arms were completely extended. Participants were instructed to perform a maximal pull for 4 seconds and the peak value was used for analysis. The force obtained (in Newtons, N) was normalized to body mass (i.e., N/kg) to allow for a better comparison among genotypes. Thereafter, participants performed a maximal handgrip strength test with both hands (dominant and non-dominant) using a handgrip dynamometer (Grip-D, Takei, Japan). Performance was expressed in N and two attempts were performed with each hand; the peak value was used for statistical analysis.

The lunge test was performed as a measure of dorsiflexion range of motion [40]. Participants placed their foot along a measuring tape on the floor with both their big toe and heel on the centerline of the measuring tape while they leaned on a wall. The weight-bearing lunge test was performed with both limbs and the maximal dorsiflexion during the test was defined as the maximum distance of the toe from the wall while maintaining contact between the wall and knee without lifting the heel. Participants were then asked to progressively move their knee forwards while they were reclined on the wall, repeating the lunge movement until the maximum distance at which they could tolerably lunge their knee to the wall without heel lift was found [41]. At this point, dorsiflexion range of motion was performed using a handheld manual goniometer (Baseline®, The Therapy Connection Inc, Windham, NY, USA) by placing the center of the goniometer just below the lateral malleolus of the ankle, with one arm lined up through the lateral aspect of the fibula and the other arm lined up with the fifth metatarsophalangeal joint [42]. The measurement was repeated three times and the maximal ankle dorsiflexion (in °) was used for analysis.

On the day of the race, participants had their usual pre-competition meal at least 3 h before the race, which was not standardized among participants to avoid affecting their individual pre-competition routine. Runners were encouraged to ingest 500 mL of water 2 h before the start of the race to increase the likelihood of being euhydrated at the start line. During the race, participants wore a race bib with a time-chip to calculate the actual amount of time that it took them from the start line of the race to the finish line (net time, in min). Participants completed the race at their own pace and drank ad libitum at the hydration stations placed at 5-km intervals with no indications about running pace or fluid and food strategies. The marathon race was held in April on a sunny day with a mean dry temperature of 21.0 ± 2.1 °C (range 15–26 °C, temperature readings at 30-min intervals from 0- to 5-h after the race onset) and a mean relative humidity of $43 \pm 2\%$ (range 40–51%).

2.4. Genetic Testing

Genomic DNA was isolated using an organic-based DNA extraction method adapted to Amicon® (Sigma-Aldrich, Madrid, Spain) Ultra 0.5-mL columns, including a final concentration step to 50 μL [43]. To avoid contamination, recommendations for molecular genetics laboratories were followed, including physically-isolated work area laboratories for each process (sample manipulation and extraction). In addition, reference samples (internal controls, blank samples and negative controls) and contamination monitoring in all steps were included. Positive controls for all genotypes were obtained from the Mexican branch of the CANDELA Consortium [44]. Genotyping of *ACTN3* rs1815739 polymorphism (c.1858C>T; p.R577X) was conducted using a TaqMan SNP Genotyping Assay (Assay ID: C___590093_1_; Applied Biosystems, Foster City, CA, USA) and the reaction was performed in an Applied Biosystems 7500 Fast Real-Time PCR System (Applied Biosystems). The results were analyzed using 7500 Software v2.0.5 (Applied Biosystems).

2.5. Statistical analysis

The difference in the distribution of men/women in each genotype group was tested with crosstab and Chi square tests, including adjusted standardized residuals. The normality of the remaining variables was initially tested with the Shapiro-Wilk test and all variables showed a normal distribution. Group comparisons (RR vs. RX vs. XX) were performed using one-way analysis of variance (ANOVA). When the ANOVA showed a significant group-effect, between-group differences were assessed using the Tukey post-hoc test. The significance level was set at 0.05. The effect size (ES) for each full ANOVA analysis was calculated using the Eta squared (η^2) by using between-groups sum of squares and the total sums of squares for all ES. The magnitude of η^2 was interpreted following the guidelines by Cohen [45] as follows: small: 0.01; medium = 0.06; large: 0.14. Data are presented as mean $\pm$ standard deviation (SD) and all the analyses were performed with the statistical package SPSS version 20.0 (SPSS Inc., Chicago, IL, USA). The ES was also calculated in all pairwise comparisons, by using the Hedges' $g \pm 95\%$ confidence intervals (CI), to assess the magnitude of the between-group differences in the phenotypes under investigation. ES were interpreted according to the following ranges: <0.2, trivial; 0.2–0.6, small; 0.6–1.2, moderate; 1.2–2.0, large; 2.0–4.0, very large; and >4.0, extremely large [46].

3. Results

The genotyping success rate was 99%. From the study sample of 136 runners, 27.2% were genotyped as *ACTN3* RR, 49.3% were RX and 23.5% were XX. Participants had similar running experience, best race time in marathon, number of completed marathons in the three previous years, and comparable training characteristics (Table 1). In addition, the net race time in the investigated marathon was similar for all three genotypes (236 ± 36, 236 ± 44, 244 ± 27 min, respectively; $p = 0.509$, $\eta^2 = 0.01$).

The ANOVA post-hoc analyses revealed that body fat percentage was higher in the XX than in the RR group (Figure 1, upper panel), although the differences between XX and the RR or RX group were small–moderate (Table 2). The ANOVA post-hoc analyses only revealed a small-to-moderate difference between RR and XX groups for thigh volume in the dominant leg (Figure 1, lower panel). By contrast, performance in the sit-and-reach-test was higher in the XX group than in the RX group (Figure 2, upper panel). Post-hoc analyses only revealed an RR vs. XX difference in the right ankle for maximal dorsiflexion during the weight-bearing lunge test (Figure 2, lower panel) with no differences between groups in the left ankle (Table 2).

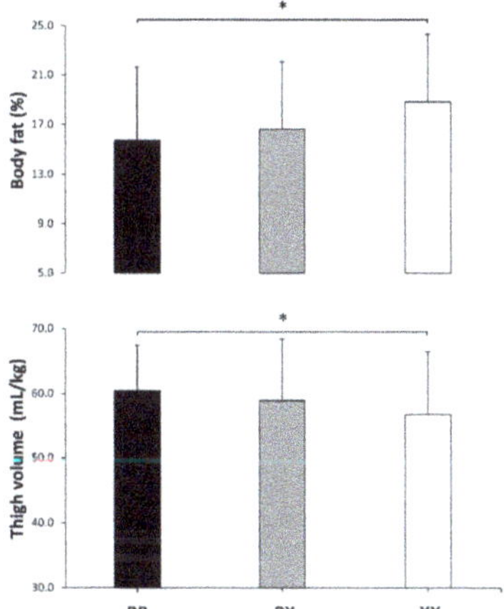

Figure 1. Body fat percentage and thigh fat-free volume in *ACTN3* RR, RX and XX recreational marathon runners. Data are mean ± standard deviation (SD) for each genotype. (*) Differences identified by a post-hoc analysis at $p < 0.05$.

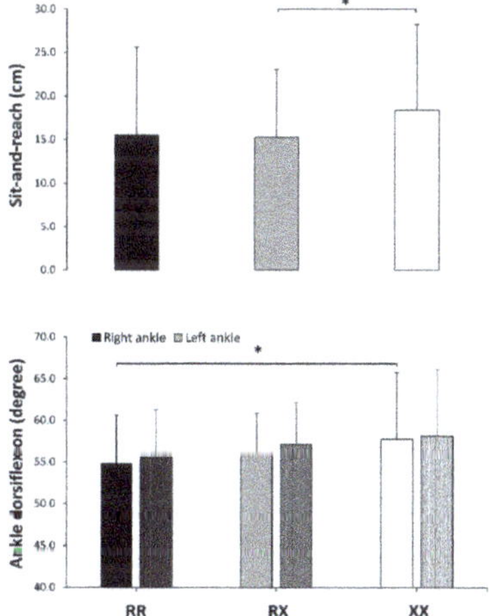

Figure 2. Distance reached in the sit-and-reach test and maximal ankle dorsiflexion in the lunge test in *ACTN3* RR, RX and XX recreational marathon runners. Data are mean ± SD for each genotype. (*) Differences identified by a post-hoc analysis at $p < 0.05$.

Table 2. Body fatness, thigh volume, distance obtained in the sit-and-reach test, and ankle dorsiflexion in *ACTN3* RR, RX and XX recreational marathon runners. Data are mean ± SD for each genotype. ES = Effect size calculated with Hedges' *g*; CI = Confidence interval. Degrees of freedom = 2, between-groups; 133, intra-group; 135, total.

Variable (Units)	RR	RX	XX	ES ± 95% CIRR vs. RX	ES ± 95% CIRR vs. XX	ES ± 95% CIRX vs. XX	*p*	η^2
Body fat (%)	15.7 ± 5.8	16.2 ± 6.1	18.8 ± 5.5	0.1 ± 0.3	0.5 ± 0.4	0.4 ± 0.3	0.024	0.06
Thigh volume (mL/kg)	60.5 ± 6.9	58.9 ± 9.5	56.8 ± 9.6	−0.2 ± 0.3	−0.4 ± 0.3	−0.2 ± 0.3	0.043	0.05
Sit-and-reach test (cm)	15.5 ± 10.1	15.3 ± 7.8	18.4 ± 9.9	0.0 ± 0.3	0.3 ± 0.4	0.4 ± 0.4	0.046	0.04
Right-ankle dorsiflexion (degrees)	54.8 ± 5.8	55.9 ± 5.0	57.7 ± 5.1	0.2 ± 0.3	0.5 ± 0.5	0.4 ± 0.4	0.044	0.05
Left-ankle dorsiflexion (degrees)	55.6 ± 4.8	57.1 ± 5.6	58.1 ± 5.1	0.3 ± 0.3	0.5 ± 0.5	0.2 ± 0.5	0.103	0.03

There were no differences in handgrip force between the genotypes (Table 3). However, the RR group had a higher isometric force relative to body mass than the XX group (Figure 3, upper panel). Jump height during a countermovement jump was similar between the genotype groups (Table 3). During the running test at 10 km/h on a treadmill, the post-hoc analyses did not reveal any difference in tidal volume or respiratory rate, although the differences in respiratory rate between XX and the other two genotype groups were small–moderate. As a result, pulmonary ventilation at 10 km/h was lower in the RR than in the XX group with a difference of moderate magnitude (Table 3). Post-hoc analyses did not reveal any between-groups difference for VO$_2$, respiratory exchange ratio or for the energy cost of running (Figure 3, lower panel and Table 3).

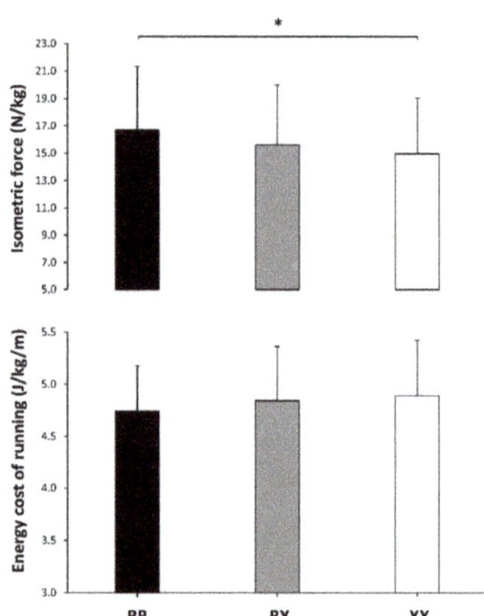

Figure 3. Maximal isometric force and energy cost of running in *ACTN3* RR, RX and XX recreational marathon runners. Data are mean ± SD for each genotype. (*) Differences identified by a post-hoc analysis at *p* < 0.05.

Table 3. Handgrip force, isometric force, countermovement jump height, and respiratory exchange data while running at 10 km/m, in *ACTN3* RR, RX, and XX marathon runners. Data are mean ± SD for each group. ES = Effect size calculated with Hedges' *g*; CI = Confidence interval; VO_2 = oxygen uptake. Degrees of freedom = 2, between-groups; 133, intra-group; 135, total.

Variable (Units)	RR	RX	XX	ES ± 95% CIRR vs. RX	ES ± 95% CIRR vs. XX	ES ± 95% CIRX vs. XX	p	η^2
Non-dominant handgrip force (N)	388 ± 73	402 ± 79	393 ± 91	0.3 ± 0.3	0.1 ± 0.4	−0.1 ± 0.4	0.663	<0.01
Dominant handgrip force (N)	414 ± 67	421 ± 81	415 ± 89	0.1 ± 0.4	0.0 ± 0.4	−0.1 ± 0.4	0.897	<0.01
Isometric force (N/kg)	16.7 ± 4.7	15.6 ± 4.4	14.7 ± 4.0	−0.2 ± 0.3	−0.5 ± 0.3	−0.2 ± 0.3	0.038	0.03
Countermovement jump height (cm)	26.9 + 4.2	26.8 + 5.3	26.2 ± 5.4	0.0 ± 0.4	−0.1 ± 0.3	−0.1 ± 0.4	0.818	<0.01
Tidal volume (L/respiration)	2.3 ± 0.7	2.5 ± 0.6	2.3 ± 0.6	−0.2 ± 0.3	0.0 ± 0.4	−0.2 ± 0.3	0.388	0.01
Respiratory rate (respirations/min)	28.7 ± 7.7	28.6 ± 8.7	31.8 ± 8.9	0.0 ± 0.3	0.4 ± 0.4	0.4 ± 0.3	0.188	0.03
Ventilation (L/min)	62.0 ± 9.5	65.8 ± 13.1	68.3 ± 12.9	0.3 ± 0.4	0.6 ± 0.3	0.2 ± 0.3	0.012	0.09
VO_2 (mL/kg/min)	38.4 ± 3.3	39.0 ± 3.9	39.3 ± 4.1	0.2 ± 0.4	0.2 ± 0.4	0.1 ± 0.4	0.389	<0.01
Respiratory exchange ratio	0.89 ± 0.06	0.89 ± 0.07	0.91 ± 0.06	0.0 ± 0.3	0.3 ± 0.4	0.3 ± 0.3	0.130	0.02
Energy cost of running (J/kg/m)	4.7 ± 0.4	4.8 ± 0.5	4.9 ± 0.5	0.2 ± 0.4	0.4 ± 0.4	0.2 ± 0.4	0.448	0.01

4. Discussion

The *ACTN3* R577X genotype is a well-characterized polymorphism that can affect physical and sports performance in elite athletes. Homozygosity for the X allele has been deemed a deleterious trait for success in elite sprint- and power-based sports, whereas the RR genotype has been considered propitious for the production of forceful and speedy contractions and thus optimal elite sport [1]. This information is being directly applied to recreational or less well-trained athletes without the knowledge of whether this polymorphism, and the consequent reduction/absence of α-actinin-3, affects muscle function in these populations as it does elite performers. Accordingly, our aim was to determine the influence of the three *ACTN3* genotypes on common exercise phenotypes in recreational marathon runners. As main outcomes, we found that the genotype frequencies of the *ACTN3* R577X polymorphism in our sample of marathoners was very similar to the ones previously reported in Spanish sedentary controls [47]. However, XX runners likely had higher body fat percentage, muscle flexibility and ankle dorsiflexion than RR or RX runners, despite similar age, running experience and training characteristics. Moreover, XX runners likely presented lower whole-body muscle force production and lower thigh fat-free mass volume than their RR counterparts. On the other hand, the *ACTN3* genotype did not affect the energy cost of running or the time necessary to complete a marathon. Overall, these data indicate that the R577X polymorphism might affect several characteristics in recreational endurance runners. Specifically, the outcomes of this investigation suggest that α-actinin-3-deficient runners might have lower strength values, but this does not translate into a poorer marathon performance. Although the magnitude of the *ACTN3* genotype effect on the exercise phenotypes investigated here was small-to-moderate in most cases, the consistency of the differences between the XX runners and the other two genotypes suggests a likely effect of α-actinin-3 deficiency on force production, muscle mass and muscle flexibility in endurance runners (Figure 4)

A myriad of case control studies has confirmed that homozygosity for the X allele hinders success in elite speed sports [2]. Nevertheless, these types of investigations do not allow for causal conclusions or explanations for this genetic influence on performance. To better understand the effect of α-actinin-3 deficiency, MacArthur et al. [6] generated an *Actn3* knock-out (KO) mouse, which recapitulates the features of human α-actinin-3 deficiency through loss of the expression of this protein in fast-type skeletal muscle fibers, while at the same time showing elevated expression of α-actinin-2. *Actn3*-KO mice have lower grip strength [29] and inferior fast force muscle production [48] than their wild-type littermates, which might explain in part the low frequency of XX in the pool of elite athletes in speed sports. In addition, *Actn3*-KO mice have less body weight due to lower lean body mass [25], with no changes in fat mass.

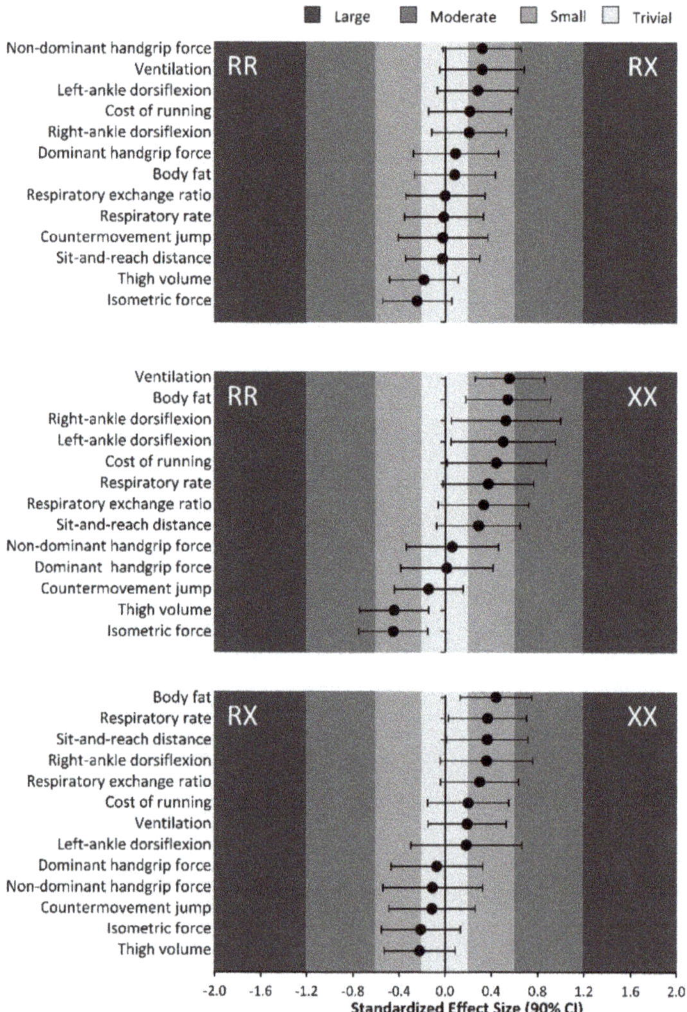

Figure 4. Magnitude of the *ACTN3* genotype effect, calculated with Hedges' $g \pm 95\%$ confidence interval, on the phenotypes investigated for the comparison of RR vs. RX (upper panel), RR vs. XX (middle panel), and RX vs. XX (lower panel) genotypes. Negative values indicate higher values of the variable for the group categorized at the left. Positive values indicate higher values of the variable for the group categorized at the right.

Interestingly, some of the main phenotypes found in the *Actn3*-KO mouse model have also been found in our sample of recreational runners, as XX runners likely presented lower values of maximal isometric force, lower fat-free volume and higher body fat percentage than RR or RX runners (in this case, without any effect on body mass). While the lack of α-actinin-3 has been related to these phenotypes in untrained individuals [8,9,49], this is the first study to show these negative effects on a sample of trained and competitive individuals, which might mirror the situation in elite athletes.

A summary of how the *ACTN3* genotype alters muscle function during exercise has been recently performed by Lee et al. [50] using the main outcomes found in the *Actn3*-KO mouse model. Briefly, α-actinin-3 interacts with a number of partner proteins, which broadly fall into three biological pathways: structural, metabolic and signaling. First, α-actinin-3 is a structural protein with a role for attaching and cross-linking actin filaments, and thus, its deficiency might negatively affect the structure of the sarcomere as well as its ability to produce force during muscle shortening [29]. As an additional cause for the lower values of force associated with the XX genotype, it has also been found that *Actn3* KO mice have a lower lactate dehydrogenase [6] and glycogen phosphorylase enzyme activity [51] in fast twitching muscle fibres. These enzymatic changes are consistent with a lower ability to catabolize glycogen into glucose and a subsequent reduction in the capacity to convert glucose into lactate in skeletal muscle fibres, which would be a limiting metabolic factor for high-intensity/fast muscle contractions. The spectrum of changes within muscle fibers associated with α-actinin-3 deficiency also includes a higher calcineurin activity, which is found in both *Actn3*-KO mice and humans with the XX genotype [52]. In this context, the muscle tissue of α-actinin-3-deficient individuals might be theoretically more prone to adapt to endurance training stimuli rather than to strength- or power-oriented programs [53]. Thus, based on the present findings, and those of the animal model, it might be assumed that the *ACTN3* XX genotype is likely associated with lower muscle strength. Whether these effects can be offset with individualized training warrants further investigation. In addition, exploring the relationship between α-actinin-3 deficiency and body fatness requires further analysis in active, untrained and obese populations, because, to date, this link is unclear [25].

It has been hypothesized that the X allele might have helped in the adaptation to environments with scarce food resources, where a more efficient muscle metabolism and lower cost of locomotion would be essential for survival [54]. Indeed, this theory was confirmed in the *Actn3*-KO mouse model, which showed a shift towards more efficient and aerobic muscle metabolism [6,29]. In humans, higher VO_{2peak} and higher running speed at the ventilatory threshold have been reported in XX versus RR counterparts [55,56]. However, a study designed to relate the XX genotype to a lower cost of locomotion found that the RX genotype was more efficient for running [31]. The present investigation also disputes this association, because the energy cost of running at 10 km/h was similar in all three genotypes groups. The lack of a lower cost of running in XX runners compared with R-allele carriers, might be related to another phenotype present in runners with α-actinin-3 deficiency, that is, a higher muscle flexibility and ankle range of motion (Table 2). Indeed, the less flexible runners might also be the most economical when running [57,58], suggesting that low muscle flexibility in certain areas of the musculoskeletal system may enhance running economy in sub-elite male runners. Thus, it is possible that the higher muscle flexibility of XX runners offsets the metabolic effect of α-actinin-3 deficiency, although this suggestion requires further investigation. In addition, XX runners needed greater pulmonary ventilation values than RR counterparts to meet the metabolic demands of running at 10 km/h. This finding is a novelty of this investigation as no previous investigation has reported changes in pulmonary ventilation across the three different *ACTN3* genotypes. Although the cause for a higher ventilation in XX is not evident from our data, this might be the result of a lower running speed at the respiratory compensation threshold. In any case, the explanation of an effect of the *ACTN3* genotype on the mechanics of ventilation during running, if any, merits further research. Overall, our findings suggest that the 577X allele does not increase running economy and, thus, the persistence of this null allele might be explained beyond its potential metabolically thrifty properties, as previously suggested [26].

An alternative hypothesis for the survival of the 577XX genotype emerges from our data and previous research. Higher values of muscle flexibility and range of motion have been found for XX individuals in different joints [20–22], which, although negative for running economy, might confer enhanced muscle function. In the present study, likely higher trunk flexibility and ankle dorsiflexion values were identified in XX runners, which might be associated with a higher ground force application [59] and lower leg stiffness during running [60]. In addition, higher muscle flexibility might produce a protective role against muscle damage during exercise [24]. However, despite the potential benefits of the enhanced muscle flexibility exhibited by our XX runners, previous evidence suggests that running economy or reduction of exercise-induced muscle damage are not improved by α-actinin-3 deficiency [1]. Lastly, it has been suggested that the *ACTN3* genotype affects α-actinin-3 expression in a dose-dependent fashion, indicating that RX individuals might have intermediate phenotypes [61]. Interestingly, this notion is reinforced by the present data because RX individuals were at the mid-point for isometric muscle strength, body fatness, thigh volume, muscle flexibility, and ankle dorsiflexion (Figures 1–3), while the magnitude for the effect with respect to XX runners was essentially lower than when comparing RR and XX genotypes (Figure 4). Thus, both the XX and RX genotype can be related to some favorable phenotypes for endurance running, although their utility should be consolidated with further research.

The present investigation does have some limitations. First, we recruited a convenience sample of 136 marathoners that were registered in a competitive marathon. After grouping by genotype, we found that groups were similar in the distribution of sexes, age, running experience, and training characteristics. However, there was high intersubject variability even within each group. Thus, further investigation with larger and more homogeneous samples and controlled training habits should be performed to confirm these outcomes. Also, we used a battery of testing to identify the effect of *ACTN3* genotypes on several exercise phenotypes of utility for endurance runners; however, other phenotypes such as VO_{2max} and running velocity at lactate threshold should be investigated because of their high association with endurance performance [62]. Several of the major outcomes of the present investigation could have been influenced by some variables that we did not assess, such as diet and previous resistance or muscle flexibility training background. The lack of control for these potential confounders therefore represents a limitation of our investigation. Finally, the present study was focused on the influence of only one polymorphism on these phenotypes, and several other candidate genes might have exerted an influence on the phenotypes investigated. Despite these limitations, we believe that the investigation contributes to the current knowledge on the effect α-actinin-3 deficiency has on muscle function, exercise traits.

5. Conclusions

In summary, compared with their RR peers, *ACTN3* XX marathon runners likely had lower values of whole-body isometric muscle force and lower fat-free mass volume in the thigh, and a higher percentage of body fatness. By contrast, *ACTN3* XX marathon runners had higher muscle flexibility and ankle range of motion, whereas no clear genotype-effect was found for running economy, handgrip force, and jump height. Thus, α-actinin-3 deficiency is associated with several physiological and anthropometric traits in recreational endurance runners. Nevertheless, the magnitude of differences among *ACTN3* genotypes was small-to-moderate and did not affect marathon performance. Future investigations should determine whether personalized endurance training based on genetics is effective to reduce the partially negative effects of α-actinin-3 deficiency in trained athletes. Although there is not yet a scientific rationale for the use of commercial genetic tests to predict sports performance [63,64], the outcomes of this investigation suggest that those recreational marathon runners with the *ACTN3* XX genotype might perhaps benefit from personalized strength training (to compensate for their lower muscle force capacity) more than their counterparts who are carriers of the R-allele.

Author Contributions: Conceptualization, J.D.C., V.M., J.G.-H., G.B.-M., C.R.-M., M.A.-N. and A.L.; methodology, J.D.C., V.M., J.G.-H., G.B.-M., C.R.-M. and M.A.-N.; data acquisition, J.D.C., V.M., J.G.-H., G.B.-M., C.R.-M., M.A.-N. and B.L.; formal analysis, J.D.C. and A.L.; writing—original draft preparation, J.D.C. and A.L. writing—review and editing, V.M., J.G.-H., G.B.-M., C.R.-M., M.A.-N. and B.L.; project administration, J.D.C.; funding acquisition, J.D.C.

Funding: The study was part of the ACTN3 project supported by a Grant-in-aid from the Vice-Rectorate of Research and Science, at the Camilo Jose Cela University. Research by A.L. in the field is funded by the Spanish Ministry of Economy and Competitiveness (Fondo de Investigaciones Sanitarias) and Fondos FEDER (grant numbers PI15/00558 and PI18/00139).

Acknowledgments: The authors wish to thank the participants in this study for their invaluable contribution. In addition, we are very grateful to the Organization of Rock'n' Roll Madrid Marathon & ½ for their help in establishing the investigation areas at the start and finish lines.

Conflicts of Interest: The authors declare that they have no conflict of interest derived from the outcomes of this study.

References

1. Del Coso, J.; Hiam, D.; Houweling, P.J.; Pérez, L.M.; Eynon, N.; Lucía, A. More than a "speed gene": *ACTN3* R577X genotype, trainability, muscle damage, and the risk for injuries. *Eur. J. Appl. Physiol.* **2019**, *119*, 49–60. [CrossRef]

2. Houweling, P.J.; Papadimitriou, I.D.; Seto, J.T.; Pérez, L.M.; Del Coso, J.; North, K.N.; Lucia, A.; Eynon, N. Is evolutionary loss our gain? The role of *ACTN3* p.Arg577Ter (R577X) genotype in athletic performance, ageing, and disease. *Hum. Mutat.* **2018**, *39*, 1774–1787. [CrossRef]

3. North, K.N.; Yang, N.; Wattanasirichaigoon, D.; Mills, M.; Easteal, S.; Beggs, A.H. A common nonsense mutation results in α-actinin-3 deficiency in the general population. *Nat. Genet.* **1999**, *21*, 353–354. [CrossRef]

4. Seto, J.T.; Lek, M.; Quinlan, K.G.R.; Houweling, P.J.; Zheng, X.F.; Garton, F.; MacArthur, D.G.; Raftery, J.M.; Garvey, S.M.; Hauser, M.A.; et al. Deficiency of α-actinin-3 is associated with increased susceptibility to contraction-induced damage and skeletal muscle remodeling. *Hum. Mol. Genet.* **2011**, *20*, 2914–2927. [CrossRef]

5. Pickering, C.; Kiely, J. *ACTN3*, morbidity, and healthy aging. *Front. Genet.* **2018**, *9*, 15. [CrossRef]

6. MacArthur, D.G.; Seto, J.T.; Raftery, J.M.; Quinlan, K.G.; Huttley, G.A.; Hook, J.W.; Lemckert, F.A.; Kee, A.J.; Edwards, M.R.; Berman, Y.; et al. Loss of *ACTN3* gene function alters mouse muscle metabolism and shows evidence of positive selection in humans. *Nat. Genet.* **2007**, *39*, 1261–1265. [CrossRef]

7. amazon.com Test de ADN 24Genetics. Available online: https://www.amazon.es/Test-24Genetics-Todo-Nutrigenetica-Farmacogenetica/dp/B071ZVNWXL/ref=sr_1_fkmr1_1?ie=UTF8&qid=1552053631&sr=8-1-fkmr1&keywords=24+genetics+deporte (accessed on 8 March 2019).

8. Walsh, S.; Liu, D.; Metter, E.J.; Ferrucci, L.; Roth, S.M. *ACTN3* genotype is associated with muscle phenotypes in women across the adult age span. *J. Appl. Physiol.* **2008**, *105*, 1486–1491. [CrossRef]

9. Clarkson, P.M.; Devaney, J.M.; Gordish-Dressman, H.; Thompson, P.D.; Hubal, M.J.; Urso, M.; Price, T.B.; Angelopoulos, T.J.; Gordon, P.M.; Moyna, N.M.; et al. *ACTN3* genotype is associated with increases in muscle strength in response to resistance training in women. *J. Appl. Physiol.* **2005**, *99*, 154–163. [CrossRef]

10. Broos, S.; Van Leemputte, M.; Deldicque, L.; Thomis, M.A. History-dependent force, angular velocity and muscular endurance in *ACTN3* genotypes. *Eur. J. Appl. Physiol.* **2015**, *115*, 1637–1643. [CrossRef]

11. Norman, B.; Esbjörnsson, M.; Rundqvist, H.; Osterlund, T.; von Walden, F.; Tesch, P.A. Strength, power, fiber types, and mRNA expression in trained men and women with different *ACTN3* R577X genotypes. *J. Appl. Physiol.* **2009**, *106*, 959–965. [CrossRef]

12. Hanson, E.D.; Ludlow, A.T.; Sheaff, A.K.; Park, J.; Roth, S.M. *ACTN3* genotype does not influence muscle power. *Int. J. Sports Med.* **2010**, *31*, 834–838. [CrossRef]

13. Kikuchi, N.; Nakazato, K.; Min, S.; Ueda, D.; Igawa, S. The *ACTN3* R577X polymorphism is associated with muscle power in male Japanese athletes. *J. Strength Cond. Res.* **2014**, *28*, 1783–1789. [CrossRef]

14. Norman, B.; Esbjörnsson, M.; Rundqvist, H.; Österlund, T.; Glenmark, B.; Jansson, E. *ACTN3* genotype and modulation of skeletal muscle response to exercise in human subjects. *J. Appl. Physiol.* **2014**, *116*, 1197–1203. [CrossRef] [PubMed]

15. Broos, S.; Malisoux, L.; Theisen, D.; van Thienen, R.; Ramaekers, M.; Jamart, C.; Deldicque, L.; Thomis, M.A.; Francaux, M. Evidence for *ACTN3* as a speed gene in isolated human muscle fibers. *PLoS ONE* **2016**, *11*, e0150594. [CrossRef]

16. Delmonico, M.J.; Kostek, M.C.; Doldo, N.A.; Hand, B.D.; Walsh, S.; Conway, J.M.; Carignan, C.R.; Roth, S.M.; Hurley, B.F. α-actinin-3 (*ACTN3*) R577X polymorphism influences knee extensor peak power response to strength training in older men and women. *J. Gerontol. A Biol. Sci. Med. Sci.* **2007**, *62*, 206–212. [CrossRef]

17. Del Coso, J.; Valero, M.; Salinero, J.J.; Lara, B.; Díaz, G.; Gallo-Salazar, C.; Ruiz-Vicente, D.; Areces, F.; Puente, C.; Carril, J.C.; et al. *ACTN3* genotype influences exercise-induced muscle damage during a marathon competition. *Eur. J. Appl. Physiol.* **2017**, *117*, 409–416. [CrossRef]

18. Del Coso, J.; Salinero, J.J.; Lara, B.; Gallo-Salazar, C.; Areces, F.; Puente, C.; Herrero, D. *ACTN3* X-allele carriers had greater levels of muscle damage during a half-ironman. *Eur. J. Appl. Physiol.* **2017**, *117*, 151–158. [CrossRef]

19. Del Coso, J.; Valero, M.; Salinero, J.J.; Lara, B.; Gallo-Salazar, C.; Areces, F. Optimum polygenic profile to resist exertional rhabdomyolysis during a marathon. *PLoS ONE* **2017**, *12*, e0172965. [CrossRef]

20. Zempo, H.; Fuku, N.; Murakami, H.; Miyachi, M. The relationship between α-Actinin-3 gene R577X polymorphism and muscle flexibility. *Juntendo Med. J.* **2016**, *62*, 118. [CrossRef]

21. Kikuchi, N.; Zempo, H.; Fuku, N.; Murakami, H.; Sakamaki-Sunaga, M.; Okamoto, T.; Nakazato, K.; Miyachi, M. Association between *ACTN3* R577X polymorphism and trunk flexibility in 2 different cohorts. *Int. J. Sports Med.* **2017**, *38*, 402–406. [CrossRef]

22. Kikuchi, N.; Tsuchiya, Y.; Nakazato, K.; Ishii, N.; Ochi, E. Effects of the *ACTN3* R577X genotype on the muscular strength and range of motion before and after eccentric contractions of the elbow flexors. *Int. J. Sports Med.* **2018**, *39*, 148–153. [CrossRef]

23. Kim, J.H.; Jung, E.S.; Kim, C.-H.; Youn, H.; Kim, H.R. Genetic associations of body composition, flexibility and injury risk with *ACE*, *ACTN3* and *COL5A1* polymorphisms in Korean ballerinas. *J. Exerc. Nutr. Biochem.* **2014**, *18*, 205–214. [CrossRef]

24. Chen, C.-H.; Nosaka, K.; Chen, H.-L.; Lin, M.-J.; Tseng, K.-W.; Chen, T.C. Effects of flexibility training on eccentric exercise-induced muscle damage. *Med. Sci. Sport. Exerc.* **2011**, *43*, 491–500. [CrossRef]

25. Houweling, P.J.; Berman, Y.D.; Turner, N.; Quinlan, K.G.R.; Seto, J.T.; Yang, N.; Lek, M.; Macarthur, D.G.; Cooney, G.; North, K.N. Exploring the relationship between α-actinin-3 deficiency and obesity in mice and humans. *Int. J. Obes.* **2017**, *41*, 1154–1157. [CrossRef]

26. Moran, C.N.; Yang, N.; Bailey, M.E.S.; Tsiokanos, A.; Jamurtas, A.; MacArthur, D.G.; North, K.; Pitsiladis, Y.P.; Wilson, R.H. Association analysis of the *ACTN3* R577X polymorphism and complex quantitative body composition and performance phenotypes in adolescent Greeks. *Eur. J. Hum. Genet.* **2007**, *15*, 88–93. [CrossRef]

27. Riedl, I.; Osler, M.E.; Benziane, B.; Chibalin, A.V.; Zierath, J.R. Association of the *ACTN3* R577X polymorphism with glucose tolerance and gene expression of sarcomeric proteins in human skeletal muscle. *Physiol. Rep.* **2015**, *3*, e12314. [CrossRef]

28. Amorim, C.E.G.; Acuña-Alonzo, V.; Salzano, F.M.; Bortolini, M.C.; Hünemeier, T. Differing evolutionary histories of the *ACTN3*R577X polymorphism among the major human geographic groups. *PLoS ONE* **2015**, *10*, e0115449. [CrossRef]

29. MacArthur, D.G.; Seto, J.T.; Chan, S.; Quinlan, K.G.R.; Raftery, J.M.; Turner, N.; Nicholson, M.D.; Kee, A.J.; Hardeman, E.C.; Gunning, P.W.; et al. An *Actn3* knockout mouse provides mechanistic insights into the association between α-actinin-3 deficiency and human athletic performance. *Hum. Mol. Genet.* **2008**, *17*, 1076–1086. [CrossRef]

30. Deschamps, C.L.; Hittel, D.S. *ACTN3*: A thrifty gene for speed? *Endocr. Pract.* **2016**, *22*, 897–898. [CrossRef]

31. Pasqua, L.A.; Bueno, S.; Matsuda, M.; Marquezini, M.V.; Lima-Silva, A.E.; Saldiva, P.H.N.; Bertuzzi, R. The genetics of human running: *ACTN3* polymorphism as an evolutionary tool improving the energy economy during locomotion. *Ann. Hum. Biol.* **2016**, *43*, 255–260. [CrossRef]

32. Papadimitriou, I.D.; Lockey, S.J.; Voisin, S.; Herbert, A.J.; Garton, F.; Houweling, P.J.; Cieszczyk, P.; Maciejewska-Skrendo, A.; Sawczuk, M.; Massidda, M.; et al. No association between *ACTN3* R577X and ACE I/D polymorphisms and endurance running times in 698 Caucasian athletes. *BMC Genom.* **2018**, *19*, 13. [CrossRef]

33. Guilherme, J.P.L.F.; Bertuzzi, R.; Lima-Silva, A.E.; Pereira, A.d.C.; Lancha Junior, A.H. Analysis of sports-relevant polymorphisms in a large Brazilian cohort of top-level athletes. *Ann. Hum. Genet.* **2018**, *82*, 254–264. [CrossRef]

34. Stewart, A.; Marfell-Jones, M. International Society for Advancement of Kinanthropometry. In *International Standards for Anthropometric Assessment*; International Society for the Advancement of Kinanthropometry: Lower Hutt, New Zealand, 2011; ISBN 0620362073.

35. Durnin, J.V.; Womersley, J. Body fat assessed from total body density and its estimation from skinfold thickness: Measurements on 481 men and women aged from 16 to 72 years. *Br. J. Nutr.* **1974**, *32*, 77–97. [CrossRef]

36. Jones, P.R.; Pearson, J. Anthropometric determination of leg fat and muscle plus bone volumes in young male and female adults. *J. Physiol.* **1969**, *204*, 63P–66P.

37. Brouwer, E. On simple formulae for calculating the heat expenditure and the quantities of carbohydrate and fat oxidized in metabolism of men and animals, from gaseous exchange (Oxygen intake and carbonic acid output) and urine-N. *Acta Physiol. Pharmacol. Neerl.* **1957**, *6*, 795–802.

38. Tam, E.; Rossi, H.; Moia, C.; Berardelli, C.; Rosa, G.; Capelli, C.; Ferretti, G. Energetics of running in top-level marathon runners from Kenya. *Eur. J. Appl. Physiol.* **2012**, *112*, 3797–3806. [CrossRef]

39. Salinero, J.J.; Soriano, M.L.; Lara, B.; Gallo-Salazar, C.; Areces, F.; Ruiz-Vicente, D.; Abián-Vicén, J.; González-Millán, C.; Del Coso, J. Predicting race time in male amateur marathon runners. *J. Sports Med. Phys. Fitness* **2017**, *57*, 1169–1177.

40. Calatayud, J.; Martin, F.; Gargallo, P.; García-Redondo, J.; Colado, J.C.; Marín, P.J. The validity and reliability of a new instrumented device for measuring ankle dorsiflexion range of motion. *Int. J. Sports Phys. Ther.* **2015**, *10*, 197–202.

41. Bennell, K.; Talbot, R.; Wajswelner, H.; Techovanich, W.; Kelly, D.; Hall, A. Intra-rater and inter-rater reliability of a weight-bearing lunge measure of ankle dorsiflexion. *Aust. J. Physiother.* **1998**, *44*, 175–180. [CrossRef]

42. Driller, M.W.; Overmayer, R.G. The effects of tissue flossing on ankle range of motion and jump performance. *Phys. Ther. Sport* **2017**, *25*, 20–24. [CrossRef]

43. Sambrook, J.; Fritsch, E.; Maniatis, T. *Molecular Cloning: A Laboratory Manual*, 2nd ed.; Cold Spring Harbor Laboratory Press: Cold Spring Harbor, NY, USA, 1989.

44. Consorcio Candela Análisis de la Diversidad y Evolución de Latinoamérica. Available online: https://www.ucl.ac.uk/candela/spanish/inicio (accessed on 10 April 2019).

45. Cohen, J. *Statistical Power Analysis for the Behavioral Sciences*, 2nd ed.; L. Erlbaum Associates: Hillsdale, NJ, USA, 1988; ISBN 9780805802832.

46. Hopkins, W.G.; Marshall, S.; Batterham, A.; Hanin, J. Progressive statistics for studies in sports medicine and exercise science. *Med. Sci. Sport. Exerc.* **2009**, *41*, 3–13. [CrossRef]

47. Eynon, N.; Ruiz, J.R.; Femia, P.; Pushkarev, V.P.; Cieszczyk, P.; Maciejewska-Karlowska, A.; Sawczuk, M.; Dyatlov, D.A.; Lekontsev, E.V.; Kulikov, L.M.; et al. The *ACTN3* R577X polymorphism across three groups of elite male European athletes. *PLoS ONE* **2012**, *7*, e43132. [CrossRef]

48. Seto, J.T.; Chan, S.; Turner, N.; MacArthur, D.G.; Raftery, J.M.; Berman, Y.D.; Quinlan, K.G.R.; Cooney, G.J.; Head, S.; Yang, N.; et al. The effect of α-actinin-3 deficiency on muscle aging. *Exp. Gerontol.* **2011**, *46*, 292–302. [CrossRef]

49. Erskine, R.M.; Williams, A.G.; Jones, D.A.; Stewart, C.E.; Degens, H. The individual and combined influence of *ACE* and *ACTN3* genotypes on muscle phenotypes before and after strength training. *Scand. J. Med. Sci. Sports* **2014**, *24*, 642–648. [CrossRef]

50. Lee, F.X.Z.; Houweling, P.J.; North, K.N.; Quinlan, K.G.R. How does α-actinin-3 deficiency alter muscle function? *Mechanistic insights into ACTN3, the 'gene for speed.' Biochim. Biophys. Acta—Mol. Cell Res.* **2016**, *1863*, 686–693. [CrossRef]

51. Quinlan, K.G.R.; Seto, J.T.; Turner, N.; Vandebrouck, A.; Floetenmeyer, M.; Macarthur, D.G.; Raftery, J.M.; Lek, M.; Yang, N.; Parton, R.G.; et al. α-actinin-3 deficiency results in reduced glycogen phosphorylase activity and altered calcium handling in skeletal muscle. *Hum. Mol. Genet.* **2010**, *19*, 1335–1346. [CrossRef]

52. Seto, J.T.; Quinlan, K.G.R.; Lek, M.; Zheng, X.F.; Garton, F.; MacArthur, D.G.; Hogarth, M.W.; Houweling, P.J.; Gregorevic, P.; Turner, N.; et al. *ACTN3* genotype influences muscle performance through the regulation of calcineurin signaling. *J. Clin. Investig.* **2013**, *123*, 4255–4263. [CrossRef]

53. Garton, F.C.; Seto, J.T.; Quinlan, K.G.R.; Yang, N.; Houweling, P.J.; North, K.N. α-Actinin-3 deficiency alters muscle adaptation in response to denervation and immobilization. *Hum. Mol. Genet.* **2014**, *23*, 1879–1893. [CrossRef]

54. Friedlander, S.M.; Herrmann, A.L.; Lowry, D.P.; Mepham, E.R.; Lek, M.; North, K.N.; Organ, C.L. *ACTN3* allele frequency in humans covaries with global latitudinal gradient. *PLoS ONE* **2013**, *8*, e52282. [CrossRef]

55. Silva, M.S.M.; Bolani, W.; Alves, C.R.; Biagi, D.G.; Lemos, J.R.; da Silva, J.L.; de Oliveira, P.A.; Alves, G.B.; de Oliveira, E.M.; Negrão, C.E.; et al. Elimination of influences of the *ACTN3* R577X variant on oxygen uptake by endurance training in healthy individuals. *Int. J. Sports Physiol. Perform.* **2015**, *10*, 636–641. [CrossRef] [PubMed]

56. Pasqua, L.A.; Bueno, S.; Artioli, G.G.; Lancha, A.H.; Matsuda, M.; Marquezini, M.V.; Lima-Silva, A.E.; Saldiva, P.H.N.; Bertuzzi, R. Influence of *ACTN3* R577X polymorphism on ventilatory thresholds related to endurance performance. *J. Sports Sci.* **2016**, *34*, 163–170. [CrossRef] [PubMed]

57. Jones, A.M. Running economy is negatively related to sit-and-reach test performance in international-standard distance runners. *Int. J. Sports Med.* **2002**, *23*, 40–43. [CrossRef]

58. Craib, M.W.; Mitchell, V.A.; Fields, K.B.; Cooper, T.R.; Hopewell, R.; Morgan, D.W. The association between flexibility and running economy in sub-elite male distance runners. *Med. Sci. Sports Exerc.* **1996**, *28*, 737–743. [CrossRef] [PubMed]

59. Schrödter, E.; Brüggemann, G.-P.; Willwacher, S. Is soleus muscle-tendon-unit behavior related to ground-force application during the sprint start? *Int. J. Sports Physiol. Perform.* **2017**, *12*, 448–454. [CrossRef]

60. Goodwin, J.S.; Blackburn, J.T.; Schwartz, T.A.; Williams, D.S.B. Clinical predictors of dynamic lower extremity stiffness during running. *J. Orthop. Sport. Phys. Ther.* **2019**, *49*, 98–104. [CrossRef]

61. Hogarth, M.W.; Garton, F.C.; Houweling, P.J.; Tukiainen, T.; Lek, M.; Macarthur, D.G.; Seto, J.T.; Quinlan, K.G.R.; Yang, N.; Head, S.I.; et al. Analysis of the *ACTN3* heterozygous genotype suggests that α-actinin-3 controls sarcomeric composition and muscle function in a dose-dependent fashion. *Hum. Mol. Genet.* **2016**, *25*, 866–877. [CrossRef] [PubMed]

62. Santos-Concejero, J.; Oliván, J.; Maté-Muñoz, J.L.; Muniesa, C.; Montil, M.; Tucker, R.; Lucia, A. Gait-cycle characteristics and running economy in elite Eritrean and European runners. *Int. J. Sports Physiol. Perform.* **2015**, *10*, 381–387. [CrossRef] [PubMed]

63. Kreutzer, A.; Martinez, C.A.; Kreutzer, M.; Stone, J.D.; Mitchell, J.B.; Oliver, J.M. Effect of *ACTN3* polymorphism on self-reported running times. *J. Strength Cond. Res.* **2019**, *33*, 80–88. [CrossRef] [PubMed]

64. Webborn, N.; Williams, A.; McNamee, M.; Bouchard, C.; Pitsiladis, Y.; Ahmetov, I.; Ashley, E.; Byrne, N.; Camporesi, S.; Collins, M.; et al. Direct-to-consumer genetic testing for predicting sports performance and talent identification: Consensus statement. *Br. J. Sports Med.* **2015**, *49*, 1486–1491. [CrossRef]

© 2019 by the authors. Licensee MDPI, Basel, Switzerland. This article is an open access article distributed under the terms and conditions of the Creative Commons Attribution (CC BY) license (http://creativecommons.org/licenses/by/4.0/).

 *genes*

Article

GCKR rs780094 Polymorphism as A Genetic Variant Involved in Physical Exercise

Isabel Espinosa-Salinas [1], Rocio de la Iglesia [1,2], Gonzalo Colmenarejo [3], Susana Molina [1], Guillermo Reglero [1,4], J. Alfredo Martinez [1,5,6], Viviana Loria-Kohen [1,*] and Ana Ramirez de Molina [1,*]

[1] Nutrition and Clinical Trials Unit, IMDEA Food CEI UAM + CSIC, 28049 Madrid, Spain
[2] Department of Pharmaceutical and Health Sciences, Faculty of Pharmacy Universidad San Pablo-CEU, CEU Universities, Urbanización Montepríncipe, 28925 Madrid, Spain
[3] Biostatistics and Bioinformatics Unit, IMDEA Food CEI UAM + CSIC, 28049 Madrid, Spain
[4] Department of Production and Characterization of New Foods, Institute of Food Science Research (CIAL) CEI UAM + CSIC, 28049 Madrid, Spain
[5] Department of Food Sciences and Physiology, University of Navarra, 31009 Pamplona, Spain
[6] CIBERobn, Instituto Carlos III, 28029 Madrid, Spain
* Correspondence: viviana.loria@imdea.org (V.L.-K.); ana.ramirez@imdea.org (A.R.d.M.); Tel.: +34-91-727-8100 (V.L.-K. & A.R.d.M.)

Received: 27 June 2019; Accepted: 25 July 2019; Published: 28 July 2019

Abstract: Exercise performance is influenced by genetics. However, there is a lack of knowledge about the role played by genetic variability in the frequency of physical exercise practice. The objective was to identify genetic variants that modulate the commitment of people to perform physical exercise and to detect those subjects with a lower frequency practice. A total of 451 subjects were genotyped for 64 genetic variants related to inflammation, circadian rhythms, vascular function as well as energy, lipid and carbohydrate metabolism. Physical exercise frequency question and a Minnesota Leisure Time Physical Activity Questionnaire (MLTPAQ) were used to qualitatively and quantitatively measure the average amount of physical exercise. Dietary intake and energy expenditure due to physical activity were also studied. Differences between genotypes were analyzed using linear and logistic models adjusted for Bonferroni. A significant association between *GCKR* rs780094 and the times the individuals performed physical exercise was observed ($p = 0.004$). The carriers of the minor allele showed a greater frequency of physical exercise in comparison to the major homozygous genotype carriers (OR: 1.86, 95% CI: 1.36–2.56). The analysis of the *GCKR* rs780094 variant suggests a possible association with the subjects that present lower frequency of physical exercise. Nevertheless, future studies are needed to confirm these findings.

Keywords: glucokinase-regulator; exercise; behavior; genotyping; obesity

---

1. Introduction

Physical exercise and energy intake are fundamental factors in the energy balance equation and in body weight management [1]. Knowledge about the factors associated with physical activity practice and energy expenditure as well as the potential mechanisms involved in fuel homeostasis is essential in order to understand human thermodynamics [1].

The energy used during physical performance is the most variable component of total energy expenditure [2]. This feature depends on different factors such as body composition, intensity and duration of the physical activity, and the individual genetic profile [3,4]. Also, the sport aptitude has a strong genetic component [5,6]. However, there is a lack of research about the role that genetic variability plays in the motivation for practicing any physical exercise, the adherence to it and the related efficiency.

Multiple metabolic and physiological processes have been related to physical training [7]. The study of genetic variants occurring in genes associated with the regulation of such metabolic processes could help determine how these processes are implicated in the practice of physical exercise. Although there are Genome-wide association studies (GWAS) linking specific locus with exercise and, also, reviews determining associations between SNPs with sports performance, there is scarce evidence on the relationship between obesity and metabolism-associated polymorphisms with physical exercise [8–11]. Further studies may be of interest to establish more precise relationships between genes related to metabolism and how they affect the practice of physical exercise.

Thus, genes related to inflammation such as those encoding interleukins, or the C-reactive protein have been linked to athletic performance [12]. In turn, genes associated with energy metabolism such as *FTO* and *POMC*, as well as with circadian rhythms like *CLOCK* and *PER2* gene, have been related to the muscular system or the response to sport and exercise [13–16]. In the case of vascular function, genes such as the *NOS3* gene, whose effect is modified by the practice of physical exercise, or the *GNB3* gene, known to be associated with elite athletes, have also been investigated [17,18]. Moreover, studies of genetic variants involved in lipid metabolism such as apolipoproteins or peroxisome proliferator-activated receptors (*PPARs*) and in carbohydrate metabolism such as *ADIPOQ*, have been associated with sports practice, which could also influence predisposition to the practice of physical exercise [19–21].

In this context, the study of the glucokinase regulatory protein (*GCKR*) gene, involved in lipid and carbohydrate metabolism, can be of interest as it might also interfere with the amount of physical exercise performed [22]. The *GCKR* gene, located on chromosomal region 2p23.2–3, modulates glucokinase (GCK) that is a key regulatory enzyme of glucose metabolism and storage, and potentially implicated in energy utilization [23,24]. Thus, a hypothetical mechanism behind a physically active or sedentary behaviour may lie on the greater or lesser availability of energy substrates, as well as the signals these substrates exert at the brain level.

Recently, the single nucleotide polymorphism rs780094, located in an intronic region of the *GCKR* gene, was found to be related with triacylglycerides (TAG) and fasting plasma glucose concentrations [23]. Specifically, the major allele (C) of *GCKR* rs7800094 is associated with decreased TAG, but increased fasting plasma glucose, while minor allele (T) is associated with lower levels of fasting plasma glucose and insulin, and higher levels of TAG [25]. In this regard, it is interesting to point out that GCKR has been identified in the same brain area than GCK, what could modulate the feeding behavior and energy balance [26]. GCKR and GCK, which act as glucose-level sensors, might interact with appetite-regulating peptides and interfere in the individual feeling of satiety [27].

Information about genetic variants that modulate energy intake and expenditure through the adherence to physical exercise performance can be useful to detect those subjects with lower frequency of exercise practice and those less prone to maintain an adequate energy balance to maintain a healthy status, to prevent overweight and obesity, and to personalize loss weight strategies. In this context, the aim of the present study was to identify genetic variants that modulate the amount of physical activity and exercise performed to detect those subjects that exercise less, with the goal to prevent overweight.

2. Materials and Methods

2.1. Subjects and Study Protocol

A total of 557 subjects (155 men and 402 women, aged from 18 to 65 years), who participated in a GENYAL Platform (Platform for Clinical Trials in Nutrition and Health) observational study, encompassed this investigation. Participants were recruited in Cantoblanco Campus (Autonomous University of Madrid, Madrid, Spain), being almost entirely of Caucasian origin. Inclusion criteria were: free-living adults aged from 18 to 70 years that gave written informed consent to be contacted to perform clinical trials and nutritional intervention studies. Exclusion criteria were: To suffer from any serious illness (kidney or liver diseases or other condition that affects lifestyle or diet), to present dementia or impaired cognitive function and to be pregnant or breastfeeding. Of the total of the participants, it was not possible to obtain all the data. Thus, only 490 were evaluated for physical activity and exercise practice while only 451 were genotyped. The study was conducted according to the guidelines laid down in the Declaration of Helsinki, and all procedures involving human subjects were approved by the Research Ethic Committee of Autonomous University of Madrid (CEI 27-666). Written informed consent was obtained from all subjects.

2.2. Anthropometric Measures and Dietary Intake

Anthropometric and body composition variables, such as height, weight, fat mass percentage and waist circumference, were measured by standard validated techniques [28]. Body weight and fat mass percentage were assessed using the body composition monitor BF511 (Omron Healthcare UK Ltd., Kyoto, Japan). BMI was calculated as the body weight divided by the squared height (kg/m^2). Waist and hip circumferences were measured using a Seca 201 non-elastic tape (Quirumed, Valencia, Spain). A 72-h food record was collected from all participants. DIAL (2.16 version, Alce Ingeniería, Las Rozas, Madrid, Spain) Software was used to analyze the energy intake, macro and micronutrients of the collected records [29].

2.3. Physical Activity Measures

The frequency of physical exercise practice on a regular basis ("exercise" variable), was quantified by a specific question in order to define how many times per week they used to practice physical exercise: 0, 1, 2, 3, 4 or 5 and > 5 times. A dichotomized version of this variable was created ("exercise classification") as follows: (1) Subjects who practiced physical exercise 0 times per week and (2) subjects who practiced physical exercise at least one day per week. This classification was made in order to better split the sample in terms of exercise practice: those who did not practice at all vs. those who practiced some exercise.

In addition, Minnesota Leisure Time Physical Activity Questionnaire (MLTPAQ) was used to quantitatively measure the average physical activity practice (kcal/day) by the volunteers [30]. Based on the Compendium of physical activities, Energy Expenditure in Physical Activity (EEPA) was estimated as follows: EEPA = I × N × T; where "I" represents the degree of intensity for each physical activity in kilocalories / minute; "N", the number of times that physical activity was developed; and "T", the time in minutes spent in each session [31].

2.4. Biochemical Measurements

Blood samples were taken early in the morning at IMDEA Food after a 12-h overnight fast and stored at 4 °C to 6 °C until analysis (always performed within 48 h) by Laboratory CQS Consulting. This laboratory integrates preanalytical, analytical and post-analytical processes. Total cholesterol (TC), high-density lipoprotein (HDL) and low-density lipoprotein cholesterol (LDL), triacylglycerols (TAG) and glucose were determined by enzymatic spectrophotometric assay using an Architect CI8200 instrument (Abbott Laboratories, Chicago, IL, USA). The triglyceride-glucose index (TyG index) was calculated as the natural logarithm (Ln) of the product of plasma glucose and TAG according to the following formula: Ln (TAG (mg/dL) × glucose (mg/dL))/2 [32].

2.5. DNA Isolation and Genotyping

A total of 64 genetic variants related to inflammation (7 genes; 9 SNPs), circadian rhythms (2 genes; 6 SNPs), vascular function (8 genes; 10 SNPs) as well as energy (11 genes; 17 SNPs, lipid (9 genes; 15 SNPs) and carbohydrate metabolism (3 genes; 7 SNPs), were analyzed. The selection was based on a previous study in accordance with the allelic frequency and TaqMan® probe (Applied Biosystems, Waltham, MA USA) availability [22]. Blood samples were taken and stored at −80 °C until DNA extraction. Genomic DNA from each participant was isolated from whole blood using the QIAamp DNA Blood Mini Kit (Qiagen Sciences, Inc., Germantown, MD, USA) and recovered in 100 mL of nuclease-free water. Its concentration and quality were then measured in a nanodrop ND-2000 spectrophotometer (ThermoScientific, Waltham, MA, USA). The mean concentration of the samples was 80 to 90 ng/mL. Genotyping was performed using the QuantStudio_ 12 K Flex Real-Time PCR System (Life Technologies Inc., Carlsbad, CA, USA) with a TaqMan OpenArray plates following the manufactured instructions (Real-Time PCR Handbook and education center of Applied Biosystem) [33]. The results were analyzed using TaqMan Genotyper software (V 1.3, Applied Biosystems, Waltham, MA USA). The proportion of genotypes not passing the quality threshold was <5%.

2.6. Statistical Analysis

Statistical analysis was performed with the R software, version 3.4.1 (R Foundation for Statistical Computing, Vienna, Austria) [34]. Deviations from Hardy-Weinberg equilibrium of genotype frequencies at individual loci were assessed using standard $\chi2$ tests. Descriptive analyses were implemented for different continuous and categorical variables by sex, SNP (single nucleotide polymorphism) rs780094 and exercise classification. *p*-Values were obtained using one-way ANOVA for continuous variables, and Fisher exact test for categorical variables. Associations between the two variables, exercise variable and energy expenditure due to physical activity per week, were assessed by computing the Pearson correlation coefficient (r) and the corresponding statistical test. Identification of significant SNPs was performed by deriving logistic model vs. SNPs, adjusted by sex, age and additional appropriate covariables when necessary. Model assumptions were checked in all the cases. Three logistic genetic models were evaluated: Additive, dominant and codominant. These genetic models refer to the dominant or recessive character of each of the alleles. Thus, the additive model indicates a multiplicative effect depending on the number of rare alleles. In the dominant model, the less frequent allele is considered dominant (common homozygous vs. heterozygous + rare homozygous), and the codominant model considers each genotype individually, with independent genetic effects. The Bonferroni method was used to correct the *p*-values for multiple test (64 SNPs). Significance level was set to $\alpha = 0.05$ adjusted after Bonferroni correction. No power analysis could be conducted as there was no idea of the variability of the responses analyzed in advance, and the sample used was the largest available.

3. Results

Out of all the genetic variants studied, only one significant statistical association between *GCKR* rs780094 polymorphism and the frequency of physical activity performed was found, when adjusted for multiple corrections. Considering the number of polymorphisms studied and the obtained results, only the related ones to this association will be discussed.

Descriptive statistics concerning gender, exercise classification and *GCKR* rs780094 data showed several differences regarding population characteristics (Table 1). This table shows anthropometric measures, biochemical variables, dietary intake and physical activity and exercise characteristics, according to gender, exercise classification (inactive vs. active subjects, see above) and genotype *GCKR* rs780094.

The analysis showed significant differences concerning gender for the total anthropometric variables. Thus, males presented significantly higher values of body weight, height, BMI, lean mass percentage, visceral fat, waist circumference and waist-hip ratio than females. On the other hand, fat mass in males resulted in significantly lower values than females. Several biochemical parameters were also significantly different concerning gender. Hence, women had higher total cholesterol and HDL cholesterol levels while triglycerides/glucose index, Total Cholesterol/HDL and LDL/HDL were of lower values. Besides, men consumed and expended statistically significant more calories, as well as performed more times of physical exercise per week than women.

Concerning exercise classification (inactive subjects who never practiced any kind of physical exercise per week vs. active subjects who did once or more times per week), significant differences for several anthropometric variables were expected (Table 1). Indeed, BMI ($p = 0.005$), fat mass ($p = 0.001$), visceral fat classification ($p = 0.049$), and waist circumference ($p = 0.042$) were more elevated in inactive individuals as compared to active subjects. Moreover, water intake resulted in statistically significant differences according to exercise classification variable, where active subjects consumed more water compared to inactive subjects ($p = 0.001$). Likewise, energy expenditure due to physical practice (calories were calculated from the METs, **metabolic equivalents of task,** results for the physical activity practice registered in the MLTPAQ) resulted in statistically significantly differences in terms of exercise classification ($p < 0.001$). No significant differences were found for biochemical parameters.

The association between energy expenditure by physical activity and exercise variable, has a Pearson correlation of $r = 0.37$ ($p = 8 \times 10^{-17}$), indicating a significant, although moderate, association between these two variables.

Concerning the genotype analyses, for *GCKR* rs7800094, genotype frequencies were in Hardy–Weinberg equilibrium ($p = 0.116$). No significant differences were found in anthropometric variables (Table 1), although there were significant differences regarding total cholesterol ($p = 0.041$). Total caloric value of fat intake resulted in statistically significant differences since rare homozygous carriers consumed more amount of fat in contrast to mayor allele carriers ($p = 0.036$). In this sense, a tendency for a different energy intake has been observed ($p = 0.074$), where rare homozygous carriers consumed more total energy than mayor allele carriers. At the same time, there were significant differences in exercise variable (Table 1). Particularly, rare homozygous carriers practiced physical exercise a greater amount of times per week compared to major allele carriers who presented lower physical exercise per week ($p = 0.016$). There were no significant differences with respect of energy expenditure attributed to physical activity practice concerning genotype ($p = 0.582$).

In this study, the association between the above described 64 genetic variants and exercise classification, were tested by developing logistic regression models adjusted by sex, age and BMI. Of the total of the polymorphisms analyzed, only *GCKR* rs780094 was found to be significantly associated with exercise classification after correction for multiple tests. Thus, Table 2 shows the distribution of our sample with regards to rs780094 genotype and exercise classification. It displays the corresponding odds ratios and *p*-values of the three logistic models: Codominant, additive and dominant. Statistically significant differences were found for the three designs. Within the exercise classification variable: active subjects, there was a higher percentage of rare homozygous carriers in comparison with mayor homozygous carriers. Thus, Figure 1 shows the sample distribution of the exercise classification variable according to the genotype. Almost 40% of common homozygote carriers did not perform any physical exercise per week, compared to the homozygous variants where the percentage of no-active people was much lower (14%). Most heterozygous carriers (86%), performed at least some type of exercise once a week ($p = 0.004$).

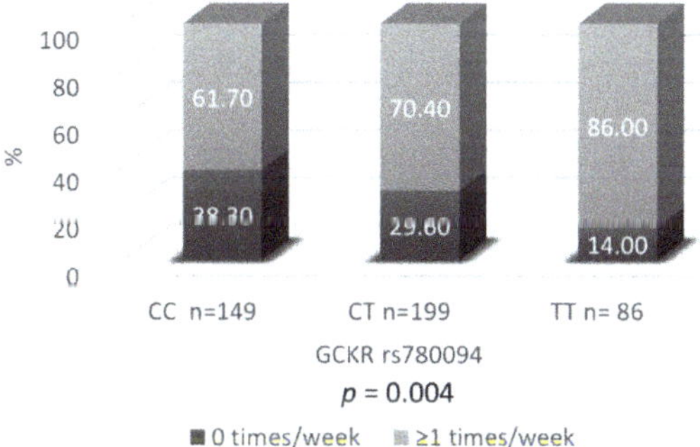

Figure 1. Percentage of the population classified according to the number of times that they perform physical exercise according to the genotype. *p*-Value of an additive model adjusted by 64 SNPs is also shown.

Table 1. Descriptive characteristics of the study participants (Mean, SD).

Ariables	Gender			Exercise Classification			Genotype (rs780094)			
	Male $n = 155$	Female emphn $= 402$	p	No Activity $n = 150$	Activity $n = 340$	p	Common Homozygous (C/C) $n = 155$	Heterozygous (C/T) $n = 204$	Rare Homozygous (T/T) $n = 92$	p
Nutritional status										
Body Weight (kg)	84.29 (14.70)	67.69 (13.45)	<0.001	73.83 (17.63)	71.23 (14.64)	0.117	73.39 (16.73)	71.99 (15.27)	70.51 (14.34)	0.375
Height (cm)	176.26 (6.44)	162.30 (6.36)	<0.001	164.69 (8.19)	166.52 (9.11)	0.029	166.87 (9.45)	165.77 (8.51)	164.54 (8.85)	0.164
BMI (kg/m^2)	27.18 (4.78)	25.72 (4.84)	0.003	27.09 (5.45)	25.65 (4.54)	0.005	26.26 (5.06)	26.18 (4.91)	26.04 (4.76)	0.949
Fat mass (%)	24.87 (7.96)	36.66 (8.34)	<0.001	35.74 (9.45)	32.66 (9.73)	0.001	32.93 (10.24)	34.01 (9.53)	34.38 (10.45)	0.495
Lean mass (%)	35.76 (4.93)	26.57 (3.58)	<0.001	27.92 (4.87)	29.39 (5.92)	0.004	29.61 (5.95)	28.55 (5.59)	28.62 (5.97)	0.214
Visceral fat classification	9.75 (5.30)	6.32 (2.80)	<0.001	7.78 (4.59)	6.93 (3.58)	0.049	7.27 (4.04)	7.26 (3.98)	6.93 (3.17)	0.715
Waist circumference (cm)	94.39 (14.07)	84.41 (13.71)	<0.001	89.12 (15.76)	86.05 (13.80)	0.042	87.07 (14.49)	87.51 (14.81)	85.98 (13.39)	0.695
Waist-Hip ratio	0.89 (0.08)	0.81 (0.08)	<0.001	0.83 (0.10)	0.82 (0.09)	0.283	0.82 (0.08)	0.83 (0.09)	0.82 (0.08)	0.493
Biochemical parameters										
Glucose (mg/dL)	88.58 (16.94)	86.77 (10.92)	0.358	88.00 (16.38)	87.76 (10.36)	0.891	86.94 (10.37)	88.92 (14.23)	89.37 (12.56)	0.311
Total-C (mg/dL)	194.77 (38.67)	202.75 (35.42)	0.048	199.06 (36.05)	200.81 (37.74)	0.659	194.93 (38.07)	205.79 (35.93)	203.17 (35.73)	0.041
HDL-C (mg/dL)	46.57 (10.84)	58.34 (14.04)	<0.001	55.65 (15.59)	55.83 (13.79)	0.912	54.59 (15.05)	57.44 (14.06)	55.00 (12.05)	0.185
LDL-C (mg/dL)	126.77 (32.68)	126.41 (29.87)	0.916	125.55 (30.67)	126.85 (31.19)	0.704	122.37 (29.96)	128.84 (31.27)	127.98 (30.70)	0.185
TAG (mg/dL)	103.39 (50.34)	95.53 (46.27)	0.138	99.06 (46.20)	96.28 (47.60)	0.585	90.92 (39.11)	98.95 (50.78)	102.33 (53.76)	0.151
TyG Index	4.56 (0.24)	4.48 (0.23)	0.009	4.51 (0.23)	4.49 (0.23)	0.543	4.46 (0.21)	4.51 (0.25)	4.54 (0.24)	0.072
Total-C/HDL-C	4.43 (1.30)	3.69 (1.03)	<0.001	3.86 (1.11)	3.83 (1.14)	0.808	3.83 (1.20)	3.81 (1.14)	3.82 (0.90)	0.988
LDL-C/HDL-C	2.87 (0.99)	2.3 (0.81)	<0.001	2.46 (0.92)	2.41 (0.86)	0.629	2.41 (0.88)	2.40 (0.89)	2.43 (0.78)	0.962
Dietary intake										
Energy (TCV: kcal/day)	2325 (650)	2037 (719)	<0.001	2121 (842)	2150 (674)	0.714	2176 (809)	2049 (580)	2260 (929)	0.074
CHO (TCV%)	37.79 (6.99)	38.48 (6.30)	0.311	38.42 (6.51)	38.34 (6.42)	0.911	39.03 (6.16)	38.30 (6.36)	37.13 (6.91)	0.108
Proteins (TCV%)	17.23 (3.27)	17.28 (3.51)	0.894	17.22 (3.29)	17.14 (3.41)	0.808	17.01 (3.12)	17.38 (3.54)	17.04 (3.30)	0.549
Fats (TCV%)	39.98 (6.38)	40.04 (6.38)	0.927	40.59 (5.99)	39.72 (6.50)	0.163	39.56 (5.92)	39.74 (6.35)	41.52 (5.90)	0.036
Water intake (mL)	1495 (622)	1454 (727)	0.522	1320 (738)	1568 (694)	0.001	1544 (798)	1450 (667)	1476 (687)	0.511
Physical activity performance										
Exercise (times/week)	2.34 (1.67)	1.73 (1.50)	<0.001	0.00 (0.00)	2.72 (1.13)	N.A.	1.79 (1.69)	1.85 (1.53)	2.31 (1.37)	0.016
Energy expenditure physical activity/week (kcal)	2701 (2434)	2038 (1676)	0.007	1504 (1656)	2533 (1948)	<0.001	2222 (1935)	2199 (1899)	1975 (1811)	0.582

BMI, Body mass index; CHO, carbohydrates; HDL-C, high density lipoprotein cholesterol; LDL-C, low density lipoprotein; N.A., not applicated; TyG Index, Triglycerides/glucose index; SD, standard deviation; TCV, total caloric value, Total-C, total cholesterol. No activity, 0 times of physical activity performance per week. Activity, one or more times of physical activity performance per week. p-values were obtained from one-way ANOVA for continuous variables, and Fisher exact test for categorical variables. Significance level $p \le 0.05$.

Table 2. Association of exercise classification performance and genotype according to genetic models.

Model	No Activity			Activity			OR (CI)	p-Value [1]
	Major Allele Homozygote	Heterozygote	Minor Allele Homozygote	Major Allele Homozygote	Heterozygote	Minor Allele Homozygote		
	n (%)	n (%)	n (%)	n (%)	n (%)	n (%)		
Codominant	57 (38.30)	59 (29.60)	12 (14.00)	92 (61.70)	140 (70.40)	74 (86.00)	1.52 (0.96–2.41)/4.02 (2.04–8.51)	0.012
Additive	57 (38.30)	59 (29.60)	12 (14.00)	92 (61.70)	140 (70.40)	74 (86.00)	1.86 (1.36–2.56)	0.004
Dominant	57 (38.30)	71 (24.90)		92 (61.70)	214 (75.10)		3.18 (1.70–6.46)	0.012

[1] Adjusted by sex, age and BMI. p-value was corrected by multiple SNP comparisons (Bonferroni). Significance level $p \leq 0.05$. No activity, 0 times of physical activity performance per week. Activity, 1 or more times of physical activity performance per week.

4. Discussion

Physical exercise provides a plethora of health benefits, whereas sedentary lifestyle is considered to be a main risk factor in the development of multiple chronic diseases [35]. Currently, there is a scarcity of knowledge about the mechanisms that trigger the practice of physical exercise and energy expenditure. However, this knowledge in the area could lead to personalized strategies focused on people prone to weight gain [36], even before overweight manifests itself [22]. Therefore, research on the mechanisms that modulate energy expenditure could be useful to reduce the overweight rates of the population.

Genetic factors could play an influential role in the behavior of individuals regarding their physical exercise [5]. In this context, the *GCKR* gene which regulates glucokinase activity in the liver, influences the regulation of lipid metabolism and hepatic glucose, so that it is strongly associated with fasting TAG and glucose levels [37]. Specifically, it has been observed that the minor T allele of *GCKR* rs780094 is associated with metabolic traits including higher levels of TAG, even though glycemia levels were adequate [38]. This involvement in metabolism may interfere in the frequency of physical exercise performance.

Our results revealed that total cholesterol levels are significantly different concerning genotype, which may lead to possible influence of this polymorphism on cholesterol metabolism. Likewise studies of other *GCKR* polymorphisms show similar outcomes [39]. Even so, it is necessary to consider that the differences observed in our study concerning lipid profile, showed significant differences depending also on gender. Although the differences observed on HDL levels between men and women are in line with the existing bibliography, minority allele carriers who presented higher levels of cholesterol, also tended to consume more dietary fat [40,41]. This fact could influence the levels of blood cholesterol so further studies are needed to be conducted in this respect.

Conversely to results found in previous researches, our analyses show no significant differences for glucose and TAG according to rs780094 *GCKR* polymorphism [38]. Understanding the regulatory mechanisms of CGK activity is a complex issue. On the one hand, GCK glucokinase activity is regulated by fructose 6-phosphate (F6P), and fructose 1-phosphate (F1P) whose presence enhances and reduces GKRP-mediated inhibition, respectively [42]. On the other hand, several studies suggested that nuclear interaction with GKRP plays an important role in establishing and regulating GCK protein concentration, a fact that is essential for the maintenance of glucose homeostasis [42].

Regarding physical exercise practice and genetic association, it has been found that around 40% of common homozygous genotype (CC) carriers do not perform any type of exercise per week, compared to 86% of homozygous variants (TT) that perform at least some physical exercise once a week. Involvement of GCKR in lipid and glucose metabolism, support the idea about the existence of an influence on the capability to perform more frequently physical exercise. However, according to Alfred et al. (2013) no evidence about cognitive and physical capability has been found at this moment concerning this genotype [43].

In this study we used the exercise classification variable as a measure of the exercise practice of the individuals. There are alternatives, and complementary, measures of physical activity like the Minnesota questionnaire. In fact, we observed a significant correlation of exercise variable and energy expenditure by this questionnaire (see above), although we did not find a significant association between the rs780094 SNP and the energy expenditure (Table 1). This could be due to the Minnesota questionnaire being a more comprehensive measure of «total» physical activity during the day, while our exercise variable would be more focused on physical exercise or sports practice, that is, extra physical activity performed after work hours. In this way, the rs780094 SNP could be associated to this fraction of the total physical exercise of the individual.

Although the genetic factor seems to be associated with a predisposal to a greater practice of physical exercise, it is crucial to take into account other multiple factors that may condition this attitude. These factors may depend on aptitudes, preferences, incentives or other aspects that might influence the level of difficulties of the physical activities performed [44]. Therefore, it is important to implement new studies to determine the genetic load on other possible factors involved in the frequency of physical exercise practice. A possible predisposition must be always validated in studies where all the conditioning variables are included. Regarding this, it is necessary to note that in the current study, all factors associated with this possible behavior have not been taken into account. Therefore, on the basis of this consideration, this study could be considered a starting point for developing future studies addressing this issue by including in the criteria these factors.

Since *GCKR* is a gene involved in the regulation of carbohydrates and lipids metabolism, it would be reasonable to assume that this gene may be also implicated in the genetic disposition to perform more frequently physical exercise. Among genes associated with adherence to physical activity practice we might also encounter the dopamine receptor 1 (*DRD1*), as well as the helix-loop-helix 2 (*NHLH2*), which are involved in eating behavior [15]. In addition, genetic variants of the *MC4R* and *LEPR* genes have also been shown to be associated with levels of physical activity according to their genotype [45]. The *GCKR* gene, is implicated in coding molecules involved in lipid and glucose turnover which refer to the fundamental pathways in body homeostasis during the practice of physical exercise.

GCK enzyme in humans is involved as a "glucose sensor" in liver, that permits to regulate the changes in plasma glucose concentrations [46]. GCK activity is potently controlled by GCKR, which is encoded by the *GCKR* gene. During fasting, GKRP is bound to GCK inhibiting its activity and locating in the nucleus. In this way, glycogenolysis and gluconeogenesis take place, and glucose is exported to the circulation for use by peripheral tissues. In postprandial state, GKRP releases GCK by glucose and F1P. GCK then binds to glucose, adopt the closed (catalytically active) conformation and exit from the nucleus to generate glucose-6-phosphate for triglycerides and glucose disposal and storage (glycogen synthesis). Alterations of the *GCKR* gene cause high levels of regulation of the metabolism of lipids and glucose and, this could determine a different frequency of physical exercise predisposition. In the case of the T/T genotype (rs780094 *GCKR*), it has been reported that insulin and blood glucose levels could remain better regulated [38], which may improve the use of substrates during physical exercise. In turn, increased levels of triglycerides in blood may favor the use of this substrate during exercise practice, and therefore preventing obesity.

Interestingly, results from this study revealed that minor genotype (TT) carriers have a higher fat consumption and tend to have greater energy intake although they are also more predisposed to perform more exercise. These outcomes would explain why anthropometric parameters were no modified, although their cholesterol levels are significantly higher. In this context, a greater motivation to perform more physical exercise, for major genotype carriers (majority of the European population), could benefit them. In the case of rare genotype carriers, recommendations could be aimed at monitoring blood cholesterol levels based on their fat intake.

5. Conclusions

The analysis of *GCKR* rs780094 variant may be useful to feature those subjects less prone to physical exercise. Thus, common allele carriers could benefit from personalized intervention strategies that would consider increasing the frequency of physical exercise. Likewise, this knowledge could contribute to prevent and manage overweight and obesity in the subjects with lower frequency of physical exercise. In the same way, in the case of minor genotype carriers, recommendations could be aimed at monitoring blood cholesterol levels based on their fat intake. Nevertheless, future studies will be needed to confirm these findings since the use of genetic information for the identification of individuals at risk of a given condition requires replication and a complete validation process.

Author Contributions: Conceptualization, I.E.-S. and J.A.M.; methodology, V.L.-K. and S.M.; validation, G.C. and S.M.; formal analysis, G.C.; investigation, I.E.-S., V.L.-K. and A.R.d.M.; resources, S.M., A.R.d.M. and G.R.; data curation, I.E.-S.; writing—original draft preparation, I.E.-S.; writing—review & editing, I.E.-S., G.C., R.d.l.I., J.A.M., V.L.-K. and A.R.d.M.; visualization, I.E.-S.; supervision, R.d.l.I., V.L.-K., A.R.d.M. and J.A.M.; project administration, G.R. and A.R.d.M.

Funding: This work was supported by IMDEA Food Institute.

Acknowledgments: We thank the members of the management team, and the participants for their management and contribution to this study.

Conflicts of Interest: The authors declare no conflict of interest. The funders had no role in the design of the study; in the collection, analyses, or interpretation of data; in the writing of the manuscript, or in the decision to publish the results.

References

1. Hill, J.O.; Wyatt, H.R.; Peters, J.C. Energy Balance and Obesity. *Circulation* **2012**, *126*, 126–132. [CrossRef] [PubMed]
2. Westerterp, K.R. Physical activity and physical activity induced energy expenditure in humans: Measurement, determinants, and effects. *Front. Physiol.* **2013**, *4*, 4. [CrossRef] [PubMed]
3. Hills, A.P.; Mokhtar, N.; Byrne, N.M. Assessment of Physical Activity and Energy Expenditure: An Overview of Objective Measures. *Front. Nutr.* **2014**, *1*, 1. [CrossRef] [PubMed]
4. Huppertz, C.; Bartels, M.; Jansen, I.E.; Boomsma, D.I.; Willemsen, G.; Moor, M.H.M.; de Geus, E.J.C. A Twin-Sibling Study on the Relationship Between Exercise Attitudes and Exercise Behavior. *Behav. Genet.* **2014**, *44*, 45–55. [CrossRef] [PubMed]
5. Herring, M.P.; Sailors, M.H.; Bray, M.S.; Herring, M. Genetic factors in exercise adoption, adherence and obesity. *Obes. Rev.* **2013**, *15*, 29–39. [CrossRef] [PubMed]
6. Santos, D.M.D.V.E.; Katzmarzyk, P.T.; Seabra, A.F.T.; Maia, J.A.R.; Santos, D.M.V.E. Genetics of Physical Activity and Physical Inactivity in Humans. *Behav. Genet.* **2012**, *42*, 559–578. [CrossRef] [PubMed]
7. Radák, Z. *The Physiology of Physical Training*, 1st ed.; Academic Press: San Diego, CA, USA, 2018; ISBN 978-0-12-815137-2.
8. Hara, M.; Hachiya, T.; Sutoh, Y.; Matsuo, K.; Nishida, Y.; Shimanoe, C.; Tanaka, K.; Shimizu, A.; Ohnaka, K.; Kawaguchi, T.; et al. Genomewide Association Study of Leisure-Time Exercise Behavior in Japanese Adults. *Med. Sci. Sports Exerc.* **2018**, *50*, 2433–2441. [CrossRef]
9. De Moor, M.H.; Liu, Y.-J.; Boomsma, D.I.; Li, J.; Hamilton, J.J.; Hottenga, J.-J.; Levy, S.; Liu, X.-G.; Pei, Y.-F.; Posthuma, D.; et al. Genome-wide Association Study of Exercise Behavior in Dutch and American Adults. *Med. Sci. Sports Exerc.* **2009**, *41*, 1887–1895. [CrossRef]
10. Eynon, N.; Hanson, E.D.; Lucia, A.; Houweling, P.J.; Garton, F.; North, K.N.; Bishop, D.J. Genes for Elite Power and Sprint Performance: ACTN3 Leads the Way. *Sports Med.* **2013**, *43*, 803–817. [CrossRef]
11. Ahmetov, I.I.; Fedotovskaya, O.N. Current Progress in Sports Genomics. *Adv. Clin. Chem.* **2015**, *70*, 247–314.
12. Jones, N.; Kiely, J.; Suraci, B.; Collins, D.J.; De Lorenzo, D.; Pickering, C.; Grimaldi, K.A. A genetic-based algorithm for personalized resistance training. *Boil. Sport* **2016**, *33*, 117–126. [CrossRef]
13. Kraemer, W.J.; Rogol, A.D. *The Endocrine System in Sports and Exercise*; John Wiley & Sons: Hoboken, NJ, USA, 2008; ISBN 978-0-470-75780-2.
14. Schroder, E.A.; Esser, K.A. Circadian Rhythms, skeletal muscle molecular clocks and exercise. *Exerc. Sport Sci. Rev.* **2013**, *41*, 224–229. [CrossRef]
15. Heffernan, S.M.; Stebbings, G.K.; Kilduff, L.P.; Erskine, R.M.; Day, S.H.; Morse, C.I.; McPhee, J.S.; Cook, C.J.; Vance, B.; Ribbans, W.J.; et al. Fat mass and obesity associated (FTO) gene influences skeletal muscle phenotypes in non-resistance trained males and elite rugby playing position. *BMC Genet.* **2017**, *18*, 4. [CrossRef]
16. Lee, H.; Ash, G.I.; Angelopoulos, T.J.; Gordon, P.M.; Moyna, N.M.; Visich, P.S.; Zoeller, R.F.; Gordish-Dressman, H.; Thompson, P.D.; Hoffman, E.P.; et al. Obesity-related Genetic Variants and Their Associations With Physical Activity. *Med. Sci. Sports Exerc.* **2015**, *47*, 79. [CrossRef]
17. Sawczuk, M.; Maciejewska-Karłowska, A.; Cięszczyk, P.; Leońska-Duniec, A. IS GNB3 C825T Polymorphism Associated with Elite Status of Polish Athletes? *Biol. Sport* **2014**, *31*, 21–25. [CrossRef]
18. Grøntved, A.; Andersen, L.B.; Franks, P.W.; Verhage, B.; Wareham, N.J.; Ekelund, U.; Loos, R.J.F.; Brage, S. NOS3 variants, physical activity, and blood pressure in the European Youth Heart Study. *Am. J. Hypertens.* **2011**, *24*, 444–450. [CrossRef]
19. Leońska-Duniec, A.; Grzywacz, A.; Jastrzębski, Z.; Jażdżewska, A.; Lulińska-Kuklik, E.; Moska, W.; Leźnicka, K.; Ficek, K.; Rzeszutko, A.; Dornowski, M.; et al. ADIPOQ polymorphisms are associated with changes in obesity-related traits in response to aerobic training programme in women. *Biol. Sport* **2018**, *35*, 165–173.

20. Petr, M.; Stastny, P.; Zajac, A.; Tufano, J.J.; Maciejewska-Skrendo, A. The Role of Peroxisome Proliferator-Activated Receptors and Their Transcriptional Coactivators Gene Variations in Human Trainability: A Systematic Review. *Int. J. Mol. Sci.* **2018**, *19*, 1472. [CrossRef]

21. Raichlen, D.A.; Alexander, G.E. Exercise, APOE genotype, and the evolution of the human lifespan. *Trends Neurosci.* **2014**, *37*, 247–255. [CrossRef]

22. Espinosa-Salinas, M.I. Identificación de SNPs Implicados en la Diferente Respuesta a Componentes de la Dieta y Asociación con Enfermedades Relacionadas con la Alimentación: Estudios Nutrigenéticos. Ph.D. Thesis, Universidad Autónoma de Madrid, Madrid, Spain, 2017.

23. Sotos-Prieto, M.; Guillén, M.; Sorli, J.V.; Portolés, O.; Guillem-Saiz, P.; Gonzalez, J.I.; Qi, L.; Corella, D. Relevant associations of the glucokinase regulatory protein/glucokinase gene variation with TAG concentrations in a high-cardiovascular risk population: Modulation by the Mediterranean diet. *Br. J. Nutr.* **2013**, *109*, 193–201. [CrossRef]

24. Irwin, D.M.; Tan, H. Evolution of glucose utilization: Glucokinase and glucokinase regulator protein. *Mol. Phylogenet. Evol.* **2014**, *70*, 195–203. [CrossRef]

25. Fesinmeyer, M.D.; Meigs, J.B.; North, K.E.; Schumacher, F.R.; Bůžková, P.; Franceschini, N.; Haessler, J.; Goodloe, R.; Spencer, K.L.; Voruganti, V.S.; et al. Genetic variants associated with fasting glucose and insulin concentrations in an ethnically diverse population: Results from the Population Architecture using Genomics and Epidemiology (PAGE) study. *BMC Med. Genet.* **2013**, *14*, 98. [CrossRef]

26. Fernández, E.B. Contribution of metabolic sensors on feeding behaviour and the control of body weight. *Anales de la Real Academia Nacional de Medicina* **2012**, *129*, 541–563.

27. Fernández, E.B. Molecular aspects of a hypothalamic glucose sensor system and their implications in the control of food intake. *Anales de la Real Academia Nacional de Medicina* **2003**, *120*, 513–522.

28. Durnin, J.V.; Fidanza, F. Evaluation of nutritional status. *Bibl. Nutr. Dieta* **1985**, *35*, 20–30.

29. Ortega, R.; López-Sobaler, A.; Andrés, P.; Requejo, A.; Molinero, L. *Programa DIAL Para Valoración de Dietas y Gestión de Datos de Alimentación*; Departamento de Nutrición (UCM) Alce Ingeniería: Madrid, Spain, 2004.

30. Elosua, R.; García, M.; Aguilar, A.; Molina, L.; Covas, M.-I.; Marrugat, J. Validation of the Minnesota Leisure Time Physical Activity Questionnaire in Spanish Women. *Med. Sci. Sports Exerc.* **2000**, *32*, 1431–1437. [CrossRef]

31. Ainsworth, B.E.; Haskell, W.L.; Whitt, M.C.; Irwin, M.L.; Swartz, A.M.; Strath, S.J.; O'brien, W.L.; Bassett, D.R.; Schmitz, K.H.; Emplaincourt, P.O.; et al. Compendium of Physical Activities: An update of activity codes and MET intensities. *Med. Sci. Sports Exerc.* **2000**, *32*, S498–S516. [CrossRef]

32. Hosseini, S.M. Triglyceride-Glucose Index Simulation. *J. Clin. Basic Res.* **2017**, *1*, 11–16. [CrossRef]

33. QuantStudio qPCR Product Portfolio. Available online: https://www.thermofisher.com/us/en/home/life-science/pcr/real-time-pcr/real-time-pcr-instruments/quantstudio-qpcr-product-portfolio.html (accessed on 21 February 2018).

34. R: What Is R? Available online: https://www.r-project.org/about.html (accessed on 13 February 2018).

35. González, K.; Fuentes, J.; Márquez, J.L. Physical Inactivity, Sedentary Behavior and Chronic Diseases. *Korean J. Fam. Med.* **2017**, *38*, 111–115. [CrossRef]

36. Horn, E.E.; Turkheimer, E.; Strachan, E.; Duncan, G.E. Behavioral and environmental modification of the genetic influence on body mass index: A twin study. *Behav. Genet.* **2015**, *45*, 409–426. [CrossRef]

37. Posthumus, M.; Collins, M. *Genetics and Sports*; Karger Medical and Scientific Publishers: Basel, Switzerland, 2016; ISBN 978 3 318 03011 2.

38. Bi, M.; Kao, W.H.L.; Boerwinkle, E.; Hoogeveen, R.C.; Rasmussen-Torvik, L.J.; Astor, B.C.; North, K.E.; Coresh, J.; Köttgen, A. Association of rs780094 in GCKR with Metabolic Traits and Incident Diabetes and Cardiovascular Disease: The ARIC Study. *PLoS ONE* **2010**, *5*, e11690. [CrossRef]

39. Teslovich, T.M.; Musunuru, K.; Smith, A.V.; Edmondson, A.C.; Stylianou, I.M.; Koseki, M.; Pirruccello, J.P.; Ripatti, S.; Chasman, D.I.; Willer, C.J.; et al. Biological, Clinical, and Population Relevance of 95 Loci for Blood Lipids. *Nature* **2010**, *466*, 707–713. [CrossRef]

40. Seidell, J.C.; Cigolini, M.; Charzewoka, J.; Ellsinger, B.-M.; Björntorp, P.; Hautvast, J.G.; Szostak, W. Fat distribution and gender differences in serum lipids in men and women from four European communities. *Atherosclerosis* **1991**, *87*, 203–210. [CrossRef]

41. Pascot, A.; Lemieux, I.; Bergeron, J.; Tremblay, A.; Nadeau, A.; Prud'Homme, D.; Couillard, C.; Lamarche, B.; Després, J.-P. HDL particle size: A marker of the gender difference in the metabolic risk profile. *Atherosclerosis* **2002**, *160*, 399–406. [CrossRef]

42. Rees, M.G.; Wincovitch, S.; Schultz, J.; Waterstradt, R.; Beer, N.L.; Baltrusch, S.; Collins, F.S.; Gloyn, A.L. Cellular characterisation of the GCKR P446L variant associated with type 2 diabetes risk. *Diabetologia* **2012**, *55*, 114–122. [CrossRef]

43. Alfred, T.; Ben-Shlomo, Y.; Cooper, R.; Hardy, R.; Deary, I.J.; Elliott, J.; Harris, S.E.; Kivimäki, M.; Kumari, M.; Power, C.; et al. Associations between a Polymorphism in the Pleiotropic GCKR and Age-Related Phenotypes: The HALCyon Programme. *PLoS ONE* **2013**, *8*, e70045. [CrossRef]

44. National Research Council (US), Committee on Physical Activity, Health, Transportation, and Land Use; Institute of Medicine. *Does the Built Environment Influence Physical Activity: Examining the Evidence—Special Report 282*; National Academies Press: Washington, DC, USA, 2005; ISBN 978-0-309-66498-1.

45. Moore-Harrison, T.; Lightfoot, J.T. Driven to Be Inactive—The Genetics of Physical Activity. *Prog. Mol. Boil. Transl. Sci.* **2010**, *94*, 271–290.

46. Matschinsky, F.M. Assessing the potential of glucokinase activators in diabetes therapy. *Nat. Rev. Drug Discov.* **2009**, *8*, 399–416. [CrossRef]

© 2019 by the authors. Licensee MDPI, Basel, Switzerland. This article is an open access article distributed under the terms and conditions of the Creative Commons Attribution (CC BY) license (http://creativecommons.org/licenses/by/4.0/).

Article

Changes in Serum Iron and Leukocyte mRNA Levels of Genes Involved in Iron Metabolism in Amateur Marathon Runners—Effect of the Running Pace

Agata Grzybkowska [1], Katarzyna Anczykowska [1], Wojciech Ratkowski [2], Piotr Aschenbrenner [3], Jędrzej Antosiewicz [1], Iwona Bonisławska [4] and Małgorzata Żychowska [3,*]

[1] Department of Biochemistry, Faculty of Physical Education, Gdansk University of Physical Education and Sport, 80-336 Gdansk, Poland; agata.p3@gmail.com (A.G.); kasia.anczykowska@gmail.com (K.A.); jant@gumed.edu.pl (J.A.)

[2] Department of Management in Tourism and Recreation, Faculty of Tourism and Recreation, University of Physical Education and Sport, 80-336 Gdansk, Poland; maraton1954@o2.pl

[3] Department of Life Sciences, Faculty of Physical Education, Gdansk University of Physical Education and Sport, 80-336 Gdansk, Poland; sqarko@gmail.com

[4] Department of Anatomy and Anthropology, Faculty of Physical Education, Gdansk University of Physical Education and Sport, 80-336 Gdansk, Poland; iwonabonislawska@gmail.com

* Correspondence: zychowska.m@gmail.com

Received: 28 April 2019; Accepted: 12 June 2019; Published: 15 June 2019

Abstract: Iron is essential for physical activity due to its role in energy production pathways and oxygen transportation via hemoglobin and myoglobin. Changes in iron-related biochemical parameters after physical exercise in athletes are of substantial research interest, but molecular mechanisms such as gene expression are still rarely tested in sports. In this paper, we evaluated the mRNA levels of genes related to iron metabolism (*PCBP1, PCBP2, FTL, FTH,* and *TFRC*) in leukocytes of 24 amateur runners at four time points: before, immediately after, 3 h after, and 24 h after a marathon. We measured blood morphology as well as serum concentrations of iron, ferritin, and C-reactive protein (CRP). Our results showed significant changes in gene expression (except for *TFRC*), serum iron, CRP, and morphology after the marathon. However, the alterations in mRNA and protein levels occurred at different time points (immediately and 3 h post-run, respectively). The levels of circulating ferritin remained stable, whereas the number of transcripts in leukocytes differed significantly. We also showed that running pace might influence mRNA expression. Our results indicated that changes in the mRNA of genes involved in iron metabolism occurred independently of serum iron and ferritin concentrations.

Keywords: iron metabolism; ferritin; gene expression; marathon runners; *PCBP1*; *PCBP2*; *TFRC*

---

1. Introduction

Iron is essential for physical activity due to its role in energy production pathways and oxygen transportation via hemoglobin and myoglobin [1]. Athletes are considered to be at greater risk of iron deficiency than the general population, although supportive data are inconclusive [2,3]. In one particular study, iron deficiency was found in 1.6% of recreational runners, but iron overload was found in 15% of the male participants [4].

Recently, research interest in iron metabolism during and after exercise has grown because physical activity can affect iron and iron-regulatory protein status in many ways, such as by inducing oxidative stress and inflammation [5–10]. In the case of intense running efforts, foot strike causes hemolysis as an additional factor that contributes to disordered iron metabolism [11,12]. Terink et al. [13] reported that most of the studies related to iron metabolism were conducted on well-trained athletes, mainly during short and intensive efforts. Additionally, the changes in iron metabolism were determined only on protein concentrations in plasma or serum and usually showed an increase in blood ferritin values, although the findings were conflicting. To the best of our knowledge, there are no data focused on the effect of endurance exercise on the mRNA of genes related to iron metabolism.

The popularity of marathon running has increased in recent years especially in amateurs of different ages, sex and physical capabilities. It seems that the runner's age, running speed and level of adaptation to training are main influences of the physiological response to physical exercise. However, published studies also have some contradictory findings. For example, Jastrzębski et al. [14] reported that during a 100 km run, muscle and liver damage was age but not pace-dependent while negative metabolic changes were independent of age.

The aim of our study was to examine the changes in serum iron and ferritin concentrations together with the changes in leukocyte mRNA levels of genes encoding proteins involved in iron metabolism i.e., *PCBP1* (poly(rC) binding protein 1), *PCBP2* (poly(rC) binding protein 2), *FTH1* (ferritin heavy chain 1), *FTL* (ferritin light chain) and *TFRC* (transferrin receptor). The expression of these genes is expected to be affected by marathon running as the proteins they encode are involved in exercise-induced oxidative stress and inflammation [15–17]. We also aimed to evaluate the relationship between changes in gene expression and baseline serum iron and ferritin concentrations, and runner pace, during a run. We hypothesized that marathon running will induce an increase in the mRNA levels of genes associated with iron metabolism similarly to serum changes and that these changes will be pace-dependent.

2. Materials and Methods

2.1. Characteristics of the Subjects and Baseline Laboratory Parameters

A total of 28 healthy young men who reported regular physical activity involving a running program participated in our study. All participants were asked to refrain from changing their diet and to avoid nicotine and alcohol use, for one month prior to undertaking the study marathon run (42.2 km) at an athletic stadium (Gdansk University of Physical Education and Sport, Gdansk, Poland). The run was completed by 26 of the 28 participants, and a further two subjects were excluded as they had mRNA levels that were far from the average. Anthropometric data for the 24 included subjects are shown in Table 1. All the subjects were informed of the purpose of the study and the possible risks involved before giving written consent. The study was approved by the Bioethics Committee for Clinical Research at the Regional Medical Chamber in Gdansk (NKBBN/448/2016). The principles of the Helsinki Declaration were respected.

Table 1. Characteristic of the participants. Data are presented as a range or means ± standard deviation (SD).

Participant's Characteristic	
	Baseline ($n = 24$)
Age (years)	48.8 ± 6.5
Body mass (kg)	80.1 ± 8.5
Height (cm)	178.7 ± 5.3
BMI (kg/m^2)	25.1 ± 2.3
PBF (%)	15.5 ± 5.0
Pace during the run (km/h)	10.9 ± 1.4
Training units per week	Between 1 and 7
Training regimen (km/week)	Between 20 and 115
Training experience (years)	Between 4 and 24
Number of finished marathons	Between 1 and 62

BMI—body mass index, PBF—percentage body fat.

2.2. Experimental Procedure

Venous blood was collected and serum was obtained from Vacutest® Clot Activator tubes (Vacutest KIMA, Arzegrande, Italy) at four time points: before the run (pre-race), immediately after finishing the run (post-race), 3 h after the run (3 h post-race) and 24 h after the run (24 h post-race). The blood samples were analyzed for blood morphology, and serum concentrations of iron, ferritin, uric acid, creatinine kinase and C-reactive protein (CRP) at an accredited laboratory (Uniwersyteckie Centrum Kliniczne, Gdansk, Poland). Right before the run, the subjects' body weight, height, body mass index (BMI) and percentage of body fat (PBF) were determined using InBody 720 (Biospace Co., Ltd., Seoul, Korea) [18].

To assess gene expression, a further 2 mL of venous blood was collected using vacutainers spray-coated with K_3EDTA as an anticoagulant at the same four time points. The collected blood was mixed within 15 min with Red Blood Cell Lysis Buffer (RBCL) (A&A Biotechnology, Gdynia, Poland) and incubated on ice for at least 15 min. The samples were then spun at 3000× g at 4 °C for 10 min. The resulting pellet was washed again with the hemolysis buffer and the remaining white blood cells lysed using Fenozol (A&A Biotechnology, Gdynia, Poland), and immediately after stored at −20 °C for up to four months, with no freeze–thaw cycles.

2.3. RNA Extraction and Reverse Transcription

Isolation of total RNA was carried out by the modified Chomczynski and Sacchi method [19]. White blood cells diluted in fenozol were thawed at 50 °C for 5 min. Then 200 μL of chloroform (POCH, Gliwice, Poland) was added and the suspension was shaken. Samples were then centrifuged at 10,000 *g* for 30 min at 4 °C. The aqueous phase was collected and mixed with 500 μL of isopropanol (POCH, Gliwice, Poland) and left for at least 30 min to precipitate RNA. Samples were again spun at 10,000 g for 15 min at 4 °C. The aqueous phase was disposed, and the remained pellet was washed 2 times in 1 mL of 75% ethanol at 7500 g at 4 °C. After drying, the pellet was resuspended in 20 μL of PCR grade water. During the optimization period for tested genes, gel electrophoresis has been performed to check for the quality and integrity of RNA. RNA concentration and purity were determined by spectrophotometer (Multiskan Sky Microplate Spectrophotometer, ThermoFisher Scientific, Warszawa, Poland) by absorbance at UV 260/280, and a ratio >1.7 was accepted as pure RNA suitable for further analysis. RNA was then reverse transcribed to cDNA in Eppendorf Mastercycler Gradient 5331, using 0.2 μM oligo(dT) and a Transcriptor First Strand cDNA Synthesis Kit as per the manufacturer's instructions (Roche, Warszawa, Poland). For the analysis 1000 ng of RNA has been used. Thermal conditions used for this step were as follows: Incubation—60 min at 50 °C—followed by inactivation— 5 min at 85 °C. Prepared samples were frozen immediately after the reverse transcription and then stored at −20 °C for up to one month, with no freeze–thaw cycles. For gene expression analysis, the obtained cDNA was diluted 10 times, just before the qRT-PCR step.

2.4. Quantitative Polymerase Chain Reaction Assay

Quantitative real-time polymerase chain reaction (qRT-PCR) analyses were carried out on six genes of particular physiological significance in the context of iron metabolism. The AriaMx Real-Time PCR System (Agilent Technologies, Warszawa Poland) and FastStart Universal SYBR® Green Master (Rox) (Roche, Warszawa, Poland) were used, according to the manufacturer's protocol, on 96-well PCR plates in triplicate for each sample. 2 μL of diluted cDNA was used for qRT-PCR. The thermal cycling conditions comprised an activation step: 95 °C for 10 min followed by 40 cycles of annealing; and an extension step: 95 °C for 15 s and 60 °C for 1 min. Additionally, the melt curve analysis was performed for each reaction. *TUBB* (tubulin beta class I, NM 001293213) was chosen experimentally and used as the reference gene. The relative mRNA expression of *PCBP1* (NM_006196), *PCBP2* (NM_001128913), *FTH1* (NM 002032), *FTL* (NM 000146), *CAT* (catalase, NM 001752) and *TFRC* (NM 001128148) was calculated using qRT-PCR. The primer sequences were designed by the authors using the Primer3 Web

tool. In silico specificity screen has been performed using USCS genome browser. The primers were then ordered from Genomed, Warszawa, Poland. Primer sequences (5′-3′), were:
TUBB
Forward primer: TCCACGGCCTTGCTCTTGTTT
Reverse primer: GACATCAAGGCGCATGTGAAC
PCBP1
Forward primer: AGAGTCATGACCATTCCGTAC
Reverse primer: TCCTTGAATCGAGTAGGCATC
PCBP2
Forward primer: TCCAGCTCTCCGGTCATCTTT
Reverse primer: ACTGAATCCGGTGTTGCCATG
FTH1
Forward primer: TCCTACGTTTACCTGTCCATG
Reverse primer: CTGCAGCTTCATCAGTTTCTC
FTL
Forward primer: GTCAATTTGTACCTGCAGGCC
Reverse primer: CTCGGCCAATTCGCGGAA
CAT
Forward primer: GATGGACATCGCCACATGAAT
Reverse primer: AAGATCCCGGATGCCATAGTC
TFRC
Forward primer: TGCAGCAGTGAGTCTCTTCA
Reverse primer: AGGCCCATCTCCTTAACGAG

2.5. Statistical Analysis

2.5.1. Serum Parameters

Whole blood measurements were corrected for plasma volume shift using the Dill and Costill equation [20]. The normality of the distributions was checked for all parameters using the Shapiro–Wilk test. Values were compared statistically using the one-way analysis of variance (ANOVA) test followed by Tukey's multiple comparisons test for parametric data, and Kruskal–Wallis test followed by Dunn's multiple comparisons test for nonparametric data.

2.5.2. mRNA Levels

Relative mRNA expression was determined using the Schmittgen and Livak delta delta C_t method [21] in Microsoft Excel (2017). The mRNA levels of the tested genes were described as the differences in the cycle threshold value normalized to the *TUBB* mRNA level, i.e., $\Delta C_T = C_T$ of gene—C_T of *TUBB*. All statistical analyses were performed using GraphPad Prism 8.0 (GraphPad Software, Inc., La Jolla, CA, USA).

Relative mRNA expression data were linearly transformed and then the normality of the distribution was checked with the Shapiro–Wilk test. Results were analyzed using the Wilcoxon matched pairs [22] test for nonparametric variables. A p-value of <0.05 was considered significant.

3. Results

3.1. Blood Morphology

Statistically significant changes were observed in all white blood cells, uric acid and creatine kinase immediately after the run. Changes in other laboratory parameters occurred 3 h post-race, except for CRP which was significantly elevated after 24 h post-race. All values were corrected for changes in plasma volume (%Delta PV). All results are shown in Table 2.

Table 2. Changes in laboratory parameters after a marathon run ($n = 24$). $p < 0.05$ is statistically significant. Data are presented as the mean $\pm$ SD for the pre-race and the three following measurements. Values are corrected for plasma volume changes (%Delta PV). Statistical analyses were undertaken using Tukey's multiple comparisons test for the parametric values and Dunn's multiple comparisons test for nonparametric values, compared to the pre-race values. * $p < 0.05$.

	Pre-Race	Post-Race	3 h Post-Race	24 h Post-Race
%Delta PV	-	1.05%	11.58% *	12.43% *
	-	$p > 0.9999$	$p < 0.0001$	$p < 0.0001$
Hemoglobin (g/dL)	14.94 ± 0.85	14.78 ± 1.36	12.63 ± 1.06 *	12.41 ± 0.96 *
	-	$p = 0.9515$	$p < 0.0001$	$p < 0.0001$
RBC ($\times 10^{12}$/L)	4.97 ± 0.33	4.91 ± 0.44	4.19 ± 0.35 *	4.13 ± 0.32 *
	-	$p = 0.9445$	$p < 0.0001$	$p < 0.0001$
Hematocrite (%)	43.33 ± 2.17	42.61 ± 3.96	35.91 ± 3.09 *	35.91 ± 2.58 *
	-	$p > 0.9999$	$p < 0.0001$	$p < 0.0001$
Reticulocytes ($\times 10^{9}$/L)	59.44 ± 11.6	61.78 ± 13.77	49.38 ± 11.67 *	46.63 ± 12.52 *
	-	$p = 0.9088$	$p = 0.0261$	$p = 0.0027$
White blood cells ($\times 10^{9}$/L)	5.27 ± 1.22	14.74 ± 3.45 *	12.02 ± 2.07 *	6.35 ± 1.54
	-	$p < 0.0001$	$p < 0.0001$	$p = 0.9970$
Neutrophils ($\times 10^{9}$/L)	2.84 ± 0.9	12.2 ± 3.11 *	9.82 ± 1.97 *	3.58 ± 1.63
	-	$p < 0.0001$	$p < 0.0001$	$p > 0.9999$
Lymphocytes ($\times 10^{9}$/L)	1.7 ± 0.31	1.48 ± 0.47 *	1.26 ± 0.28	1.94 ± 0.45
	-	$p = 0.1485$	$p = 0.0007$	$p = 0.9166$
Monocytes ($\times 10^{9}$/L)	0.47 ± 0.14	0.95 ± 0.26 *	0.89 ± 0.21 *	0.56 ± 0.13
	-	$p < 0.0001$	$p < 0.0001$	$p > 0.9999$
Eosinophils ($\times 10^{9}$/L)	0.22 ± 0.14	0.04 ± 0.04 *	0.01 ± 0.01 *	0.19 ± 0.12
	-	$p < 0.0001$	$p < 0.0001$	$p > 0.9999$
Basophils ($\times 10^{9}$/L)	0.04 ± 0.02	0.06 ± 0.02 *	0.03 ± 0.01	0.04 ± 0.01
	-	$p = 0.0032$	$p > 0.9999$	$p > 0.9999$
CRP (mg/L)	1.4 ± 3.7	1.31 ± 3.26	1.87 ± 3.12	9.79 ± 7.28 *
	-	$p > 0.9999$	$p = 0.0807$	$p < 0.0001$
Uric acid (mg/dL)	5.26 ± 1.08	5.9 ± 1.07 *	5.71 ± 1.03 *	4.86 ± 0.96 *
	-	$p < 0.0001$	$p = 0.0007$	$p = 0.0113$
Creatine kinase (U/L)	171.56 ± 68.52	569.55 ± 490.71 *	871.04 ± 900.02 *	1410.66 ± 1444.06 *
	-	$p < 0.0001$	$p < 0.0001$	$p < 0.0001$

3.2. Serum Ferritin and Iron Concentrations

No significant changes in ferritin concentrations were observed immediately after the run or during the recovery period, compared with baseline. There was a slight tendency to an increase in ferritin immediately after the run compared with the pre-race rest value (113.1 versus 93 ng/mL, respectively). However, ferritin at 24 h after the run was essentially unchanged from the pre-race value (97 versus 93 ng/mL, respectively). The same direction of changes was observed in serum iron, but at 3 h after the run there was a significant decrease compared with baseline values (66.1 versus 102.3 μg/dL, respectively; $p = 0.002$). Between 3 h and 24 h post race, serum iron increased and had returned to baseline by 24 h (Figure 1).

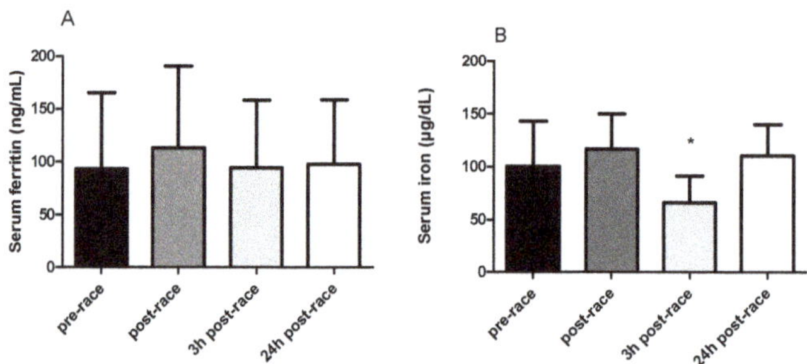

Figure 1. Serum ferritin (**A**) and iron (**B**) concentrations at different time points in the 24 participants. Values are presented as mean ± SD. * $p < 0.05$ compared to the pre-race value.

3.3. Effect of Exercise on mRNA Levels of Selected Genes

Out of six genes tested, five were down-regulated at the end of the race compared with baseline, with the differences for *PCBP1*, *PCBP2*, *FTH* and *CAT* achieving statistical significance ($p = 0.0359$, $p = 0.0443$, $p = 0.0158$ and $p = 0.0182$, respectively) (Figure 3). There was a trend for up-regulation in *PCBP1* and *PCBP2* ($p = 0.0826$ and $p = 0.2435$, respectively), and a significant up-regulation in *FTH* and *FTL* genes ($p = 0.0056$ and $p = 0.0064$, respectively) at 3 h after the marathon run. The mRNA levels of all genes except for *TFRC*, which remained insignificantly decreased, returned to baseline levels at 24 h after the run.

3.4. Relationship Between Baseline Levels of Serum Iron and Ferritin, and Changes in mRNA Levels with Exercise

There were no statistically significant differences in mRNA levels at any time point in participants with baseline serum iron concentrations below (serum iron ≤ 105 µg/dL) and above (serum iron >105 µg/dL) the median baseline value (data not shown). There were also no statistically significant differences in mRNA levels in participants with baseline serum ferritin concentrations below (serum ferritin ≤ 78.08 ng/mL) and above (serum ferritin > 78.08 ng/mL) the median baseline serum ferritin value (data not shown).

3.5. Effect of Running Pace on mRNA Levels of Selected Genes

To determine if the running pace had any effect on gene expression, the participants were divided into two groups (slow and fast) by the median split. The characteristic of two groups is shown in Table 3. A significant difference between groups was observed for pace ($p = 0.0001$), BMI ($p = 0.006$) and age of the participants ($p = 0.0001$). The mean ± SD pace value in the slow group was 10.0 ± 0.5 km/h and in the fast group was 12.2 ± 0.7 km/h ($p < 0.0001$). The mRNA levels of the genes tested in these two groups are shown in Figure 2.

Table 3. Characteristics of slow and fast groups. Data are presented as means ± standard deviation (SD). * $p < 0.05$ for comparison between two groups.

Slow and Fast Groups Characteristics		
	Slow ($n = 12$)	Fast ($n = 12$)
Pace during the run (km/h)	10.04 ± 0.52	12.18 ± 0.71 *
Age (years)	53.58 ± 5.45	44.25 ± 3.49 *
BMI (kg/m^2)	26.28 ± 1.88	23.83 ± 1.95 *
Training units per week	3.00 ± 1.04	4.58 ± 1.26
Training regimen (km/week)	43.58 ± 16.53	71.27 ± 25.22
Training experience (years)	10.83 ± 6.90	8.08 ± 5.02
Number of finished marathons	16.83 ± 20.85	13.58 ± 9.18
Baseline iron level (µg/dL)	112.75 ± 40.54	92.92 ± 42.44
Baseline ferritin level (ng/mL)	93.41 ± 36.58	88.18 ± 98.29

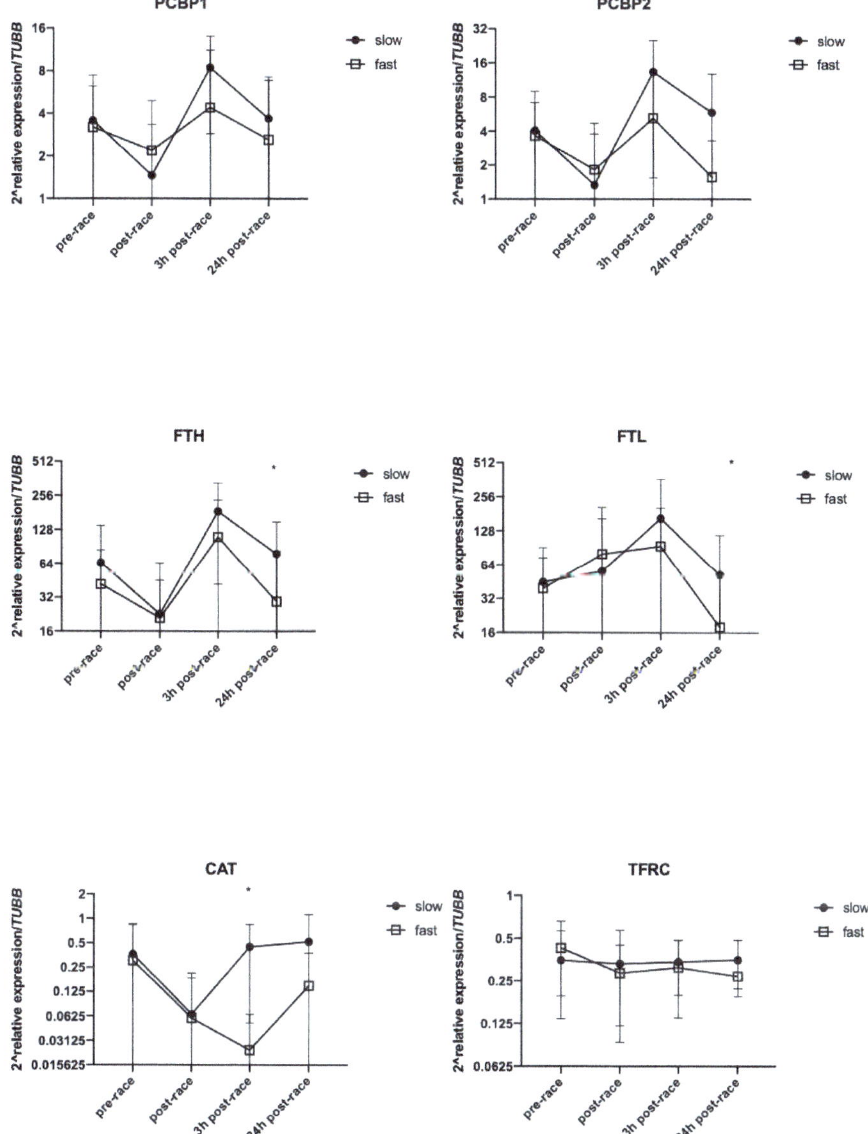

Figure 2. Changes in mRNA levels between slow and fast runners ($n = 12$ per group). Values are mean ± SD. * $p < 0.05$ for comparison between two groups.

The direction of change in *PCBP1*, *PCBP2* and *FTH* gene expression were the same i.e., a decrease immediately after the race and a statistically significant increase 3 h post-race. At 24 h post-race, the values returned almost to baseline (Figure 2). *FTL* mRNA levels were more stable than *FTH* mRNA levels between the end of the run and 3 h post-race. However, similarly to *FTH*, a significant difference in *FTL* mRNA levels between groups was observed 24 h after the race ($p = 0.0245$ and $p = 0.0128$, respectively). The slow group presented with higher levels than the fast group. The opposite changes were observed in *CAT* mRNA levels at 3 h post-race. In the slow group *CAT* mRNA levels dropped, while they increased in the fast pace group ($p = 0.0017$).

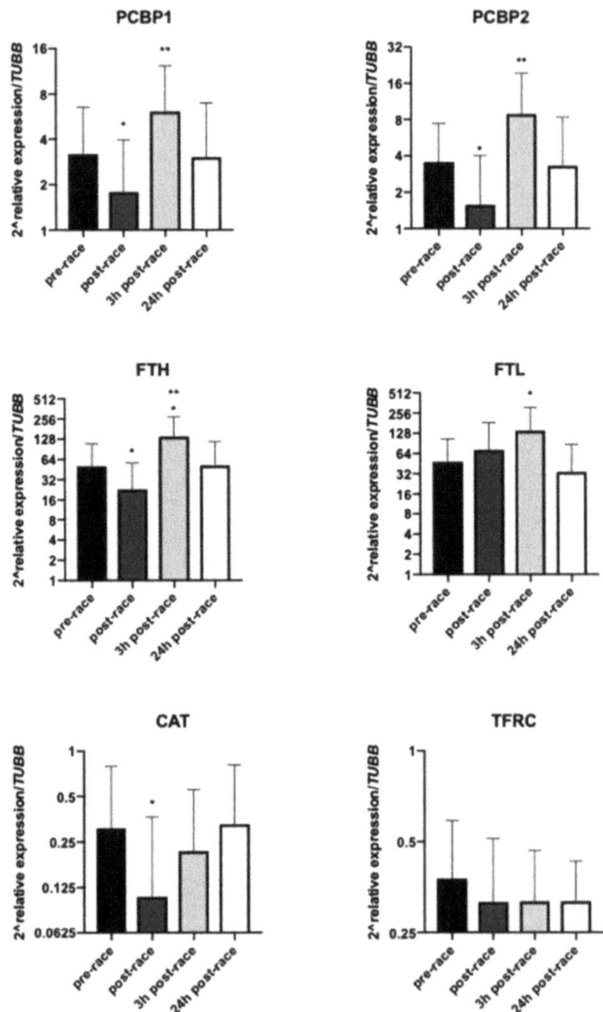

Figure 3. mRNA levels of selected genes at different time points ($n = 24$). Values are presented as mean ± SD. * $p < 0.05$ compared to the pre-race value and ** $p < 0.05$ compared to the post-race value.

4. Discussion

The results of this study did not confirm our hypothesis associated with serum iron, ferritin and expression of genes involved in iron metabolism. Serum ferritin concentrations remained almost unchanged at all time points. Iron status immediately and 24 h after completion of a marathon also did not differ from baseline but there was a significant decrease 3 h after the run. Moreover, changes in iron and ferritin did not correlate with each other (data not shown). Interestingly, a significant decrease in *FTH*, *PCBP1*, *PCBP2* and *CAT* mRNA was observed immediately after the run, and a significant increase in *PCBP1*, *PCBP2*, *FTH* and *FTL* mRNA was seen at 3 h after the run. *TFRC* mRNA remained unchanged. Furthermore, changes in serum indicators and gene expression in leukocytes occurred at different time points.

4.1. Changes in Serum Iron and Ferritin Concentrations

Interindividual variability was observed in baseline serum iron (39–196 µg/L) and ferritin (8.2–367.9 ng/mL) concentrations. The literature on the changes in iron status induced by endurance exercise is equivocal. A decrease in serum iron concentrations 24 h after a marathon was reported by Roecker et al. [23], by Terink et al. [13] after repeated walking and by Chiu et al. [24] after an ultramarathon. On the other hand, an increase in iron concentrations was reported by Peeling et al. [25] after a triathlon and by Buchman et al. [26] after a marathon. According to Terink et al. [13], these differences could be associated with changes in plasma volume and whether this parameter was taken into consideration before the analysis of the results. We corrected for changes in plasma volume and our findings were similar to those reported by Duca et al. [27]. These authors found no change in serum iron or ferritin at 24 h after a half-marathon. Similar findings in serum iron and ferritin concentrations at 24 and 48 h after a marathon were also reported by Weight et al. [28]. Unfortunately, there appear to be no studies in which ferritin and iron concentrations were determined at 3 h after exercise. It is important to note that at this time point increased serum hepcidin was observed [25], and it can be assumed that this would be accompanied by a drop in serum iron, which is consistent with our data. At 24 h after the marathon, basal values had been attained in the participants of our study. In contrast to our results, at the same time point (1 day after prolonged walking) Terink et al. [13] reported decreased iron concentrations. These authors also corrected their results for the change in plasma volume. Lack of a significant correlation between serum ferritin and iron was observed earlier by Galanello et al. [29]. These authors reported that after a stressful event such as a marathon run, the serum ferritin concentrations could not accurately reflect body iron status. Moreover, the observed nonsignificant changes in ferritin concentrations at 24 h after a marathon are in agreement with data reported by Terink et al. [13]. Indirectly, the nonsignificant changes in ferritin in our study might indicate low or no inflammation in the study participants (since Peeling et al. [16] reported an increase in ferritin during exercise-induced inflammation), low or no oxidative stress [15] and minimal damage including damaged blood cells [8].

4.2. Changes in the mRNA of Genes Involved in Iron Metabolism

The genes related to iron metabolism that were selected for analysis are easily induced by stressful conditions, and sensitive to intracellular iron concentrations, oxidative stress and hypoxia [30,31]. To the best of our knowledge, this is the first study in which changes in the mRNA of these genes were examined after a marathon run. The significant decrease in mRNA of *PCBP1* and *PCBP2* (expression partners) and *FTH* was observed post-race while at 3 h after the race an up-regulation occurred in *PCBP1* and *PCBP2* as well as in *FTH* and *FTL*. Furthermore, 24 h after the run the gene mRNA levels returned to baseline values. Unfortunately, discussion about these changes is hard since, as mentioned before, there are no data on this topic in the current literature. We assumed that the increase in the mRNA of genes involved in apoptosis and inflammatory response reported earlier [32] would cause long-term up-regulation in our tested genes i.e., that remained up-regulated 24 h after a marathon run. Unfortunately, this was not confirmed by our results. According to the literature, PCBP1 and PCBP2 proteins are iron chaperones that deliver iron to ferritin, the iron storage protein [17,33]. Thus, it is expected that an increase in the expression of these genes might play a protective role against iron toxicity. The mRNA levels of *TFRC* remained unchanged during the marathon run and in the recovery period (with a slight tendency to decrease compared to basal values) suggesting that the intracellular labile iron pool was kept under control. In turn, the *CAT* mRNA level decreased after the marathon run but also returned to baseline after 24 h. The results at 3 h after the marathon showed a significant increase in mRNA levels. It is established that during exercise, changes in many plasma or serum parameters influence intracellular homeostasis. Oxidative stress is another indicator of tested gene expression; thus, we evaluated the mRNA levels of C*AT* for additional information on changes in intracellular oxidative stress. One of the functions of catalase is an increase in antioxidative capacity (Sureda et al. [34]), thus its expression indirectly shows the level of oxidative stress in the cell.

4.3. Relationship Between the mRNA of Genes Involved in Iron Metabolism and Running Speed

Generally, the same direction of changes in *PCBP1*, *PCBP2*, *FTH*, *FTL* and *TFRC* mRNA was observed in both groups, indicating a tendency to decrease immediately after the run and increase 3 h post-race. However, significant differences in *FTH* and *FTL* mRNA were observed between the slow and fast groups at 24 h after finishing the marathon. In faster participants, the mRNA levels of these genes were significantly lower compared to slower participants. According to Jastrzębski et al. [14], based on organ damage indicators, our findings could be caused by a better adaptation to a long-lasting effort in the faster group. The cited authors concluded that participants choose their running speed to individual possibilities determined by changes in tested parameters. In our opinion, the results obtained in our experiment, regarding changes in gene expression, indicating that this hypothesis could be true. Additionally, significant differences between groups (slow and fast) indicated that faster runners were significantly younger than slower runners. This finding indicated possibilities of influence of age to obtained results during marathon run. However, after dividing the participants of the run into two groups by the median split of age (younger and older), no significant differences between groups have been observed.

5. Conclusions

We concluded that marathon running induced changes in biochemical parameters and the expression of genes involved in iron metabolism, but these changes occurred at different time points. Interestingly, in faster runners, the return to basal values occurred faster than in slower runners. Generally, the amateurs could adjust the pace of the run to their capabilities.

Author Contributions: Conceptualization, J.A, W.R., A.G. and M.Ż.; methodology, M.Ż., A.G. and K.A.; validation, M.Ż., K.A., I.B. and A.G.; formal analysis, A.G.; investigation, I.B., W.R. and P.A.; resources, W.R., M.Ż. and P.A.; writing—original draft preparation, M.Ż. and A.G.; writing—review and editing, J.A.; visualization, K.A. and A.G.; supervision, J.A. and M.Ż.; project administration, J.A. and M.Ż.; funding acquisition, W.R and J.A.

Funding: This research was funded by the knowledge grant of the Polish Ministry of Science and Higher Education N RSA4 06754.

Conflicts of Interest: The authors declare no conflict of interest.

References

1. Alaunyte, I.; Stojceska, V.; Plunkett, A. Iron and the Female Athlete: A Review of Dietary Treatment Methods for Improving Iron Status and Exercise Performance. *J. Int. Soc. Sports Nutr.* **2015**, *12*, 38. [CrossRef] [PubMed]

2. Clénin, G.E.; Cordes, M.; Huber, A.; Schumacher, Y.; Noack, P.; Scales, J.; Kriemler, S. Iron Deficiency in Sports—Definition, Influence on Performance and Therapy. *Schweizerische Zeitschrift für Sportmedizin und Sporttraumatologie* **2016**, *64*, 6–18. [CrossRef] [PubMed]

3. Sandström, G.; Börjesson, M.; Rödjer, S. Iron Deficiency in Adolescent Female Athletes-Is Iron Status Affected by Regular Sporting Activity? *Clin. J. Sport Med.* **2012**, *22*, 495–500. [CrossRef] [PubMed]

4. Mettler, S.; Zimmermann, M.B. Iron Excess in Recreational Marathon Runners. *Eur. J. Clin. Nutr.* **2010**, *64*, 490–494. [CrossRef] [PubMed]

5. Bárány, P. Inflammation, Serum C-Reactive Protein, and Erythropoietin Resistance. *Nephrol. Dial. Transplant.* **2001**, *16*, 224–227. [CrossRef] [PubMed]

6. Borghini, A.; Giardini, G.; Tonacci, A.; Mastorci, F.; Mercuri, A.; Mrakic-Sposta, S.; Moretti, S.; Andreassi, M.G.; Pratali, L. Chronic and Acute Effects of Endurance Training on Telomere Length. *Mutagenesis* **2015**, *30*, 711–716. [CrossRef] [PubMed]

7. Dalle Carbonare, L.; Manfredi, M.; Caviglia, G.; Conte, E.; Robotti, E.; Marengo, E.; Cheri, S.; Zamboni, F.; Gabbiani, D.; Deiana, M.; et al. Can Half-Marathon Affect Overall Health? The Yin-Yang of Sport. *J. Proteomics* **2018**, *170*, 80–87. [CrossRef] [PubMed]

8. Kell, D.B.; Pretorius, E. Serum Ferritin Is an Important Inflammatory Disease Marker, as It Is Mainly a Leakage Product from Damaged Cells. *Metallomics* **2014**, *6*, 748–773. [CrossRef]

9. Peeling, P.; Dawson, B.; Goodman, C.; Landers, G.; Wiegerinck, E.T.; Swinkels, D.W.; Trinder, D. Cumulative Effects of Consecutive Running Sessions on Hemolysis, Inflammation and Hepcidin Activity. *Eur. J. Appl. Physiol.* **2009**, *106*, 51–59. [CrossRef]

10. Spanidis, Y.; Priftis, A.; Stagos, D.; Stravodimos, G.A.; Leonidas, D.D.; Spandidos, D.A.; Tsatsakis, A.M.; Kouretas, D. Oxidation of Human Serum Albumin Exhibits Inter-Individual Variability after an Ultra-Marathon Mountain Race. *Exp. Ther. Med.* **2017**, *13*, 2382–2390. [CrossRef]

11. Tan, D.; Dawson, B.; Peeling, P. Hemolytic Effects of a Football-Specific Training Session in Elite Female Players. *Int. J. Sports Physiol. Perform.* **2012**, *7*, 271–276. [CrossRef] [PubMed]

12. Telford, R.D.; Sly, G.J.; Hahn, A.G.; Cunningham, R.B.; Bryant, C.; Smith, J.A. Footstrike Is the Major Cause of Hemolysis during Running. *J. Appl. Physiol.* **2015**, *94*, 38–42. [CrossRef] [PubMed]

13. Terink, R.; ten Haaf, D.; Bongers, C.W.G.; Balvers, M.G.J.; Witkamp, R.F.; Mensink, M.; Eijsvogels, T.M.H.; Gunnewiek, J.M.T.K.; Hopman, M.T.E. Changes in Iron Metabolism during Prolonged Repeated Walking Exercise in Middle-Aged Men and Women. *Eur. J. Appl. Physiol.* **2018**, *118*, 2349–2357. [CrossRef] [PubMed]

14. Jastrzębski, Z.; Zychowska, M.; Radzimiński, Ł.; Konieczna, A.; Kortas, J. Damage to Liver and Skeletal Muscles in Marathon Runners during a 100 Km Run with Regard to Age and Running Speed. *J. Hum. Kinet.* **2015**, *45*, 93–102. [CrossRef] [PubMed]

15. Orino, K.; Lehman, L.; Tsuji, Y.; Ayaki, H.; Torti, S.V.; Torti, F.M. Ferritin and the Response to Oxidative Stress. *Biochem. J.* **2015**, *357*, 241–247. [CrossRef]

16. Peeling, P.; Dawson, B.; Goodman, C.; Landers, G.; Trinder, D. Athletic Induced Iron Deficiency: New Insights into the Role of Inflammation, Cytokines and Hormones. *Eur. J. Appl. Physiol.* **2008**, *103*, 381–391. [CrossRef] [PubMed]

17. Shi, H.; Bencze, K.Z.; Stemmler, T.L.; Philpott, C.C. A Cytosolic Iron Chaperone That Delivers Iron to Ferritin. *Science* **2008**, *320*, 1207–1210. [CrossRef]

18. Völgyi, E.; Tylavsky, F.A.; Lyytikäinen, A.; Suominen, H.; Alén, M.; Cheng, S. Assessing Body Composition with DXA and Bioimpedance: Effects of Obesity, Physical Activity, and Age. *Obesity* **2008**, *16*, 700–705. [CrossRef]

19. Chomczynski, P.; Sacchi, N. Single-Step Method of RNA Isolation by Acid Guanidinium Extraction. *Anal. Biochem.* **1987**, *162*, 156–159. [CrossRef]

20. Dill, D.B.; Costill, D.L. Calculation of Percentage Changes in Volumes of Blood, Plasma, and Red Cells in Dehydration. *J. Appl. Physiol.* **1974**, *37*, 247–248. [CrossRef]

21. Schmittgen, T.D.; Livak, K.J. Analyzing Real-Time PCR Data by the Comparative CT method. *Nat. Protoc.* **2008**, *3*, 1101–1108. [CrossRef] [PubMed]

22. Atamaniuk, J.; Stuhlmeier, K.M.; Vidotto, C.; Tschan, H.; Dossenbach-Glaninger, A.; Mueller, M.M. Effects of Ultra-Marathon on Circulating DNA and MRNA Expression of pro- and Anti-Apoptotic Genes in Mononuclear Cells. *Eur. J. Appl. Physiol.* **2008**, *104*, 711–717. [CrossRef] [PubMed]

23. Roecker, L.; Meier-Buttermilch, R.; Brechtel, L.; Nemeth, E.; Ganz, T. Iron-Regulatory Protein Hepcidin Is Increased in Female Athletes after a Marathon. *Eur. J. Appl. Physiol.* **2005**, *95*, 569–571. [CrossRef] [PubMed]

24. Chiu, Y.H.; Lai, J.I.; Wang, S.H.; How, C.K.; Li, L.H.; Kao, W.F.; Yang, C.C.; Chen, R.J. Early Changes of the Anemia Phenomenon in Male 100-Km Ultramarathoners. *J. Chin. Med. Assoc.* **2015**, *78*, 108–113. [CrossRef] [PubMed]

25. Peeling, P.; Sim, M.; Badenhorst, C.E.; Dawson, B.; Govus, A.D.; Abbiss, C.R.; Swinkels, D.W.; Trinder, D. Iron Status and the Acute Post-Exercise Hepcidin Response in Athletes. *PLoS ONE* **2014**, *9*. [CrossRef]

26. Buchman, A.L.; Dunn, J.K.; Keen, C.; Commisso, J.; Dunn, J.K.; Killip, D.; Dunn, J.K.; Ou, C.N.; Rognerud, C.L.; Dunn, J.K.; et al. The Effect of a Marathon Run on Plasma and Urine Mineral and Metal Concentrations *J. Am. Coll. Nutr.* **1998**, *17*, 124–127. [CrossRef] [PubMed]

27. Duca, L.; Da Ponte, A.; Cozzi, M.; Carbone, A.; Pomati, M.; Nava, I.; Cappellini, M.D.; Fiorelli, G. Changes in Erythropoiesis, Iron Metabolism and Oxidative Stress after Half-Marathon. *Intern. Emerg. Med.* **2006**, *1*, 30–34. [CrossRef]

28. Weight, L.M.; Byrne, M.J.; Jacobs, P. Haemolytic Effects of Exercise. *Clin. J. Sport Med.* **2006**, *2*, 74. [CrossRef]
29. Galanello, R.; Giagu, N.; Barella, S.; Maccioni, L.; Origa, R. Lack of Correlation between Serum Ferritin and Liver Iron Concentration in Beta-Zero Thalassemia Intermedia. *Blood* **2004**, *104*, 3620.
30. Huang, B.W.; Miyazawa, M.; Tsuji, Y. Distinct Regulatory Mechanisms of the Human Ferritin Gene by Hypoxia and Hypoxia Mimetic Cobalt Chloride at the Transcriptional and Post-Transcriptional Levels. *Cell. Signal.* **2014**, *26*, 2702–2709. [CrossRef]
31. Mieszkowski, J.; Kochanowicz, M.; Ma, B.; Kochanowicz, A.; Grzybkowska, A.; Anczykowska, K.; Sawicki, P.; Borkowska, A.; Niespodzinski, B.B.; Antosiewicz, J. Ferritin Genes Overexpression in PBMC and a Rise in Exercise Performance as an Adaptive Response to Ischaemic Preconditioning in Young Men. *BioMed Res. Int.* **2019**, *2019*, 9576876. [CrossRef] [PubMed]
32. Maqueda, M.; Roca, E.; Brotons, D.; Soria, J.M.; Perera, A. Affected Pathways and Transcriptional Regulators in Gene Expression Response to an Ultra-Marathon Trail: Global and Independent Activity Approaches. *PLoS ONE* **2017**, *12*, 1–26. [CrossRef] [PubMed]
33. Leidgens, S.; Bullough, K.Z.; Shi, H.; Li, F.; Shakoury-Elizeh, M.; Yabe, T.; Subramanian, P.; Hsu, E.; Natarajan, N.; Nandal, A.; et al. Each Member of the Poly-r(C)-Binding Protein 1 (PCBP) Family Exhibits Iron Chaperone Activity toward Ferritin. *J. Biol. Chem.* **2013**, *288*, 17791–17802. [CrossRef] [PubMed]
34. Sureda, A.; del Mar Bibiloni, M.; Julibert, A.; Bouzas, C.; Argelich, E.; Llompart, I.; Pons, A.; Tur, J.A. Adherence to the Mediterranean Diet and Inflammatory Markers. *Nutrients* **2018**, *10*, 62. [CrossRef] [PubMed]

© 2019 by the authors. Licensee MDPI, Basel, Switzerland. This article is an open access article distributed under the terms and conditions of the Creative Commons Attribution (CC BY) license (http://creativecommons.org/licenses/by/4.0/).

 *genes*

Review

Physical Activity and Brain Health

Carlo Maria Di Liegro [1], Gabriella Schiera [1], Patrizia Proia [2] and Italia Di Liegro [3,*]

1   Department of Biological, Chemical and Pharmaceutical Sciences and Technologies (Dipartimento di Scienze e Tecnologie Biologiche, Chimiche e Farmaceutiche) (STEBICEF), University of Palermo, 90128 Palermo, Italy; carlomaria.diliegro@unipa.it (C.M.D.L.); gabriella.schiera@unipa.it (G.S.)
2   Department of Psychology, Educational Science and Human Movement (Dipartimento di Scienze Psicologiche, Pedagogiche, dell'Esercizio fisico e della Formazione), University of Palermo, 90128 Palermo, Italy; patrizia.proia@unipa.it
3   Department of Biomedicine, Neurosciences and Advanced Diagnostics (Dipartimento di Biomedicina, Neuroscienze e Diagnostica avanzata) (Bi.N.D.), University of Palermo, 90127 Palermo, Italy
*   Correspondence: italia.diliegro@unipa.it; Tel.: +39-091-238-97415 (ext. 446)

Received: 7 August 2019; Accepted: 12 September 2019; Published: 17 September 2019

Abstract: Physical activity (PA) has been central in the life of our species for most of its history, and thus shaped our physiology during evolution. However, only recently the health consequences of a sedentary lifestyle, and of highly energetic diets, are becoming clear. It has been also acknowledged that lifestyle and diet can induce epigenetic modifications which modify chromatin structure and gene expression, thus causing even heritable metabolic outcomes. Many studies have shown that PA can reverse at least some of the unwanted effects of sedentary lifestyle, and can also contribute in delaying brain aging and degenerative pathologies such as Alzheimer's Disease, diabetes, and multiple sclerosis. Most importantly, PA improves cognitive processes and memory, has analgesic and antidepressant effects, and even induces a sense of wellbeing, giving strength to the ancient principle of "*mens sana in corpore sano*" (i.e., a sound mind in a sound body). In this review we will discuss the potential mechanisms underlying the effects of PA on brain health, focusing on hormones, neurotrophins, and neurotransmitters, the release of which is modulated by PA, as well as on the intra- and extra-cellular pathways that regulate the expression of some of the genes involved.

Keywords: physical activity; brain health; myokines; BDNF; Irisin; lactate; exercise and neurodegeneration; exercise and aging

---

1. Introduction

The discovery of the nervous system dates back to the ancient Greek physicians-philosophers Alcmaeon, Praxagoras, Herophilus [1,2], and Erasistratus [2]. Herophilus (c335–c280 B.C.), in particular, by dissecting human cadavers, was able to describe the structure of the brain and nerves, and to realize that motor nerves were joined to muscles, while other nerves (the sensory ones) went to organs, and were responsible for sensation. He promoted a cerebrocentric view of mind [1–5] and, interestingly, believed that exercise and a healthy diet were fundamental for maintaining a healthy body, and a healthy mind [3]. Over the centuries this idea has recurred many times. However, we have only recently begun to understand the cellular and molecular reasons why sedentary life is detrimental for human health, and to realize that physical activity (PA) can be a powerful medicine to counteract its effects. Actually, this is not surprising since the ability of our species to survive in many different environments, to escape predators, and to look around for food has depended on, and still depends on the ability to perform PA, and PA has thus shaped our physiology [6]. Starting from the consideration that modern humans have not only a very large brain but also a remarkable endurance capacity, it was suggested that PA also shaped our brains: It was reported, for example, that the appearance in evolution of skeletal properties related to endurance capacity correlated with the increase of brain size in hominins such as *Homo erectus* [7–9]. As reported by Hill and Polk [9], aerobic fitness (required for successful endurance activity), and aerobic capacity (measured as maximal oxygen consumption during exercise, VO_2 max) correlate with brain size, both in humans and other animals; moreover, selective breeding in rodents for endurance running capacity affects both their general physiology and their brain, and also potentiates their cognitive abilities [9,10]. A further aspect of humans that might correlate with PA concerns the integumentary system: Our hairless skin indeed enhances evaporation, thus allowing dispersion of excess heat produced during endurance activity [9,11–13]; at the same time, a hairless skin facilitates production of vasodilatory factors, such as nitric oxide (NO), with different mechanisms [14,15].

In this context, it is important to underline that, when the importance of PA during the evolution of our species is discussed, the focus is on every movement that requires activity of our skeletal muscles, and energy expenditure. On the other hand, any planned and structured activity that is voluntarily aimed at improving and/or maintaining our physical fitness should be better defined as exercise [16]. Thus, most of the experimental work cited in this review actually concerns "exercise" since the observations reported rely on a specific series of structured, planned, and repetitive activities. Exercise is, however, only a subset of physical activity; accordingly, we will use the term "exercise" when describing the results of programmed sets of experiments, and the expression "physical activity" (PA) when discussing the effects on health of either programmed or not programmed skeletal muscle movements, in daily life.

There are clear indications that PA also has important effects on human brain health at any age and have been included, for example, in the Physical Activity Guidelines for Americans, issued by the U.S. Department of Health and Human Services (HHS) in 2018 [17–19]. Interestingly, in these guidelines, four classes of age, with different PA requirements, have been set: 1. Preschool-Aged Children (3–5 years)—they should be physically active throughout the day to enhance growth and development, it is also important to underline that playing develops mental capacities and social interactions in many ways; 2. Children and Adolescents (6–17 years)—they should do 60 min or more per day of moderate-to-vigorous physical activity, most of which should be aerobic, with vigorous activity for at least 3 days per week, including muscle- and bone-strengthening physical activity; 3. Adults—according to the Guidelines "Adults should move more and sit less throughout the day". They should do at least 150–300 min of moderate-intensity PA, or 75–150 min of vigorous aerobic PA per week, together with muscle-strengthening activities of moderate-high intensity, at least 2 days a week; 4. Old Adults—they should do as much aerobic and muscle-strengthening activities as they can, on the basis of their individual health conditions. In addition, the guidelines suggest physical training for women during pregnancy and post-partum period and for adults with chronic diseases and/or disabilities [17].

PA is thus recommended as a non-pharmacologic therapy for different pathological affections as well as for the maintenance of general health status. Habitual exercise improves cardiorespiratory fitness and cardiovascular health [20–24], helps reducing body mass index [25,26], and can represent a natural, anti-inflammatory "drug" in chronic diseases, such as type 2 diabetes mellitus (T2DM) and cardiovascular disease (CVD) [27,28]. Moreover, given the strong association of pathologic conditions such as high blood pressure with blood–brain barrier alterations and brain dysfunctions, PA can also have beneficial effects on cerebrovascular and cognitive functions [23]. In addition, anti-depressive- [29], and analgesic-PA effects have been reported [30]. However, it has also been suggested that the anti-inflammatory effects can differ among different training programs [31], and that, while regular exercise can increase immune competence and reduce the risk of infection with respect to a sedentary lifestyle, acute and heavy bouts of activity can even have the opposite effect [27], and, in general, negative effects on health [32,33].

As discussed below, both endurance activity (i.e., long-lasting aerobic activity, such as running) and resistance exercise (i.e., exercise in which the predominant activity involves pushing against a force) have been shown to induce an increase of circulating growth factors (such as insulin-like growth factor 1, IGF-1), and neurotrophins (such as the brain-derived neurotrophic factor, BDNF) which have an effect on the brain both during development and in the adult. The same factors might have had an impact during hominin brain evolution [9], and can affect brain plasticity in the young as well as in the adult, under many different conditions, such as physiologic aging, neurodegenerative pathologies, and recovery after acute brain damage.

In this review we will discuss the putative cellular and molecular mechanisms underlying the mentioned effects of PA on the nervous system, focusing on genes known to be involved, as well as on epigenetic effects due to DNA methylation, histone post-translational modifications and exchange, and on the possible role of non-coding RNAs.

2. Brain Plasticity, Adult Neurogenesis, and Physical Activity

The brain capacity to adapt to ever-changing conditions, known as brain plasticity, depends on the ability of neurons to modify the strength and composition of their connections in response to both external and internal stimuli. The long-term potentiation (LTP) in synaptic efficacy constitutes the physiologic base for learning and memory. An important way for regulating neuronal function is the activity-dependent synapse-to-nucleus signalling, that can arise both in the post-synaptic and in the presynaptic element [34–38]. These signals are generated through different mechanisms, such as: (i) Calcium waves due to calcium-induced calcium release (CIRC) from the endoplasmic reticulum (ER) [35,39,40]; (ii) retrograde transport of proteins (e.g., **Jacob,** CREB Regulated Transcriptional Coactivator 1, **CRTC1**); Abelson-interacting protein 1, **Abi1**; the amyloid precursor protein intracellular domain associated-1 protein, **AIDA-1**; and the nuclear factor kappa-light-chain-enhancer of activated B cells, **NF-κB**); these proteins are post-translationally modified following synaptic activity, and transported to the nucleus, where they act on gene transcription, and thereafter on synaptic plasticity [34–38,41,42]; (iii) formation and microtubule-dependent trafficking of mRNA-protein complexes, that, after exiting the nucleus, move to neuronal periphery, where the mature transcripts localize in a repressed state, in response to local signalling, through activity-dependent activation of specific enzymes, the regulatory proteins can be then modified, for example, by phosphorylation, and the mRNAs can be translated; some of the newly synthesized proteins can accumulate at the synapse, while others can shuttle back to the nucleus to modify chromatin structure and expression [43].

By regulating synapse-to-nucleus signalling, all these events are crucial for allowing synapse activity to result in the specific gene expression programs necessary for learning and memory. In agreement with this idea, the impaired function of these signalling proteins brings about intellectual disability, psychiatric disorders, or neurodegeneration [37,38,42]. On the other hand, we can hypothesize that an increase of their function, for example as a response to PA, could also enhance brain functions and plasticity.

In the past, it was generally accepted that new neurons could not be generated in the adult to replace dying cells, and this limitation was also considered to be the main cause of neurodegeneration as well as of cognitive decline in the elderly population. However, since the 1960s, many researchers presented data suggesting that, in all the mammals analysed, new neurons could be generated in the sub granular zone (SGZ) of the dentate gyrus of the hippocampus, and in the sub-ventricular zone (SVZ) of the lateral ventricles, in the postnatal and adult life [44–50]. In particular, neurons born in the SGZ were shown to differentiate and integrate into the local neural network of the hippocampus. These findings are extremely important since the hippocampus is fundamental for the formation of certain types of memory, such as episodic memory and spatial memory [51–54]. In addition, hippocampus-dependent learning is one of the major regulators of hippocampal neurogenesis [55]: living in environments which stimulate learning enhances, in rats, the survival of neurons, born in the adult from neural stem cells (NSCs) [52].

Now, increasing evidence suggests that PA, largely due to factors released by contracting muscles (Section 3; Figure 1), can improve brain functions, such as memory and attention, in both children and adults [56–64]. A few examples of single studies (first three rows) and reviews/meta-analyses (second three rows), aimed at ascertaining any relationship between PA and learning/memory, are given in Table 1.

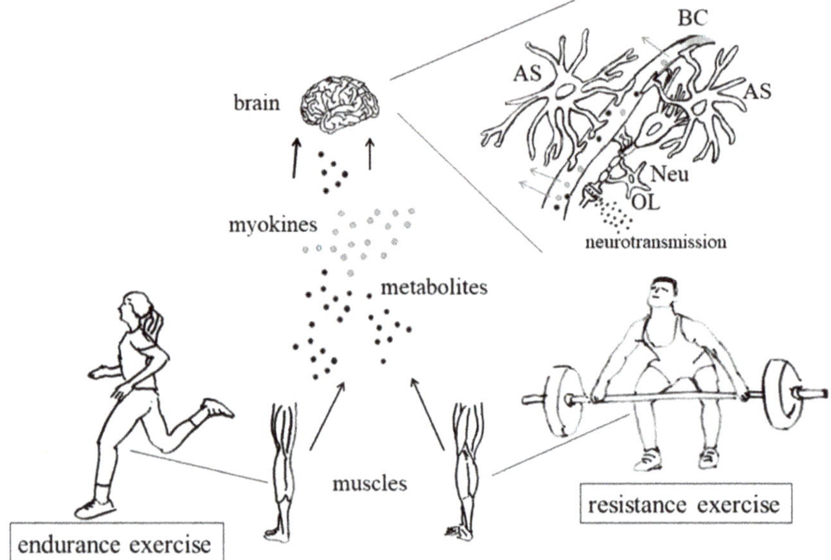

Figure 1. Hypothetical pathway for the exercise-mediated effects on brain functions: both endurance and resistance exercise, even if with different kinetics and properties, allow muscle synthesis, and release myokines (e.g., brain-derived neurotrophic factor, BDNF), as well as of metabolites (such as lactate) into the circulation; these molecules can cross the blood–brain barrier (BBB) at the level of the brain capillaries (grey arrows) and affect the functions of both neurons and glial cells, thus modifying neurotransmission in different regions of the brain. As explained in the text, neurotransmission can then activate pathways leading to modifications of gene expression. AS: astrocytes; BC: brain capillaries; Neu: neurons; OL: oligodendrocytes.

Table 1. Effects of physical activity (PA) on learning and memory in children and adolescents. In the first three rows single studies are reported, while the second three rows refer to reviews/meta-analyses. In the "Conclusions" column, the main results of the analyses, as well as a few comments on them, are given.

Protocol/Aims [Ref]	Subjects/Studies Included	Methods of Analysis	Conclusions
Analysis based on a randomized controlled trial (Ballabeina Study: [65]) aimed at evidencing any relationship between aerobic fitness/motor skills and working memory and attention in pre-school children [59]	245 ethnically diverse pre-school children (49% girl, mean age 5.2 years) were analysed at the beginning of the activity and 9 months later.	Physical tests: 1. Aerobic fitness, assessed according to the 20 m shuttle run [66], 2. Agility, assessed by an obstacle course, 3. Dynamic balance on a beam. In order to evaluate spatial memory and attention, each child was tested individually by focused tests.	**Higher baseline aerobic fitness and motor skills were related to higher levels of working memory and attention.** A further improvement of these latter abilities was noticed in the following 9 months.

Table 1. *Cont.*

Protocol/Aims [Ref]	Subjects/Studies Included	Methods of Analysis	Conclusions
The aim of the study was to ascertain whether very low-intensity exercise (i.e., walking), practiced during foreign-language (Polish) vocabulary encoding, improves subsequent recall, in comparison with encoding during physical rest [62]	49 right-handed, monolingual, Germans, healthy subjects (aged 18–30 years). Criteria of exclusion: a history of psychiatric or neurological disorders, smoking, obesity, and any knowledge of Polish or other Slavic languages.	In the first session, participants learned 40 Polish words while walking on the motor-driven treadmill, at their previously determined preferred rate. In the second session, the participants learned a further group of 40 words, while sitting in a chair. Each session lasted 30 min. The order of sessions was different for different subjects, in a balanced way, and the experiments were repeated twice.	In both experiments, participants' **performance was better when they exercised during learning** compared to learning when sedentary. Serum BDNF levels and salivary cortisol concentration were also measured: serum BDNF was unrelated to memory performance; on the other hand, a positive correlation between the salivary cortisol and the number of correctly recalled words was found.
The aim of the study was to clarify whether mnemonic discrimination is improved by an acute bout of moderate-intensity aerobic exercise [63]	21 healthy young adults (mean age 20.5 ± 1.4 years, 10 females), without histories of neurological or psychiatric disorders. All participants had normal or corrected-to-normal vision, and normal colour vision.	In this study moderate intensity is defined as 40–59% of $\dot{V}O_2$ peak, as established by the American College of Sports Medicine (ACSM) [67]. The activity was performed by a recumbent ergometer. Mnemonic task: the participants were first shown 196 pictures of everyday objects and asked, for each of them, whether it was an indoor or an outdoor item. Then they were asked to identify by pressing a button, in the second group of 256 items, which were 'previously seen', 'similar but not identical' or 'not previously seen'.	The **lure discrimination index (LDI) for high-similarity items was higher after 10 min of moderate aerobic exercise** than in resting controls, thus suggesting that a bout of acute aerobic exercise could improve pattern separation, that seems to rely on the dentate gyrus (DG) in humans.

Table 1. *Cont.*

Protocol/Aims [Ref]	Subjects/Studies Included	Methods of Analysis	Conclusions
The aim of the analysis was to search the literature, looking for evidence of chronic PA effects on mental health in children and adolescents [58].	Review articles reporting chronic physical activity and at least one mental health outcome (i.e., depression, anxiety/stress, self-esteem and cognitive functioning) in children/adolescents. Reviews chosen: 4 papers on the evidence concerning PA and depression; 4 for anxiety; 3 for self-esteem; 7 for cognitive functions.	Analysis based on data collected from PubMed, SPORTDiscus, PsychINFO, Web of Science, Medline, Cochrane Library, and ISI Science Citation Index, by using search terms related to the variables of interest (e.g., sport, exercise, physical activity) and mental health outcome variables (e.g., depression, anxiety, self-esteem, cognitive functioning).	**Associations between PA and mental health in young people (Tables 1–4 in Ref. [58]) is evident, but the effects are small-to-moderate, probably because of weakness of the research designs.** Small but consistent association between sedentary time and poorer mental health is also evident.
The aim of this systematic review was to find out studies elucidating the relationship between aerobic PA and children's cognition, academic achievement, and psychosocial function [60]	Studies analysed concerned interventions of aerobic PA in children younger than 19 years. Only randomized control trials that measured psychological, behavioural, cognitive, or academic outcomes were included.	The review was performed using MEDLINE, Cochrane, PsycINFO, SPORTDiscus, and EMBASE. Additional studies were identified through back-searching bibliographies.	**Aerobic PA is positively associated with cognition, academic achievement, behaviour, and psychosocial functioning outcomes. More rigorous trials, however, required for deducing detailed relationships.**
Systematic review and meta-analysis of studies concerning associations between PA/sedentary lifestyle and mental health. Meta-analyses were performed in randomized controlled trials (RCTs) and non-RCTs (i.e., quasi-experimental studies) [64]	Studies published from January 2013 to April 2018. Studies were included if they comprehended PA or sedentary behaviour data and at least one psychological ill-being (i.e., depression, anxiety, stress, etc.) or psychological well-being (i.e., self-esteem, optimism, happiness, etc.) outcome in pre-schoolers (2–5 years of age), children (6–11 years of age) or adolescents (12–18 years of age).	Analysis based on data collected through a systematic search of the PubMed and Web of Science databases by two independent researchers. A narrative synthesis of observational studies was conducted.	**PA improves adolescents' mental health, but additional studies are needed to confirm the effects of PA on children.** Findings from observational studies, however, suggest that promoting PA and decreasing sedentary behaviour might have a protecting effect on mental health in both children and adolescents.

The data reported in Table 1 clearly indicate that PA has a positive effect on mental health and abilities, especially in adolescents; however, as reported in the "Conclusions" column (sentences in bold letters), most authors agree on the fact that the previous studies do not yet give uniform indications on the relationships between the type/intensity/frequency of exercise and the brain health outcomes; these limitations derive, on one hand, from the wide range of conditions set in the exercise programs, and on the other hand, the differences from study to study also depend on the variability of the parameters chosen to evaluate mental health. We also have to add to these considerations the poor knowledge we still have of 'mind' and of 'mental health'. Thus, many laboratories are now focusing on exercise-dependent cellular and molecular modifications of brain cells activity, in the attempt to uncover the mechanisms underlying PA–mental health biochemical relationships.

At the cellular level, it was found that treadmill exercise can increase hippocampal neurogenesis in aged mice [68]. Interestingly, exercise can also affect the proliferation [69,70], as well as size and function, of astrocytes [71]. These latter events regulate, in turn, the number and localization of neuronal synapses, and might influence LTP and episodic memory formation [72].

Many researchers suggested that all these effects are also regulated by the brain capillaries (BC, Figure 1) that reach the neurogenic niche, supplying angiogenetic growth factors, such as the growth and differentiation factor 11 (**GDF11**), the vascular endothelial growth factor (**VEGF**) [59], and **BDNF**, that activates a cellular survival pathway involving the serine-threonine kinase **AKT** and **CREB**, thus inducing the transcription of genes responsible for almost all the aspects of neuroplasticity [59,72]. The neurogenic niche also receives axonal inputs from both local and distant neurons, which release a variety of neurotransmitters, such as serotonin, glutamate, and GABA [59]. For example, glutamate, through interaction with NMDARs, is thought to regulate LTP in response to exercise [73]. Many epidemiologic studies, mostly in the last two decades, also revealed a link between PA, human brain health (and longevity) and epigenetic modifications of the genome, even leading, on one hand, to the concept of "epigenetic age" or "DNA methylation age" (essentially measured, however, as blood cells DNA methylation) [74–78], and, on the other hand, to the acknowledgment that epigenetic mechanisms induced by PA can build up an "epigenetic memory" that affects long-term brain plasticity, neurogenesis, and function [79–82]. Intriguingly, it has been proposed that epigenetic modifications caused by lifestyle and diet, as well as the effects of PA can be heritable (discussed in [83]).

Epigenetic processes modify eukaryotic chromatin structure, and hence gene expression, without changing the underlying DNA sequence, through at least three mechanisms: (i) DNA methylation/demethylation, and post-translational modifications (such as methylation/demethylation and acetylation/deacetylation), of histones on specific residues of their N-terminal tails; (ii) substitution of some histone isotypes with other histone variants; (iii) sliding and/or removal of the basic chromatin structural organization elements (nucleosomes), due to specific ATP-dependent chromatin remodelling complexes [84–87]. Specific proteins are then able to "read" and bind DNA and histone tail modifications, thus creating synergic complexes which can activate or depress transcription [88–92]. Importantly, in some of these remodelling events, long noncoding RNAs (lncRNAs) also play a role [93]. Finally, gene expression can be regulated by short noncoding RNAs, called microRNAs (miRNAs), which are able to pair with sequences mainly present in the 3'-UTR of their target mRNAs, thus inducing inhibition of their translation or even their degradation [94–96].

In summary, while the genome of an organism is relatively stable over the lifespan, its expression (i.e., the phenotype) is influenced by many epigenetic factors. Most important, we now know that inactivity is epigenetically deleterious: for example, it has been reported that nine days of bed rest can induce insulin resistance in otherwise healthy subjects. The analysis of the pathways affected revealed a significant downregulation of 34 pathways, mainly involving genes associated with the mitochondrial function, including the peroxisome proliferator-activated receptor γ co-activator 1α (PPARGC1A, or **PGC-1α**). An increase of PPARGC1A DNA methylation was also reported, and this epigenetic modification was not completely reversed after four weeks of retraining, thus highlighting the importance of daily physical activity [76,97].

2.1. Brain-Derived Neurotrophic Factor (BDNF)

BDNF is a neurotrophin involved in all the most important aspects of neuroplasticity, from neurogenesis to neuronal survival, from synaptogenesis to cognition, as well as in the regulation of energy homeostasis.

Both in humans and rodents, the *BDNF* gene contains nine exons, each of which has its own promoter. As a result of this gene structure, many species of mature transcripts are known, even if the final translation product is the same for all of them [98,99]. The existence of different promoters, however, is important in terms of temporal and spatial regulation, including the possibility that different promoters are used in different cell types and brain regions [99].

In the published literature, a generalized exercise-dependent increase of BDNF has been reported. A few examples of both single studies (first six rows) and reviews/meta-analyses (last two rows) aimed at ascertaining PA effects on BDNF levels are reported in Table 2.

The BDNF increase seems to correlate with the exercise volume (given by "intensity + duration + frequency" of activity) [100]. However, it was also reported that the greatest responses are given by well-trained individuals, while mainly sedentary subjects show lower or even no response [100,101]. Interestingly, open-skill exercise (e.g., badminton) increases BDNF levels more than closed-skill exercise (e.g., running), probably because open-skill activities require additional attention to face ever-changing situations [102], and possibly also because they are more enjoyable.

Table 2. Effects of PA on circulating BDNF levels. In the first six rows, single studies have been reported, while the last two rows refer to reviews/meta-analyses. In the "Conclusion" column the main results of the analyses, as well as a few comments on them, are given.

Protocol/Aims [Ref]	Subjects/Studies Included	Methods of Analysis	Conclusions
The aim of the study was to test the effects of two high-intensity exercise protocols, already known to improve cardiovascular health, to also affect BDNF levels [103]	Experiment 1: 8 men (average age: 28 years) Experiment 2: 21 men (average age: 27 years) Both experiments included: -high-intensity interval-training (HIT), at 90% of maximal work rate for 1 min, alternating with 1 min of rest; -continuous exercise (CON), at 70% of maximal work rate. Both protocols lasted 20 min.	Experiment 1: serum [BDNF] was measured at 30 min before starting the exercise, at 0, 6, 10, 14, and 18 min during the exercise, and at the end of the exercise (20 min). Experiment2: Serum BDNF was measured only at the beginning (0 min) and at the end (20 min) of the experiment. BDNF was evaluated by an enzyme-linked immunoassay (ELISA).	**-Similar BDNF kinetics were observed in both protocols, with maximal BDNF level reached toward the end of training;** **-Both protocols (CON and HIT) significantly increased BDNF, with HIT more effective Shorter bouts of high-intensity exercise are slightly more effective** than continuous high-intensity exercise for elevating serum BDNF. Moreover, 73% of the participants preferred the HIT protocol Thus, the authors suggest that the HIT is an effective and preferred intervention for elevating BDNF and potentially promoting brain health.

Table 2. *Cont.*

Protocol/Aims [Ref]	Subjects/Studies Included	Methods of Analysis	Conclusions
The aim of this analysis was to study the possible relationship between exercise intensity, memory, and BDNF [104]	16 young subjects (average age: 23 years): 9 men and 7 women	3 exercise sessions at different intensities relative to ventilator threshold (Vt) (VO2max, Vt − 20%, Vt + 20%). Each session lasted approximately 30 min. Following exercise, the Rey Auditory Verbal Learning Test (RAVLT) was performed to assess short-term memory, learning, and long-term memory recall. 24 h later, the participants completed the RAVLT recognition trial, to evaluate another measure of long-term memory. Blood was drawn before exercise, immediately post-exercise, and after the 30-min recall test. Serum BDNF was evaluated by ELISA.	Long-term memory as assessed after the 24-h delay differed as a function of exercise intensity: the largest benefits were observed with the maximal intensity exercise. **BDNF significantly increased in response to exercise.** However, **no difference was noticed in relation to exercise intensity**. Similarly, no significant association was found with memory. The authors suggest that "future research is warranted so that we can better understand how to use exercise to benefit cognitive performance".
The aim of the study was to compare basal- and post-exercise- levels of circulating BDNF, in comparison with cognitive training and mindfulness practice [105]	19 healthy subjects (age: 65–85 years)	Exercises: (1) physical aerobic exercise at a moderate level, using a Swedish version of the EA Sports Active 2™ program on a Microsoft Xbox360™ game console connected to a Microsoft Kinect™ accessory and an ordinary TV set; (2) cognitive training through a computerized working memory training program; (3) mindfulness practice through the use of the Mindfulness App (http://www.mindapps.se/themindfulnessapp/). Each program lasted 35 min. All the participants went through all the three training programs, in a random sequence. Serum BDNF was evaluated by ELISA.	**Exercise caused a significant increase in BDNF levels.** Moreover, in the same subject, **a single bout of exercise had a significantly higher impact on serum BDNF levels** than cognitive training and mindfulness practice. However, considerable variability of BDNF responses was found when comparing different subjects.

Table 2. *Cont.*

Protocol/Aims [Ref]	Subjects/Studies Included	Methods of Analysis	Conclusions
The aim of the study was to compare the effect of 'open-skill' with 'closed-skill' exercise (as defined in terms of predictability of context situations) on BDNF production [102]	20 adult males: all subjects participated in both closed (running) and open (badminton) skill exercise sessions, in counterbalanced order on separate days. Exclusion criteria: - cardiovascular disease, diabetes, history of neurological problems, pre-existing injuries, smoking or intake of recreational drugs; hearing or vision problems.	Exercise sessions: −5 min of warm-up exercises, −30 min of running or badminton. Exercise intensity: 60% of the heart rate reserve level (HRR) During each session, venous blood **samples were obtained immediately before and after exercise.** Serum BDNF was evaluated by ELISA. Cognitive performance was also evaluated by a modified form of the task-switching paradigm, and controlled via the Neuroscan Stim software.	**Badminton exercise resulted in significantly higher serum BDNF levels relative to running.** This study provides interesting evidence in support of the benefits of open-skills exercise on BDNF production and executive function.
The aim of the study was to analyse the effect of aquarobic exercise on serum irisin and BDNF levels [106]	26 elderly women: Control group: 12 subjects Exercise group: 14	Exercise sessions: 16-week aquarobic exercise program, including two sessions a week. Each session lasted for 60 min: −10 min of warm-up, −40 min of exercise, −10 min of cool. Serum irisin and BDNF levels were evaluated (three times in the exercise group and two times in the control group) by ELISA.	**Aquarobic exercises improve the serum irisin and BDNF levels.**

Table 2. *Cont.*

Protocol/Aims [Ref]	Subjects/Studies Included	Methods of Analysis	Conclusions
The aim of this study was to evaluate the effect of long-term exercise on memory and biomarkers related to cognition and oxidative stress, in healthy middle-aged subjects [107]	68 healthy men: Group 1: 21 young sedentary subjects (age: 17–25 years); Group 2: 16 young trained subjects (age: 18–25 years), Group 3: 25 middle-aged sedentary subjects (age: 47–67 years) Group 4: 24 middle-aged trained subjects (age: 46–68 years). Exclusion criteria: -history of severe disease, pain, cognitive deficiencies, head trauma. -use of neuroactive or psychoactive drugs or antioxidants.	Comparison of the BDNF levels in the four groups was performed by a two-way ANOVA. The effect of PA on cognitive abilities was evaluated by a combination of neuropsychological tests, among which: the Trail Making Test, Part A and Part B, the Wechsler Adult Intelligence Scale IV Digit Span Subtest32, the Stroop Interference Test31, the Computerized tests from Cambridge Neuropsychological Test Automated Battery (CANTAB software, Cambridge Cognition, UK), and the Free and Cued Selective Reminding Test (FCSRT)33 Serum BDNF levels were measured by ELISA.	The Free and Cued Immediate Recall tests showed significant improvements in memory in the middle-aged trained individuals when compared to the sedentary ones. **A significantly lower resting level of serum BDNF (and plasma Cathepsin B) was observed in both trained groups. In particular, BDNF and CTSB levels were inversely correlated with weekly hours of exercise.**
The aim of the analysis was to find out any exercise-dependent correlation between BDNF concentration and aerobic metabolism in healthy subjects [100]	Studies were included when they reported BDNF analysis before and after at least one session of exercise. Total studied included: 20	Analysis based on papers collected from PubMed, Scopus, and Medline databases.	**PA-induced BDNF increase is related to the amount of aerobic energy required in the exercise, in a dose-dependent manner.**
Protocols: -Preferred Reporting Items for Systematic Reviews and Meta-Analyses Protocols (PRISMA-P) -Cochrane Handbook of Systematic Reviews of Interventions [108]	Inclusion criteria: studied conducted on adolescents trained with different exercise protocols, and including evaluations of pre- and post-intervention BDNF levels.	Data derived from PubMed, EMBASE, Scopus, ScienceDirect, Web of Science, SPORTDiscus, the Cochrane Central Register of Controlled Trials (CENTRAL), and CINAHL.	**The results show that BDNF levels increase after interventions, regardless of whether the aerobic exercises were acute or chronic.**

As a whole, data reported in Table 2 indicate an exercise-dependent BDNF increase. Again, as evident in the "Conclusions" column (sentences in bold letters), however, a great variability emerges from the different studies.

In general, BDNF increase seems to correlate with increased catabolic requirements, and with a higher production of reactive oxygen species (ROS), as a consequence of the increased mitochondrial activity. Then, in the brain, BDNF stimulates mitochondrial biogenesis, and acts as a metabotrophin to mediate the effects of exercise on cognition [109,110].

Actually, *BDNF* gene transcription does not depend on a single regulatory pathway: it is synergistically stimulated by a complex array of factors, some of which, as discussed above, reach the nucleus only when neurons are active. In addition, the already mentioned transcription factor coactivator **PGC-1α** increases sharply under energy-requiring conditions, both in muscles (see Section 3) and neurons, and contributes to raising BDNF levels [100].

Notably, the expression of the *BDNF* gene is also controlled at the epigenetic level. In 2006, Tsankova et al. [111] analysed the effects of a chronic social defeat stress on the *BDNF* gene chromatin organization in the mouse hippocampus, and found that stress induced a lasting downregulation of BDNF transcripts III and IV, as well as an increase in both histone and promoter methylation. The stressing protocol was followed by treatment with an antidepressant that reversed these effects, also inducing histone acetylation and downregulation of histone deacetylase (HDAC) 5 [111]. Starting from these results, in 2011, Gomez-Pinilla et al. [112] studied the epigenetic effects of exercise on BDNF chromatin regulation, and they found that, like an antidepressant, exercise induced, in the rat hippocampus, DNA demethylation of the BDNF promoter IV, as well as an increase in the levels of phosphorylated MeCP2 (that, in this form, is released from the *BDNF* gene promoter), thus stimulating BDNF mRNA and protein synthesis [112]. By chromatin immunoprecipitation assay, they also found an increase in the levels of histone H3 (but not H4) acetylation, and a decrease of histone deacetylase 5. In parallel, the levels of CaMKII and CREB increased. Similarly, Ieraci et al. [113] showed that BDNF mRNA (transcripts 1–4, 6, and 7) levels decreased immediately after an acute stress in the hippocampus of mice, then returning to the basal level within 24 h. On the other hand, PA caused an increase in BDNF mRNA and was also able to counteract the stress effect, by inducing an increase in histone H3 acetylation at the level of specific BDNF promoters [113]. Since then, a growing body of studies has shown that PA stimulates an activity-dependent cascade of events, involving phosphorylation and other post-translational modifications of signalling proteins, which arrives at the nucleus, where structural organization and function of the chromatin (which includes, among others, the *BDNF* gene) will be targeted [114].

In conclusion, although all these findings clearly demonstrate a role of PA in regulating the levels of circulating BDNF, the analysis in Table 2 shows that there is no precise exercise protocol that can be favoured in order to obtain a maximal effect on BDNF production and, possibly, on mental health. The authors of these studies/meta-analyses all agree on the need for further research in order to better understand how to use exercise to obtain cognitive improvements.

It is also important to highlight that BDNF circulates in the blood as at least two different pools: BDNF in platelets and platelet-free, plasmatic BDNF. This latter form is probably the only one able to cross the blood–brain barrier (BBB). Thus, the method used to measure the circulating neurotrophin can introduce bias from one study to another. Serum preparations that allow clotting and BDNF release from platelets retrieve a much higher amount of BDNF, in comparison with measurements of BDNF from blood samples containing anti-coagulants [115].

These findings suggest that further experiments based on standardized methods are necessary to understand the real relationship between exercise, BDNF production, and brain health.

2.2. microRNAs and Exercise

Recently Zhao et al. [116] obtained, by deep sequencing, a genome-wide identification of miRNAs, the concentration of which is modified in the rat brain, in response to high-intensity intermittent swimming training (HIST), as compared with normal controls (NC). The authors identified a large collection of miRNAs, among which 34 were expressed at significantly different levels in the two conditions; 16 out of these latter species were upregulated, and 18 downregulated in HIST rats [116]. Among the miRNAs that underwent a significant expression modification, some had already been reported by other researchers to be important for brain functions: in particular, the **miR-200 family** had been described to regulate postnatal forebrain neurogenesis [117], differentiation and proliferation of neurons [118], plasticity during neural development [119], and olfactory neurogenesis [120]. Moreover, miR-200b and miR200c seem to have a neuroprotective effect [121]. Actually, most of the predicted targets of PA-controlled miRNAs are genes related to brain/nerve function and already mentioned above, such as *BDNF*, *Igf-1*, *ngf*, and *c-fos*. Some of these genes are also targeted by **miR-483**, another miRNA downregulated in HIST rats [116]. Interestingly, exercise seems to mitigate the effects on cognition of traumatic brain injury and aging by modulating the expression in the hippocampus of **miR-21** [122] and **miR-34a** [123].

In summary, many differentially expressed miRNAs have been evidenced, when comparing the brain of exercising and non-exercising rodents, in a variety of brain areas, including the brain cortex and hippocampus. We have to remember, however, that each miRNA can target a multiplicity of mRNAs, and each mRNA can be targeted by many different miRNAs, thus it is not yet immediately evident how exercise-induced modifications in the miRNA population fit into the general regulation of brain functions by PA.

2.3. Genes Involved in Mitochondrial and Lysosomal Biogenesis

Since the 1950s, the decline of mitochondrial oxidative functions has been considered one of the main causes of cell aging [124]. The respiratory complexes (and in particular, the Nicotinamide adenine dinucleotide, NADH, dehydrogenase and the cytochrome C oxidase complexes) decrease with aging in many tissues, including the brain—relying mostly on the oxidative metabolism— that is particularly sensitive to this decline [125,126]. Moreover, mitochondrial DNA (mtDNA) accumulates mutations with age, and this is a further reason for an aberrant functioning of mitochondria [127]. Fission arrest [128] and abnormal donut-shaped mitochondria [129] have been noticed in the prefrontal cortex of aged animals. Mitochondrial alterations of different kinds have been also noticed in a variety of brain pathologies [130–132].

On the other hand, PA has been reported to have anti-aging effects and can have a positive effect on mitochondrial biogenesis due to the increase of BDNF levels [133]. Recently, it has been reported that, in old mice, exercise can improve brain cortex mitochondrial function by selectively increasing the activity of complex I, and the levels of the mitochondrial dynamin-related protein 1 (**DRP1**), a large GTPase that controls the final part of mitochondrial fission. This finding suggests that, in the brain of old mice, exercise improves mitochondrial function by inducing a shift in the mitochondrial fission–fusion balance toward fission, even in the absence of modifications in the levels of proteins that regulate metabolism or transport, such as BDNF, HSP60, or phosphorylated mTOR [134].

Autophagy is a physiological process which requires functional lysosomes, and that is involved in recycling proteins as well as in eliminating potentially toxic protein aggregates and dysfunctional organelles [135]. It has been suggested that autophagy is essential in skeletal muscle plasticity and that it is regulated by exercise [135–138]. Recently, it has been reported that, in the brain cortex, exercise promotes nuclear translocation of the transcription factor EB (**TFEB**), a master factor in lysosomal biogenesis and autophagy [139]. The authors found that activation of TFEB depends on the NAD-dependent deacetylase sirtuin-1 (**SIRT-1**), that deacetylates it at K116, allowing its nuclear translocation. In turn, SIRT-1 is activated by the pathway induced by activation of the AMP-dependent kinase (**AMPK**) [135]. Interestingly, mitophagy (autophagy of mitochondria) declines with age, thus leading to a progressive accumulation of damaged mitochondria [140]. Thus, the autophagy increase, induced by exercise, not only contributes to the elimination of toxic protein aggregates accumulating in the brain, but also produces a specific increase of mitophagy [141].

3. Muscle Contraction and Production of Myokines

Skeletal muscle is the most abundant tissue in the body and plays a fundamental role in the maintenance of the correct posture and movement. In addition, it has a central metabolic function, since, in response to post-prandial insulin, picks up glucose from the blood and accumulates it as glycogen. As a consequence, age-related loss of skeletal muscle (known as sarcopenia) not only affects body stability and movement, but might also be a cause of hyperglycaemia. On the other hand, exercise improves glucose uptake in skeletal muscles of patients with type 2 diabetes by activating GLU4 translocation to the plasma membrane, partially independent of insulin [142,143].

Different kinds of fibres exist in skeletal muscle, which differs for both metabolic and contractile properties: slow-twitch oxidative (SO) fibres have a high content of mitochondria, and myoglobin, and are more vascularized, fast-twitch glycolytic (FG) fibres have a glycolysis-based metabolism, and finally fast-twitch oxidative glycolytic (FOG) fibres have intermediate properties [144]. Skeletal muscle fibres are also classified according to the myosin heavy chain (MHC) isotypes that they produce: type-I fibres, type-IIA fibres, and type-IIX/IIB fibres, roughly corresponding to SO-, FOG-, and FG-fibres, respectively. Other types of MHC are expressed during embryogenesis or during muscle regeneration [144]. Notably, it seems that also the type of input received from the motor nerve is different for different fibres: type I seems to receive a high amount of inputs at low frequency, while type II seems to receive short inputs at high frequency [145]. Moreover, the contractility properties of muscle fibres do not depend only on the isoforms of contractile proteins expressed, but also on the isotypes of many other proteins, such as those involved in calcium trafficking, and basal metabolism. These differences also depend on epigenetic differences that also influence the transcription rate of the active genes. For example, it has been reported that the mobility of the RNA polymerase II (Pol II) during transcription of the gene encoding PGC-1α differ between fast- and slow-twitch skeletal muscles, thus affecting the gene expression efficiency [146].

Similar to neurons, skeletal muscle cells are post-mitotic, but dynamic, and have the ability to change their structure and physiology in response to long-lasting stimuli, a property called "muscle plasticity" [145]. Thus, for example, fast, fatigable muscles could change to slower, fatigue-resistant ones following chronic electrical stimulation. This remodelling involves an overall change of the structure and metabolism of the fibres, due to modifications of myofibrillar proteins, proteins regulating Ca^{2+} homeostasis, and enzymes involved in glycolysis and in mitochondrial metabolism. All these modifications are time- and intensity-dependent, and imply both transcriptional and post-transcriptional changes of gene expression [145,147]. Adult skeletal muscle can also undergo modifications in response to a more natural way of causing electrical stimulation in the muscles: exercise [148]. One of the factors controlling fibre phenotypes is myoblast determination protein (**MyoD**), a basic helix-loop-helix transcription factor with a critical function in muscle development, that is more highly expressed in fast fibres—in *Myod1*-null mice, indeed, fast fibres shift to a slower phenotype, whereas MyoD overexpression induces the opposite shift [149,150]. A reduction of slow fibres is also observed in calcineurin knock-out mice [151] and in mice overexpressing the calcineurin inhibitor regulator of calcineurin 1 (**RCAN1**) [152]. On the other hand, it has been shown that the nuclear factor of activated T cells (**NFAT**) functions as a repressor of fast properties in slow muscles [153], and is involved in the fast-to-slow phenotype switch induced by aerobic exercise. This effect is due to the NFATs ability to inhibit MyoD action, by binding to its N-terminal transcription activating domain and blocking the recruitment of the histone acetyltransferase **p300** [154]. Interestingly, NFAT is one of the targets of calcineurin-mediated dephosphorylation. It is also worth noting that calcineurin is activated by calcium, and hence by conditions that also trigger muscle contraction.

3.1. Muscle Contraction and Gene Regulation

A large body of evidence suggests that muscle contraction per se regulates gene expression and muscle plasticity. Fluctuation in the intracellular $[Ca^{2+}]$ is certainly the most important signal during muscle contraction; thus, it is highly probable that the mentioned fibre phenotype modifications and, in general, muscle adaptation to PA are initiated by Ca^{2+}. Actually, the molecular basis for contractility depends on the mechanism known as excitation-contraction coupling (ECC), and on the complex interplay between voltage-gated and ligand-gated channels, contractile proteins (such as myosin), calcium-binding buffer proteins (such as calreticulin, parvalbumin, and calsequestrin), calcium-sensor proteins (such as calmodulin and calcineurin), and calcium-dependent ATPases [155].

Ca^{2+} **ions** are also able to regulate glycolysis by making glucose available through glycogen degradation—in muscle cells, glycogen phosphorylase kinase (PhK), the enzyme that phosphorylates and activates the glycogen breaking enzyme phosphorylase (GP), is activated by the calcium/calmodulin

(CaM) complex, that constitutes its δ subunit [156,157]. Moreover, CaM can also interact with the muscle-specific isoform of phosphofructokinase (PFK-M), the pacemaker of glycolysis [158]. Ca^{2+} influx into mitochondria also induces an increase in the energy conversion potential, and ATP production [155].

It is also important to highlight that, during muscle contraction, AMP concentration increases, thus activating **AMPK**.

Another important signal due to PA is **hypoxia**; in resting muscle cells, prolyl hydroxylases (**PHDs**) use molecular oxygen to hydroxylate the hypoxia-inducible factor 1α (**HIF-1α**), thus allowing its pVHL (von-Hippel-Lindau) E3 ligase-dependent ubiquitination, and proteasomal degradation [159]. HIF-1α activity is also modulated by the hydroxylation of an asparagine residue (Asn803) by another oxygen-dependent hydroxylase, the factor inhibiting HIF-1 (**FIH-1**); under normoxic conditions, asparagine is hydroxylated, and this modification prevents interaction of HIF-1α with CBP/p300 [160].

In the hypoxic conditions initially induced by exercise, PHDs undergo a decrease of activity, due to shortage of the oxygen substrate, thus hydroxylation of HIF-1α, and hence its ubiquitination and degradation are limited. The stabilized factor translocates to the nucleus, heterodimerizes with aryl hydrocarbon nuclear receptor translocator (ARNT)/HIF-1β, binds to DNA and induces target gene transcription [159]. Genes important for the adaptation of cells to hypoxic conditions and targets of HIFs are, for example, those encoding glucose transporters, glycolytic enzymes, and angiogenic growth factors [161,162].

A further interesting aspect of muscle activity on muscle function depends on **mechanosensing mechanisms**, that depend on forces transmitted to the cells by the extracellular matrix (ECM) or by neighbouring cells during muscle contraction; these forces are simultaneously translated into changes of cytoskeletal dynamics, contributing at the same time to elicit signal transduction pathways [163]. Increasing evidence suggests that a key role in mechanotransduction is played by yes associated protein (**YAP**), a transcriptional coactivator that can be regulated by ECM stiffness and rigidity, and by cell stretching [163,164]. This protein interacts with different signal transduction pathways, such as the one involving **Wnt/β-catenin** [165], and the one involving **Hippo** [163]. Recently, it has been reported that mechanical stress also activates the c-Jun N-terminal kinase (**JNK**), that then triggers phosphorylation of the transcription factor **SMAD** in a specific linker region. SMAD phosphorylation inhibits its nuclear translocation, thus resulting in a negative regulation of the growth suppressor **myostatin**, and induction of muscle growth [166]. This pathway is activated only by resistance exercise [166]. Interestingly, by using one-legged activity protocols, it was also found that JNK activity increased only in the exercising leg [167]. It is worth noting that global transcriptome analysis, done on muscle biopsies of young men undertaking resistance exercise, revealed that, in the initial exercises, the stress imposed by muscle contraction induced the expression of heat shock proteins (**HSPs**), as well as of muscle damage-, protein turnover-, and inflammation-markers [168]; trained muscles show instead an increase of proteins related to a more oxidative metabolism, and to anti-oxidant functions, as well as of proteins involved in cytoskeletal and ECM structures, and in muscle contraction and growth [168]. Acute resistance exercise also affects the expression of genes encoding components of the ECM, such as matrix metalloproteases, enzymes involved in ECM remodelling [169].

As in the brain, PA-dependent modification of gene expression in muscle mainly depends on epigenetic events. For example, after 60 min of cycling, **HDAC4** and **HDAC5** are exported from the nucleus, thus removing their repressive function [170], and, in general, regular aerobic exercise induces decreased DNA methylation of a number of genes [171–173]. Two of the most important epigenetically regulated genes are the above-mentioned **AMPK** and **CaMK** [142,143].

Moreover, exercise induces rapid and transient changes in the muscle miRNAs (also called **myomiRNAs**) [174,175]—for example, after an acute activity bout (cycle ergometer, 60 min, 70% VO$_2$ peak), has-miR-1, has-miR-133a, has-miR-133-b, and has-miR-181a increase, while has-miR-9, has-miR-23a, has-miR-23b, and has-miR-31 decrease in the skeletal muscle [175]. Intriguingly, has-miR-1, has-miR-133a, and has-miR-133-b have been instead shown to decrease following an endurance training (cycle ergometer, 60–120 min/section, for 12 weeks, 5 times/week) [176]. As in the case of BDNF, further research is necessary in order to understand the real relationship between PA and miRNA production. Again, the analytic methods used might cause the observed differences, thus, in addition to further studies, it will be necessary to standardize miRNA purification from muscle and blood.

Finally, muscle contraction results in a transient increase of both oxygen and nitrogen reactive species (ROS and NOS, respectively) that, by interacting with redox state-sensing pathways (such as, among others, P-38/MAPK, NFkB, and AMPK), induce cyto-protective, antioxidant responses. Activation of these pathways relies in part on post-translational oxidation of cysteines on critical enzymes/regulatory proteins by glutathionylation, that is by reversibly adding glutathione to their thiol groups; in addition to stimulating protective cell responses, this modification probably prevents further irreversible oxidation of cysteines [177,178].

3.2. Release of Myokines and Metabolites by Contracting Muscles

As a whole, the data reported indicate that PA has several effects on the nervous system—it acts as an antidepressant and an anxiolytic, and can improve mood, self-esteem, and cognition. The benefits induced by PA on the brain (as well as in other organs, such as the heart) are in part mediated by peptides (myokines) and metabolites released into the blood by the endocrine activity of contracting muscles (Figure 1) [25,179–182].

3.2.1. BDNF and Cathepsin-B (CTSB)

Contracting muscles release BDNF, that seems to be involved in autocrine signalling to the muscle itself [182–184]. In addition, BDNF probably serves as a retrograde signal to the motor neurons of the spinal cord.

It is also possible that muscle-derived BDNF has an effect on the brain, as intact BDNF was reported to cross the blood–brain barrier (BBB) in both directions by a high-capacity, saturable transport system [185].

In response to exercise, muscles also release into the plasma high levels of cathepsin-B (CTSB), an abundant, calcium-dependent cysteine protease of the calpain family, produced in all human tissues. Enzymatically active CTSB is secreted through exocytosis and can degrade components of the ECM in both physiological and pathological conditions [186,187]. Although the mechanism of action of CTSB in the brain is still a matter of debate, it was acknowledged that, after exercise-dependent release from muscles, it can cross the BBB and promote BDNF expression in the hippocampus, neurogenesis, and promote the improvement of spatial memory abilities [188]. It has been reported, for example, that in CTSB knockout mice, running did not have any effect on hippocampal neurogenesis and spatial memory [188]. Similarly, in humans, changes in CTSB levels correlate with hippocampus-dependent memory functions [188].

Intriguingly, it was recently reported that resting serum levels of both BDNF and CTSB were significantly lower in long-term trained middle-aged men in comparison with sedentary controls, even if trained men showed a significant improvement in memory, based on the Free and Cued Immediate Recall tests [107]. Thus, it seems that both BDNF and CTSB molecules increase immediately after exercise, but then decrease to levels lower than in untrained individuals, showing an inverse correlation to the intensity/duration of exercise ([107]; Table 2). It is tempting to speculate that, as proposed years ago by Ji et al. [189], and as discussed by De La Rosa et al. [107] for BDNF and CTSB, most of the regulatory molecules produced in response to PA behave in a hormetic manner: in other words, their concentration should increase at the beginning of the activities, when they play an immediate role in repair processes, at the sites of the traumatic injuries, where oxidative stress is initially induced by exercise. Then, in well-trained individuals, given the better adaptation to stress, the levels of these molecules could/should decrease. Thus, their concentrations over time, if put in a graph, should give rise to a curve with the shape of an upside-down "U". Such behaviour might also represent one of the variability sources in results found in different studies—analyses performed at different time intervals during and after the exercise might give rise to very different evaluations and interpretations.

3.2.2. FGF21 and Irisin/FNDC5

FGF21 is primarily produced by the liver, but also by skeletal muscles [179,190]—it is a critical regulator of nutrient homeostasis [191]—in response to PA, it improves thermogenesis in adipose tissue and skeletal muscle, and even induces differentiation of brown adipocytes [192]. FGF21 also crosses the BBB [193] and, in association with the co-receptor βKlotho [194], binds to its receptors in the hypothalamus, where it modulates sympathetic input to brown adipose tissue, circadian rhythms, and neuroprotection [182]. Recently, it has been shown that, although all exercise types induce an increase of FGF21, the increase is greater after resistance training than after high-intensity interval (HIIT) sessions [195].

Irisin, the proteolytic cleaved extracellular part of fibronectin type III domain-containing protein 5 (FNDC5), is a myokine the expression of which depends on PGC-1α [196], and that is positively regulated by muscle contraction [197]. Like FGF21, upon its release into the systemic circulation, irisin may contribute to the browning of white adipose tissue [196]. FNDC5 has been detected in different areas of the brain, where it seems to associate with neural differentiation; moreover, irisin can cross the BBB [198], and increased levels of circulating irisin correlate with increased levels of BDNF in the mouse hippocampus [197].

3.2.3. Cytokines Released by Muscles

Contracting muscles also release cytokines, such as **IL-6**, **IL-8**, and **IL-15**. As interleukin passage across the BBB has been reported [199], these molecules are putatively able to act on the brain too. Their effects on the brain are, however, still debated. For example, both neurodegenerative and neuroprotective properties have been attributed to IL-6 [200]; interestingly, it seems that these different effects depend on the receptors engaged, and the specific signalling pathway triggered: (1) The anti-inflammatory pathway (the classical one) involves the membrane-bound IL-6 receptor (IL-6R), expressed for example on microglia, and (2) the pro-inflammatory one (also called the trans-signalling pathway), mediates neurodegeneration in mice, and depends on a soluble form of IL-6R, able to stimulate a response on distal cells [200]. Similarly, IL-8 seems to have both neurogenic and neurotoxic effects [201]. IL-15 receptors are expressed by both glial cells and neurons, with developmental and regional differences. A neuroprotective role of IL-15 is suggested by the increased motor neuron death in knockout mice lacking the IL-15 receptor α (IL15Rα), and by the ability of IL-15 treatment to ameliorate the symptoms of the experimental autoimmune encephalomyelitis (EAE). On the other hand, increased blood levels of IL-15 have been observed in inflammation of several origins [202]. In summary, as in the case of BDNF and CTSB, the levels of the muscle-derived interleukins probably have a hormetic behaviour, and their changes depend on the general adaptation to stress.

3.2.4. Lactate

It is now widely acknowledged that **lactate**, produced in large amounts during anaerobic exercise, shuttles among cells and, inside the cells, among organelles, through specific monocarboxylate carriers (MCTs); interestingly, lactate behaves as a fuel for many cells, including neurons, in conditions of oxygen shortage [203]. Moreover, the hydroxycarboxylic acid receptor 1 (HCAR1), a G protein-coupled lactate receptor, is highly enriched in the endothelial- as well as in the pericyte-like-cells of the intracerebral microvessels. Activation of HCAR1 enhances production of the cerebral vascular endothelial growth factor A (VEGFA) and cerebral angiogenesis [204,205]. More recently, it was also found that, by signalling through HCAR1, lactate can activate responses that involve both α and βγ subunits of HCAR1 and is synergic with the activity of other receptors, such as adenosine A1, GABAB, and α2-adrenergic receptors. As a consequence, not only neurons can use lactate as a substrate during exercise but, in addition, neuronal activity might be finely tuned by this molecule [203,206]. These findings highlight the important role that lactate can play in the PA-dependent muscle–brain crosstalk.

3.2.5. Extracellular Vesicles (EVs)

In the last two decades, many laboratories have demonstrated that cells can communicate at long distances by releasing EVs (mainly exosomes and/or small membrane vesicles/ectosomes) that contain many species of proteins, nucleic acids, lipids, and metabolites. Since they are membrane-bound, EVs can also fuse with the plasma membranes of other cells, thus delivering their content into them and inducing epigenetic modification of the recipient cell functions [207,208]. Central protagonists of EV-mediated trafficking are different species of RNA, and especially miRNAs. Although many obstacles are still encountered in the identification and purification of EV-carried circulating miRNAs [209], many laboratories have reported that PA induces a significant modification of many miRNAs. Among these latter species, some (for example, miR-21 and miR-132) have a role in brain functions as critical as regulation of synaptic plasticity, memory formation, and neuronal survival [82].

It is thus possible that one of the ways through which PA and muscle activity influence brain function is by delivering into the blood different species of regulatory molecules that are protected during the trip to other tissues (and to the brain, in particular) because they are packaged into EVs, and these membrane-bound vehicles might finally deliver their cargoes to the brain across the brain capillary endothelial cells. Interestingly, indeed, exercise stimulates the release of exosomes and small vesicles into circulation [210].

4. A Few Examples of Exercise Effects on Neurodegeneration: Studies on Alzheimer's Disease, Parkinson's Disease, Huntington's Disease, and Multiple Sclerosis

The evidence that regular exercise can help to prevent and even treat neurological disorders has become stronger in recent years. At the same time, a lot of research is focusing on the mechanisms underlying the ability of PA to improve the symptomatology of neurodegenerative diseases, in the attempt to find out the best protocols to be applied to the patients.

4.1. Alzheimer's Disease (AD)

PA improves cognition in a mouse model of Alzheimer's disease (AD), stimulating neurogenesis and the simultaneous increase of both BDNF and FNDC5 [211]. For example, intracerebroventricular- or tail vein-injection of FNDC5 allowed the recovery of memory impairments and synaptic plasticity in a mouse model of AD [212]. Since PA, as already discussed in Section 3, stimulates irisin release from muscle, it is possible that the beneficial role of PA is, at least in part, due to this myokine [213]. Interestingly, the effects of irisin could be also attributed to a lower release of inflammatory cytokines by astrocytes—it was shown, that irisin has protective effects on cultures of hippocampal neurons treated with Aβ peptide, only when co-administered with astrocyte-conditioned medium [214].

In the hippocampus, PA effects include: (i) enhancement of **c-Fos-** and **Wnt3-** and inhibition of glycogen synthase kinase-3β (*GSK-3β*)-gene expression; (ii) an increase of glial fibrillary acidic protein (**GFAP**) and a decrease of the **S100B** protein levels, in astrocytes; (iii) an increase of the blood–brain barrier integrity; (iv) an increase of BDNF and tropomyosin receptor kinase B; (v) enhancement of glycogen levels; and (vi) normalization of MCT2 expression [215].

Another AD progression-slowing factor, known to be produced during physical activity and able to cross BBB is the insulin-like growth factor 1 (IGF-1) [216]. This factor acts by activating the expression of BDNF. If antibodies against its receptor are used, the PA-induced increase of BDNF mRNA, protein, and precursor does not occur anymore [217].

Moreover, as already discussed (Section 3), both lactate and BDNF produced during physical exercise seems to have stimulating effects on learning and memory processes [205]. As mentioned, CTSB also crosses BBB, and should be able to stimulate hippocampal neurogenesis, and to improve learning and memory, however, AD patients have high levels of this enzyme in the blood, thus, the real role of CTSB in AD remains controversial [205].

Some miRNAs, such as miR-124 and miR-134, have been also suggested to be involved in memory formation and maintenance [218,219]. The relationship between the role of these molecules and BDNF in AD is, however, still debated [220]. Interestingly, in a mouse model of AD, **miR-34a** is upregulated [221]; it has been hypothesized that swimming training, by inhibiting miR-34a expression, might attenuate age-related autophagy dysfunction and abnormal mitochondrial dynamics, thus delaying both physiological brain aging and AD [123].

An altered metabolic process in AD is that involving demolition of the L-tryptophan, which leads to the formation of kynurenine (**KYN**); catabolism of this latter molecule generates, in turn, neurotoxic metabolites related to AD pathogenesis. KYN is able to cross freely the BBB and, in AD patients, it is found in excess both in the plasma and in the brain. PA might be protective for neurons because it stimulates the formation of an aminotransferase (**KAT**) in the muscle, KAT then catalyses the peripheral transformation of KYN into kynurenic acid (**KYNA**), and is less harmful because it is unable to cross the BBB [182,205].

Recently, it has been also reported that 4 weeks of exercise can revert the induction of gene encoding proteins involved in inflammation and apoptosis in the hypothalamus in a mouse model of AD. After 6 weeks, an improvement in glucose metabolism was also observed, and after 8 weeks there was an evident reduction of apoptosis in some populations of hypothalamic neurons [222]. Finally, it has been suggested that the benefits noticed in early AD patients following aerobic exercise are due to the exercise-dependent enhancement of the cardiorespiratory fitness, which is in turn associated with improved memory performance and reduced hippocampal atrophy [223].

4.2. Parkinson's Disease (PD)

Parkinson's disease (PD) is the second most common neurodegenerative disorder and involves a massive degeneration of the dopaminergic neurons in the substantia nigra, in the midbrain [224]. Although the priming cause is still unknown, both genetic and environmental cues could play a role. Some of the familial cases show mutations in the gene encoding α-synuclein, a protein mainly found in the presynaptic terminals; the mutated protein is prone to aggregation and tends to form the so-called Lewi bodies, which contribute to the degeneration of neurons [225,226].

At present, pharmacological therapies able to remarkably modify or delay the disease progression, are still lacking. Thus, alternative approaches not entirely based on pharmacotherapy, and able to slow down the dopaminergic neuron degeneration, are needed. Also, in the case of PD, it has been reported that association of the pharmacological therapy with exercise can help in managing the physical and cognitive decline typically associated with PD [227]. Several studies investigated the effects of various types of exercise on both motor- and non-motor-features of PD and reported positive results: 19 systematic reviews and meta-analyses, from 2005 to 2017, were published from which an increased interest in non-pharmacologic therapies is evident [228,229].

In PD animal models, exercise induces neuroprotective effects through the expression of some brain neurotrophic factors, including BDNF and glial-derived neurotrophic factor (GDNF) [230,231]. In particular, it was demonstrated that the promoter IV of *BDNF* gene shows a reduced CpG methylation in rat, after regular enrolment in physical exercise [112]. Moreover, free-wheel running (from 1.6 to 7 km/day) could improve histone H3 phospho/acetylation and c-Fos induction in dentate granule neurons [232]. These observations suggest that the positive effects of PA depend on epigenetic regulation of genes encoding neurotrophins.

Other exercise effects in PD animal models include enhanced cell proliferation and migration of neural progenitors, and an overthrow of age-related deterioration in substantia nigra vascularization, that seems to be mediated by VEGF expression [233]. Moreover, a study on PD mice models highlighted that, after a 6-weeks treadmill training exercise, a nigrostriatal Nrf2-ARE (antioxidant response element)-dependent signalling pathway was activated, which was protective against the development of parkinsonism [234].

Treadmill running also enhanced coordination and motor balance by preventing loss of Purkinje cells in the rat cerebellum. Moreover, repression of PD-induced GFAP-positive reactive astrocytes and Iba-1-positive microglia was found, showing that PA can help in suppressing astrogliosis and microglia activation. These cellular effects were accompanied by a decreased expression of the pro-apoptotic protein Bax, and enhanced expression of the anti-apoptotic protein Bcl-2 [235].

In humans, the effects of PA have been studied on the basis of correlations found among acute effects of exercise on specific clinical variables (as emerging, for example, from the PD Questionnaire-39 on quality of life) and the amplitude of low frequency fluctuations (ALFF) that may reflect the functions of the brain before and after a single bout of exercise. The results of these analyses showed, for example, an increase of ALFF signals within the right ventromedial prefrontal cortex (PFC) and the left ventrolateral PFC, as well as a bilateral increase in the substantia nigra [236].

Another study demonstrated that 4 weeks of aerobic exercise elicited a long-lasting improvement on both motor and non-motor functions of PD patients. The principal result of this study was an increase of BDNF signalling through its TrkB receptor in the patient's lymphocytes [237]. Similar results had been also reported by Wang et al. [238], who found that repetitive transcranial magnetic stimulation enhanced BDNF-TrkB signalling in both brain and lymphocytes [238]. It is thus possible that BDNF-TrkB signalling in lymphocytes can be indicative of what happens in the cortical TrkB signalling.

On the basis of these studies, we can conclude that PA can give PD-specific clinical benefits, but only if repeated habitually over time (i.e., exercise training) [239].

4.3. Huntington's Disease (HD)

HD is a fatal genetic disorder, due to an autosomal dominant mutation that determines the expansion of poly-glutamine repeats in the huntingtin (HTT) coding region [240]. Clinical features of HD include significant motor defects together with non-motor changes, like cognitive, psychological, and behavioural disabilities, that may progressively get worse before diagnosis, and that results in limitations of daily activities [241]. Physical therapy and exercise interventions were integrated into the treatment decades ago, in order to maintain patient's independence in daily life activities, while attenuating the damages in the motor function. It is indeed known that a passive lifestyle might lead to an earlier HD onset; while, as in other neurodegenerative diseases, exercise exerts a positive effect [242,243]. Recent studies have focused on both resistance and endurance exercise training modalities, based on the suggestion that both could be of help in HD patients. All the results showed a significant increase in grey matter volume and significant improvements in verbal learning and memory, after long-training exercise [243–248].

Interestingly, it was highlighted that voluntary exercise in a rat model of HD induces DNA hypomethylation at specific CpG sites, located within an Sp1/Sp3 transcription factor recognition element of the *vegfA* gene promoter. In parallel, a significant reduction of the mRNA encoding DNA methyltransferase 3b (DNMT3B) in the hippocampus of exercised rats was also found [249].

4.4. Multiple Sclerosis (MS)

Patients with Multiple Sclerosis (MS) who perform regular physical activity have a better quality of life with less fatigue and less depression than those who are sedentary [250].

A pilot study with relapsing-remitting MS patients demonstrated that exercise may also attenuate inflammation and neurodegeneration by an increase of **erythropoietin** [251].

Mulero et al. [252] analysed gene expression in MS patients who improved their fatigue status after an aerobic exercise program and compared them with healthy controls (HC). It revealed that in patients before exercise, genes that respond to interferon were more active than in the HC. On the other hand, after training, a decrease in the expression of a group of interferon-related genes was evidenced at the transcriptomic level [252]. These results are encouraging because the expression of genes activated in response to interferon also correlates with the increase in fatigue [252]. Exercise also induced a reduction of the levels of the IL-6 receptor, that went back to normal values [252]. Moreover, in the hippocampus of an animal model of MS, both high- and low-intensity training programs induced an increase of the mRNAs encoding three important neurotrophins: BDNF, the glial-derived neurotrophic factor (GDNF), and the nerve growth factor (NGF) [253].

In addition to the PA-dependent increase of BDNF, VEGF, and IGF-1, in the context of MS, a specific increase in the expression of tight junction proteins, critical for the reestablishment of the BBB function, was also evidenced [254]. Moreover, using a mouse model of MS with overexpressed ATP-binding cassette transporter 1 (**ABCA1**), Houdebine and colleagues [255] demonstrated a PA-dependent normalization of ABCA1 mRNA levels both in the brain and the cerebellum, with an improvement of myelin status.

Actually, it has been also found that different training protocols act differently on gene expression; for example, while IGF1-R expression level decreases in the brain of MS mice subjected to forced-swimming protocol, IGF1-R mRNA level increases in the cerebellum of MS mice of a running group. In parallel, a different pattern of myelin gene stimulation was also observed—in the mice that had performed running exercise, a smaller decrease of myelin was found in the brain, whereas swimming induced greater benefits in the cerebellum [255].

In summary, these few examples of PA benefits in different neurodegenerative diseases reinforce the idea of a neuroprotective effect of exercise. Exercise increases expression of genes involved in enzymatic antioxidant responses, improves cognitive functions and memory, and can counteract the progression of diseases, or at least help patients to better perform daily life activities.

There are probably multiple cellular and molecular pathways involved and act in synergy. Moreover, specific differences in the responses of individual patients can be expected depending on genetic and epigenetic variability as well as even slight differences in the grade of the pathology.

Finally, the protocols used in different studies are highly heterogeneous and to set ideal exercises for the different neurodegenerative pathologies is at the moment impossible. Further research is still necessary, and, as already noticed above, standardized methods for analysing the results and the biomarkers are compelling.

In spite of the mentioned uncertainties and variability, the current results are of real interest and encouraging. Moreover, the understanding that many biochemical pathways are involved has been stimulating a lot of new studies, aimed at finding out the best combinations of exercise and drugs to slow down the pathology while improving the life quality of the patients.

5. Exercise-Dependent Production of Dopamine, Endocannabinoids, and Opioids: Effects on Mood, Analgesia, and Happiness

In addition to an improvement of body fitness and learning and memory skills, it is well documented that PA can induce changes in the mental status, reducing anxiety and producing a general sense of wellbeing. Moreover, it can induce analgesia. The precise mechanisms involved are not yet completely understood but a few molecules, probably acting in synergy, have been identified and are currently studied as possible mediators of these further effects of PA.

5.1. Dopamine

Dopamine (DA) producing neurons are present in distinct areas of the cerebral cortex, but are mostly concentrated in the ventral midbrain, where they are arranged in different nuclei. The two main groups constitute the pars compacta of the substantia nigra (SNc), and the ventral tegmental area (VTA). The latter neurons send projections to the nucleus accumbens of the ventromedial striatum, but also to the limbic system and the prefrontal cortex, being mostly involved in the regulation of emotional, reward-related, and cognitive functions. Dopaminergic neurons of the SNc, which regulate mainly motor function, form the nigrostriatal pathway, innervating neurons located in the caudate nucleus and in the dorsolateral striatum [256,257].

This subdivision is probably an oversimplification because different subgroups of DA neurons have recently been described in human and murine midbrain, which show distinct gene expression profiles [258,259]. Even though it is not yet known whether these DA neurons have specific roles, in some instances it was demonstrated that they have unique projection patterns, connecting them to distinctive areas such as the nucleus accumbens and amygdala [260]. Using single-cell RNA sequencing, and PITX3 protein and tyrosine hydroxylase (TH) as markers for DA neurons, Tiklovà et al. [261] identified seven different populations of neurons in the mouse developing midbrain, that could be distinguished thanks to the differential expression of other genes [261].

DA neurons appear to form a brain network regulating the motivational behaviour of animals, allowing them to learn the difference between useful and harmful things, and consequently to choose proper actions [262]. DA also seems to be necessary for performing motivated actions to achieve goals, as demonstrated by the unsuitable behaviour of dopamine deficient mice [263]. In the mammalian central nervous system, DA controls many processes [262,264], such as feeding and locomotion [265]; it is also involved in the mechanisms of cognition and 'adaptive' memory formation, influencing the hippocampal long term potentiation (LTP) [266], and upregulates BDNF in the prefrontal cortex [267]. DA most probably interacts with other neurotransmitters and neuromodulators and, for example, it has been recently demonstrated that midbrain mice DA neurons also release IGF-1 that modulates DA release and concentration as well as neuronal firing [268].

As mentioned, a lot of different evidence demonstrates that the mammalian brain is capable of changing its functional and structural characteristics to adapt to the ever-changing surrounding world. This is achieved by learning and acquiring skills, thus improving cognitive functions. Neuroplasticity is orchestrated by several neurotransmitters and neurotrophins, and many cues indicate that exercise has an important role in its regulation [269]. In particular, DA regulates emotion and reward-related brain functions, and many authors have postulated that the positive properties of PA may be due to its ability to increase DA concentration [270–272]. Interestingly, PA increases the concentration of the same neurotransmitters, including DA, also activated by some drugs and alcohol [273], and this could be the reason why it improves mood in humans [274,275].

Moreover, PA, and specifically voluntary exercise, creates a sharp increase in DA concentration, especially in the nigrostriatal pathway, and has a strong positive effect in overcoming aversion. Being a molecule involved in the regulation of movement, emotions, and learning, DA could be a key component in the mechanism. Nevertheless, even though many proofs about DA's involvement in the beneficial effects of exercise have been accumulating, to date a clear explanation of the underlying mechanism is still missing [276].

One interesting aspect of DA function is that it appears as one of the factors that distinguish physically active organisms from inactive ones, influencing the locomotory activity and even the tendency of the individuals to engage in PA [277]. Voluntary exercise is genetically controlled and depends on different neuromodulators, including DA itself. Given the enhancing effects of PA on DA production and release in the brain, it can be hypothesized that an auto-sustaining circuit exists by which DA and PA positively interact—the more DA an individual animal produces, the more it is prone to live actively, and the more DA will be consequently released in this feed-forward system [278].

Even though the mechanisms by which exercise, through dopamine, creates positive effects on brain functions are yet to be elucidated, a few hypotheses have been proposed. For example, it has been demonstrated that voluntary wheel running (VWR) activates latero-dorsal tegmental (LDT) and lateral hypothalamic area (LHA) murine neurons and these, in turn, could be responsible for the activation of the DA neurons of the lateral ventral tegmental area (lVTA) [279].

DA increase in the brain can derive from a higher activity of the tyrosine hydroxylase (TH) enzyme, most probably due to a rise in calcium concentration. Enhancement of the enzyme activity depends indeed on its phosphorylation by CaMKII, the activity of which is regulated by calcium [280,281]. Actually, wheel running in rodents causes a doubling of the TH mRNA level in the VTA [282], and an increase also in the substantia nigra [283], and in the locus coeruleus [284]. A chronic exercise-dependent high level of DA, but not of other neurotransmitters (such as noradrenalin, serotonin, or glutamate), in the rat medial prefrontal cortex (mPFC), was found by Chen et al. [285], and the effect could be reduced by a glucocorticoid receptor inhibitor—the authors suggest that the local DA increase is due to the high level of cortisol, induced by PA in mPFC [285].

In summary, PA may cause an increase in serum calcium levels, and calcium can stimulate dopamine synthesis in the brain by stimulating the activity of the CaMKII, and the consequent activation of the TH enzyme by phosphorylation. In particular, it has been shown that mice forced to physical activity have a DA level sharply higher in the neostriatum and nucleus accumbens, and that a similar effect, i.e., a specific increase of DA level in these brain regions, can be obtained by intracerebroventricular injection of calcium chloride. Moreover, following physical activity, a significant amount of TH and CaM was found in mouse neostriatum and nucleus accumbens, and in human, was found in the caudate nucleus and putamen. A possible mechanism leading to calcium increase in the brain could be the release of lactate following exercise. This may induce, in turn, an increase of blood acidity that could activate parathyroid hormone, or directly increase calcium concentration by favouring bone resorption [286].

On the other hand, PA-dependent DA increase might also be a consequence of a decrease in the activity of catabolic enzymes, such as the mitochondrial monoamine oxidase (MAO) and the catechol-O-methyltransferase (COMT) [287]. In a study aimed at associating genetic background to happiness, Chen et al. [288] found that women bearing the low expression MAO-A alleles are statistically happier than those bearing the high expressed variant. Surprisingly, no difference in happiness was found when comparing men bearing the two different type of alleles [288]. Similar results have been reported regarding the COMT gene—women bearing a particular allele, containing the COMT Val158Met polymorphism, and presenting as a consequence a higher DA concentration show an emotionally healthier behaviour [289].

5.2. Opioids, Endocannabinoids, Analgesia, and the "Runner's High"

The endogenous opioid system includes different peptides (i.e., endorphins, enkephalins, and dynorphins) that derive from larger precursors and bind to G protein-coupled receptors. Three main receptors (μ, κ, and δ) mediate analgesic effects of these molecules [30]. Several studies have demonstrated PA-dependent increase of circulating opioids, and in particular of β-endorphin, in relationship with the intensity of exercise, and this β-endorphin increase correlates with analgesic effects both in humans and in rodents. Many studies, however, suggest that opioids are not the only molecules involved in analgesia induced by exercise [30]. For example, activation by exercise of the mesolimbic system in rodents has been also related to analgesic effects [290].

The endocannabinoid system (ECS) includes two G protein-coupled cannabinoid receptors (CB1 and CB2), widely expressed all over the body, and their endogenous ligands, the most well-studied of which are two derivatives of the arachidonic acid: N-arachidonoylethanolamine (AEA, also known as anandamide) and 2-arachidonoylglycerol (2-AG). ECS also includes the enzymes necessary for synthesizing and degrading the ligands [291]. In addition to CB1 and CB2 receptors, 2-AG and AEA can bind to the vanilloid receptor (TRPV1); moreover, AEA also functions as an agonist of some subtypes of the peroxisome proliferator-activated receptor (PPAR) family of nuclear receptors [292]. ECS is critically involved in the modulation of several aspects of metabolism, and, in the hypothalamus, endocannabinoids signalling seems to function in maintaining appetite, in contrast with leptin. In particular, CB1 is probably involved in reward circuits related to food (i.e., it is responsible for the hedonic aspect of eating) [292]. An expected consequence of these ECS functions is increased production of endocannabinoids in response to exercise that induces higher energy utilization. A variety of studies have indeed shown PA-dependent increase of circulating endocannabinoids, even if the results significantly differ from one study to another. It seems, for example, that the relationship between the increase of AEA and the exercise intensity, as in the case of other already mentioned molecules, is described by an "upside-down U"-shaped curve [292,293]. On the other hand, 2-AG was found significantly elevated in response to short and intense bursts of activity. It is thus possible that different endocannabinoids (or different mixes of them) are secreted in response to different types, intensities, and durations of exercise. Moreover, "preferred" exercises significantly activate ECS, and this response may also contribute to the effects on the mood [294].

Interestingly, it was also reported that hypoxia potentiates ECS activation, and it was suggested that the muscles can be the main source of the exercise-induced increase of circulating endocannabinoids, that then can cross the BBB [292]. Overall, the levels of circulating endocannabinoids are inversely related to anxiety and depression, and positively related to BDNF concentration and, thus, to the beneficial effects on mood and to a sense of vigour and wellbeing. However, 2-AG and/or AEA levels can be higher in patients with schizophrenia or other cognitive disorders, such as borderline personality disorder [292]. Moreover, these observations are consistent with the evidence that the use of cannabinoid drugs increases the risk of developing psychotic disorders, probably also in relation to alteration of the dopamine signalling [295]. In summary, it is highly probable that these molecules can also have hormetic behaviour (see Section 3.2.2).

Since the 1960s, it was known that long-running could cause what was called the "runner's high", a sudden sense of euphoria and wellbeing, accompanied by analgesia. For a long time, exercise-dependent production of endorphins was considered responsible for at least the analgesic component of the runner's high. More recently, as mentioned, the involvement of both opioids and endocannabinoids in this aspect of the response to PA has been consistently reported [30], and, in addition, it was found that cannabinoid-agonists can enhance the release of endogenous opioids in the brain [295]. We can thus infer that the two systems act in synergy in the anti-nociceptive effects of exercise. It has been also reported that, at the molecular level, a mediator of endocannabinoid action in response to exercise is AMPK [296].

On the other hand, most other aspects of the runner's high seem to depend more directly on the endocannabinoid receptors, in mice [297], even if it is not so easy to evaluate euphoria in mice.

It was also suggested that mood improvement could relate to PA-dependent increase of the levels of neurosteroids, and in particular of dehydroepiandrosterone (DHEA) [298], a molecule with a variety of effects on different neurotransmitter receptors, such as the GABA$_A$ receptor, and the NMDA as well as the AMPA receptors for glutamate. DHEA can also bind to nuclear receptors, can contribute to regulating the mitochondrial function in response to stress, and, through activation of G-protein coupled receptors of the plasma membrane, can increase transcription of miR-21, at least in a cell line of hepatocytes [299].

6. Conclusions and Perspectives

In conclusion, habitual exercise has a variety of positive effects on the human body, from regulating cardiorespiratory and cardiovascular fitness, to improving glycaemia and insulin response. In addition, as discussed, it is a way of maintaining not only a healthy body, but also a healthy mind, at any age. In particular, it can represent a non-pharmacological (and sometimes enjoyable) strategy to delay the effects of both physiological ageing and pathological neurodegeneration on brain health. However, although exercise prescriptions (including frequency, intensity, type, and time) were given, for example, for individuals with hypertension ([20], Table 1 in [23]), we cannot yet refer to specific exercise prescriptions for maximizing the positive effects of PA on cognition [23]; the protocols used in the experiments reported in this review, as well as the subjects and the markers studied (Tables 1 and 2) are indeed quite different, many informative studies relied on rodents, and not yet on humans. Further studies are thus necessary to evaluate more precisely how the factors which influence brain functioning change in response to the type, intensity, and timing of exercise. Further studies are also required to understand the interplay among the many molecules the levels of which change during/after exercise, even in opposite directions. PA induces, indeed, a variety of cellular and molecular effects, both in the periphery and in the brain. As we have seen, every molecule/group of molecules probably affects different aspects of brain function, but their synergic effects contribute to brain health as a whole.

Among all these factors a key role seems to be played by BDNF—as a PA effect, this latter molecule is produced in the periphery and can also cross the BBB. In addition, some BDNF is directly produced in the brain due to the effect of other molecules, some of which are similarly released in the periphery, in a PA-dependent manner, and then cross the BBB, where they affect the function of resident proteins either at the transcriptional or the post-transcriptional level.

Notably, all these effects also depend on the physical pre-exercise conditions of each person.

In this context, an additional issue arises from the actual difficulties of old people and patients with neurodegeneration to perform voluntary exercise. Interestingly, a recent paper reported that neuromuscular electrical stimulation (NMES) can increase BDNF and lactate serum concentration even more than voluntary exercise, and might thus represent a solution for individuals who cannot engage in high-intensity exercise or are even unable to perform any exercise at all [300].

It should also be considered that a significant percentage of people with a sedentary, computer-dependent lifestyle consider physical activity not only hard but also boring, and thus lacks motivation to exercise. From this point of view, a challenging proposal was made in a recent paper, in which the authors reported that a virtual reality-based exercise can be of help for people who cannot and/or do not like to move [301]. Namely, they refer to a particular type of dual-task video-games (exergames) that require, on behalf of the player, also a certain degree of movement, with variable physical components. The authors suggest that exergames can be also useful for children with impaired motor functions, for people who undergo rehabilitation, and for the elders [301]. Perhaps, the possibility to perform PA in the context of a virtual game can help also in the case of children and adolescents with intellectual disabilities (ID); it has been reported that children with moderate to severe ID also suffer from low physical fitness [302].

Notably, we are now aware that pharmacological therapies should be ideally shaped on individual patients because of genetic and epigenetic differences affecting responses to the drugs. When talking of physical activity, tailoring prescriptions to individuals is even more difficult since the ability to perform exercise as well as the exercise outcomes likely depend on a wider set of genes (and their epigenetic setting). For example, single nucleotide polymorphisms have been observed in a number of genes that encode proteins involved in PA/fitness relationship, such as, for example, the genes encoding BDNF and the e4 allele of apolipoprotein apoE [56], or the genes encoding muscle proteins, such as actinin [303]. These considerations become much more stringent when we focus on the nervous system—every brain is unique because of genetic and epigenetic peculiarities that accumulate throughout our lives, as an effect of learning and experiences that sculpt our mind [304].

Now, although these concepts are clear at the theoretical level, practical applications are still in their infancy, even if rapidly progressing. In the future, animal models will certainly be helpful to study correlations among specific genes and exercise outcomes. On the other hand, further studies on humans will be helpful, provided that more homogeneous interventions and standardized measurement methods are used to evaluate exercise-dependent modifications of key parameters.

Finally, as life expectancy is increasing all over the world, it is of the utmost importance for all of us to maintain independence in the daily life activities and a sense of wellbeing as long as possible. Since PA can clearly contribute in ameliorating physical fitness as well as the mental status, it should be a social and political task to promote the conditions that allow the realization of physical exercise programs for the entire population, and especially for the elders and for children. In particular, we suggest that both healthy people and patients are encouraged by physicians to perform physical activity, underlining the higher impact and efficacy of moderate and regular exercise, in comparison with acute and heavy bouts of activity.

Author Contributions: Conceptualization and artwork, I.D.L.; writing and editing, all the authors.

Funding: The authors did not receive any external funding.

Acknowledgments. The authors are supported by the Università degli Studi di Palermo (University of Palermo), Palermo, Italy.

Conflicts of Interest: The authors declare no conflict of interest.

References

1. Panegyres, K.P.; Panegyres, P.K. The Ancient Greek discovery of the nervous system: Alcmaeon, Praxagoras and Herophilus. *J. Clin. Neurosci.* **2016**, *29*, 21–24. [CrossRef] [PubMed]
2. Wills, A. Herophilus, Erasistratus, and the birth of neuroscience. *Lancet* **1999**, *354*, 1719–1720. [CrossRef]
3. Russo, L. *The Forgotten Revolution*; Springer: Berlin/Heidelberg, Germany; New York, NY, USA, 2003; ISBN 3-540-20068-1.
4. Von Staden, H. *Herophilus: The Art of Medicine in Early Alexandria*, 1st ed.; Cambridge University Press: Cambridge, UK, 2008; ISBN 9780521041782.
5. Reveron, R.R. Herophilus and Erasistratus, pioneers of human anatomical dissection. *Vesalius* **2014**, *20*, 55–58. [PubMed]
6. Neufer, P.D.; Bamman, M.M.; Muoio, D.M.; Bouchard, C.; Cooper, D.M.; Goodpaster, B.H.; Booth, F.W.; Kohrt, W.M.; Gerszten, R.E.; Mattson, M.P.; et al. Understanding the Cellular and Molecular Mechanisms of Physical Activity-Induced Health Benefits. *Cell Metab.* **2015**, *22*, 4–11. [CrossRef] [PubMed]
7. Bramble, D.; Lieberman, D.E. Endurance running and the evolution of Homo. *Nature* **2004**, *432*, 345–352. [CrossRef] [PubMed]
8. Raichlen, D.A.; Polk, J.D. Linking brains and brawn: Exercise and the evolution of human neurobiology. *Proc. Biol. Sci.* **2013**, *280*, 20122250. [CrossRef] [PubMed]
9. Hill, T.; Polk, J.D. BDNF, endurance activity, and mechanisms underlying the evolution of hominin brains. *Am. J. Phys. Anthropol.* **2019**, *168* (Suppl. 67), 47–62. [CrossRef] [PubMed]
10. Wikgren, J.; Mertikas, G.G.; Raussi, P.; Tirkkonen, R.; Äyräväinen, L.; Pelto-Huikko, M.; Koch, L.G.; Britton, S.L.; Kainulainen, H. Selective breeding for endurance running capacity affects cognitive but not motor learning in rats. *Physiol. Behav.* **2012**, *106*, 95–100. [CrossRef]
11. Carrier, D.R. The Energetic Paradox of Human Running and Hominid Evolution. *Curr. Anthropol.* **1984**, *25*, 483–495. [CrossRef]
12. Wheeler, P.E. The thermoregulatory advantages of hominid bipedalism in open equatorial environments: The contribution of increased convective heat loss and cutaneous evaporative cooling. *J. Hum. Evol.* **1991**, *21*, 107–115. [CrossRef]
13. Ruxton, G.D.; Wilkinson, D.M. Avoidance of overheating and selection for both hair loss and bipedality in hominins. *Proc. Natl. Acad. Sci. USA* **2011**, *108*, 20965–20969. [CrossRef] [PubMed]
14. Kellogg, D.L., Jr.; Zhao, J.L.; Wu, Y. Roles of nitric oxide synthase isoforms in cutaneous vasodilation induced by local warming of the skin and whole body heat stress in humans. *J. Appl. Physiol.* **2009**, *107*, 1438–1444. [CrossRef] [PubMed]
15. Liu, D.; Fernandez, B.O.; Hamilton, A.; Lang, N.N.; Gallagher, J.M.C.; Newby, D.E.; Feelisch, M.; Weller, R.B. UVA irradiation of human skin vasodilates arterial vasculature and lowers blood pressure independently of nitric oxide synthase. *J. Invest. Dermatol.* **2014**, *134*, 1839–1846. [CrossRef] [PubMed]
16. Caspersen, C.J.; Powell, K.E.; Christenson, G.M. Physical activity, exercise, and physical fitness: Definitions and distinctions for health-related research. *Public Health Rep.* **1985**, *100*, 126–131. [PubMed]
17. *Physical Activity Guidelines for Americans*, 2nd ed.; U.S. Department of Health and Human Services: Washington, DC, USA, 2018. Available online: https://health.gov/paguidelines/second-edition/ (accessed on 1 June 2019).
18. Piercy, K.L.; Troiano, R.P.; Ballard, R.M.; Carlson, S.A.; Fulton, J.E.; Galuska, D.A.; George, S.M.; Olson, R.D. The Physical Activity Guidelines for Americans. *JAMA* **2018**, *320*, 2020–2028. [CrossRef] [PubMed]
19. Erickson, K.I.; Hillman, C.; Stillman, C.M.; Ballard, R.M.; Bloodgood, B.; Conroy, D.E.; Macko, R.; Marquez, D.X.; Petruzzello, S.J.; Powell, K.E. Physical Activity, Cognition, and Brain Outcomes: A Review of the 2018 Physical Activity Guidelines. *Med. Sci. Sports Exerc.* **2019**, *51*, 1242–1251. [CrossRef] [PubMed]
20. Pescatello, L.S.; MacDonald, H.V.; Lamberti, L.; Johnson, B.T. Exercise for Hypertension: A Prescription Update Integrating Existing Recommendations with Emerging Research. *Curr. Hypertens. Rep.* **2015**, *17*, 87. [CrossRef] [PubMed]
21. Sosner, P.; Guiraud, T.; Gremeaux, V.; Arvisais, D.; Herpin, D.; Bosquet, L. The ambulatory hypotensive effect of aerobic training: A reappraisal through a meta-analysis of selected moderators. *Scand. J. Med. Sci. Sports* **2017**, *27*, 327–341. [CrossRef]

22. McTiernan, A.; Friedenreich, C.M.; Katzmarzyk, P.T.; Powell, K.E.; Macko, R.; Buchner, D.; Pescatello, L.S.; Bloodgood, B.; Tennant, B.; Vaux-Bjerke, A.; et al. 2018 Physical Activity Guidelines Advisory Committee. *Med. Sci. Sports Exerc.* **2019**, *51*, 1252–1261. [CrossRef]

23. Rêgo, M.L.; Cabral, D.A.; Costa, E.C.; Fontes, E.B. Physical Exercise for Individuals with Hypertension: It Is Time to Emphasize its Benefits on the Brain and Cognition. *Clin. Med. Insights Cardiol.* **2019**, *13*. [CrossRef]

24. Pescatello, L.S.; Parducci, P.; Livingston, J.; Taylor, B.A. A Systematically Assembled Signature of Genes to be Deep-Sequenced for Their Associations with the Blood Pressure Response to Exercise. *Genes* **2019**, *10*, 295. [CrossRef] [PubMed]

25. Bajer, B.; Vlcek, M.; Galusova, A.; Imrich, R.; Penesova, A. Exercise associated hormonal signals as powerful determinants of an effective fat mass loss. *Endocr. Regul.* **2015**, *49*, 151–163. [CrossRef] [PubMed]

26. Liaw, Y.-C.; Liaw, Y.-P.; Lan, T.H. Physical Activity Might Reduce the Adverse Impacts of the FTO Gene Variant rs3751812 on the Body Mass Index of Adults in Taiwan. *Genes* **2019**, *10*, 354. [CrossRef] [PubMed]

27. Krüger, K.; Mooren, F.C.; Pilat, C. The Immunomodulatory Effects of Physical Activity. *Curr. Pharm. Des.* **2016**, *22*, 3730–3748. [CrossRef] [PubMed]

28. Pedersen, B.K. Anti-inflammatory effects of exercise: Role in diabetes and cardiovascular disease. *Eur. J. Clin. Investig.* **2017**, *47*, 600–611. [CrossRef] [PubMed]

29. Cooney, G.M.; Dwan, K.; Greig, C.A.; Lawlor, D.A.; Rimer, J.; Waugh, F.R.; McMurdo, M.; Mead, G.E. Exercise for depression. *Cochrane Database Syst. Rev.* **2013**, *9*, CD004366. [CrossRef]

30. Da Silva Santos, R.; Galdino, G. Endogenous systems involved in exercise-induced analgesia. *J. Physiol. Pharmacol.* **2018**, *69*, 3–13. [CrossRef]

31. Cavalcante, P.A.M.; Gregnani, M.F.; Henrique, J.S.; Ornellas, F.H.; Araújo, R.C. Aerobic but not Resistance Exercise Can Induce Inflammatory Pathways via Toll-Like 2 and 4: A Systematic Review. *Sports Med. Open* **2017**, *3*, 42. [CrossRef] [PubMed]

32. Schwellnus, M.; Soligard, T.; Alonso, J.M.; Bahr, R.; Clarsen, B.; Dijkstra, H.P.; Gabbett, T.J.; Gleeson, M.; Hägglund, M.; Hutchinson, M.R.; et al. How much is too much? (Part 2) International Olympic Committee consensus statement on load in sport and risk of illness. *Br. J. Sports Med.* **2016**, *50*, 1043–1052. [CrossRef]

33. Peake, J.M.; Neubauer, O.; Walsh, N.P.; Simpson, R.J. Recovery of the immune system after exercise. *J. Appl. Physiol.* **2017**, *122*, 1077–1087. [CrossRef]

34. Ch'ng, T.H.; Uzgil, B.; Lin, P.; Avliyakulov, N.K.; O'Dell, T.J.; Martin, K.C. Activity-dependent transport of the transcriptional coactivator CRTC1 from synapse to nucleus. *Cell* **2012**, *150*, 207–221. [CrossRef] [PubMed]

35. Lim, A.F.; Lim, W.L.; Ch'ng, T.H. Activity-dependent synapse to nucleus signaling. *Neurobiol. Learn. Mem.* **2017**, *138*, 78–84. [CrossRef] [PubMed]

36. Herbst, W.A.; Martin, K.C. Regulated transport of signaling proteins from synapse to nucleus. *Curr. Opin. Neurobiol.* **2017**, *45*, 78–84. [CrossRef] [PubMed]

37. Marcello, E.; Di Luca, M.; Gardoni, F. Synapse-to-nucleus communication: From developmental disorders to Alzheimer's disease. *Curr. Opin. Neurobiol.* **2018**, *48*, 160–166. [CrossRef] [PubMed]

38. Parra-Damas, A.; Saura, C.A. Synapse-to-Nucleus Signaling in Neurodegenerative and Neuropsychiatric Disorders. *Biol. Psychiatry* **2019**, *86*, 87–96. [CrossRef] [PubMed]

39. Ross, W.N. Understanding calcium waves and sparks in central neurons. *Nat. Rev. Neurosci.* **2012**, *13*, 157–168. [CrossRef] [PubMed]

40. Bading, H. Nuclear calcium signalling in the regulation of brain function. *Nat. Rev. Neurosci.* **2013**, *14*, 593–608. [CrossRef] [PubMed]

41. Proepper, C.; Johannsen, S.; Liebau, S.; Dahl, J.; Vaida, B.; Bockmann, J.; Kreutz, M.R.; Gundelfinger, E.D.; Boeckers, T.M. Abelson interacting protein 1 (Abi-1) is essential for dendrite morphogenesis and synapse formation. *EMBO J.* **2007**, *26*, 1397–1409. [CrossRef]

42. Spilker, C.; Nullmeier, S.; Grochowska, K.M.; Schumacher, A.; Butnaru, I.; Macharadze, T.; Gomes, G.M.; Yuanxiang, P.; Bayraktar, G.; Rodenstein, C.; et al. A Jacob/Nsmf Gene Knockout Results in Hippocampal Dysplasia and Impaired BDNF Signaling in Dendritogenesis. *PLoS Genet.* **2016**, *12*, e1005907. [CrossRef]

43. Di Liegro, C.M.; Schiera, G.; Di Liegro, I. Regulation of mRNA transport, localization and translation in the nervous system of mammals (Review). *Int. J. Mol. Med.* **2014**, *33*, 747–762. [CrossRef]

44. Altman, J.; Das, G.D. Autoradiographic and histological evidence of postnatal hippocampal neurogenesis in rats. *J. Comp. Neurol.* **1965**, *124*, 319–335. [CrossRef] [PubMed]

45. Altman, J.; Das, G.D. Postnatal neurogenesis in the guinea-pig. *Nature* **1967**, *214*, 1098–1101. [CrossRef] [PubMed]

46. Kaplan, M.S.; Hinds, J.W. Neurogenesis in the adult rat: Electron microscopic analysis of light radioautographs. *Science* **1977**, *197*, 1092–1094. [CrossRef] [PubMed]

47. Kaplan, M.S.; Bell, D.H. Mitotic neuroblasts in the 9-day-old and 11-month-old rodent hippocampus. *J. Neurosci.* **1984**, *4*, 1429–1441. [CrossRef] [PubMed]

48. Stanfield, B.B.; Trice, J.E. Evidence that granule cells generated in the dentate gyrus of adult rats extend axonal projections. *Exp. Brain Res.* **1988**, *72*, 399–406. [CrossRef] [PubMed]

49. Cameron, H.A.; Woolley, C.S.; McEwen, B.S.; Gould, E. Differentiation of newly born neurons and glia in the dentate gyrus of the adult rat. *Neuroscience* **1993**, *56*, 337–344. [CrossRef]

50. Eriksson, P.S.; Perfilieva, E.; Björk-Eriksson, T.; Alborn, A.M.; Nordborg, C.; Peterson, D.A.; Gage, F.H. Neurogenesis in the adult human hippocampus. *Nat. Med.* **1998**, *4*, 1313–1317. [CrossRef] [PubMed]

51. Squire, L.R. Memory and the hippocampus: A synthesis from findings with rats, monkeys, and humans. *Psychol. Rev.* **1992**, *99*, 195–231. [CrossRef] [PubMed]

52. Deng, W.; Aimone, J.B.; Gage, F.H. New neurons and new memories: How does adult hippocampal neurogenesis affect learning and memory? *Nat. Rev. Neurosci.* **2010**, *11*, 339–350. [CrossRef]

53. Squire, L.R.; Genzel, L.; Wixted, J.T.; Morris, R.G. Memory consolidation. *Cold Spring Harb. Perspect. Biol.* **2015**, *7*, a021766. [CrossRef]

54. Kim, S.; Dede, A.J.; Hopkins, R.O.; Squire, L.R. Memory, scene construction, and the human hippocampus. *Proc. Natl. Acad. Sci. USA* **2015**, *112*, 4767–4772. [CrossRef]

55. Gould, E.; Beylin, A.; Tanapat, P.; Reeves, A.; Shors, T.J. Learning enhances adult neurogenesis in the hippocampal formation. *Nat. Neurosci.* **1999**, *2*, 260–265. [CrossRef] [PubMed]

56. Hillman, C.H.; Erickson, K.I.; Kramer, A.F. Be smart, exercise your heart: Exercise effects on brain and cognition. *Nat. Rev. Neurosci.* **2008**, *9*, 58–65. [CrossRef] [PubMed]

57. Van Praag, H. Exercise and the brain: Something to chew on. *Trends Neurosci.* **2009**, *32*, 283–290. [CrossRef] [PubMed]

58. Biddle, S.J.; Asare, M. Physical activity and mental health in children and adolescents: A review of reviews. *Br. J. Sports Med.* **2011**, *45*, 886–895. [CrossRef] [PubMed]

59. Niederer, I.; Kriemler, S.; Gut, J.; Hartmann, T.; Schindler, C.; Barral, J.; Puder, J.J. Relationship of aerobic fitness and motor skills with memory and attention in preschoolers (Ballabeina): A cross-sectional and longitudinal study. *BMC Pediatr.* **2011**, *11*, 34. [CrossRef] [PubMed]

60. Lees, C.; Hopkins, J. Effect of Aerobic Exercise on Cognition, Academic Achievement, and Psychosocial Function in Children: A Systematic Review of Randomized Control Trials. *Prev. Chronic Dis.* **2013**, *10*, E174. [CrossRef] [PubMed]

61. Ma, C.L.; Ma, X.T.; Wang, J.J.; Liu, H.; Chen, Y.F.; Yang, Y. Physical exercise induces hippocampal neurogenesis and prevents cognitive decline. *Behav. Brain Res.* **2017**, *317*, 332–339. [CrossRef]

62. Schmidt-Kassow, M.; Zink, N.; Mock, J.; Thiel, C.; Vogt, L.; Abel, C.; Kaiser, J. Treadmill walking during vocabulary encoding improves verbal long-term memory. *Behav. Brain Funct.* **2014**, *10*, 24. [CrossRef]

63. Suwabe, K.; Hyodo, K.; Byun, K.; Ochi, G.; Yassa, M.A.; Soya, H. Acute moderate exercise improves mnemonic discrimination in young adults. *Hippocampus* **2017**, *27*, 229–234. [CrossRef]

64. Rodriguez-Ayllon, M.; Cadenas-Sánchez, C.; Estévez-López, F.; Muñoz, N.E.; Mora-Gonzalez, J.; Migueles, J.H.; Molina-García, P.; Henriksson, H.; Mena-Molina, A.; Martínez-Vizcaíno, V.; et al. Role of Physical Activity and Sedentary Behavior in the Mental Health of Preschoolers, Children and Adolescents: A Systematic Review and Meta-Analysis. *Sports Med.* **2019**, *49*, 1383–1410. [CrossRef] [PubMed]

65. Niederer, I.; Kriemler, S.; Zahner, L.; Bürgi, F.; Ebenegger, V.; Hartmann, T.; Meyer, U.; Schindler, C.; Nydegger, A.; Marques-Vidal, P.; et al. Influence of a lifestyle intervention in preschool children on physiological and psychological parameters (Ballabeina): Study design of a cluster randomized controlled trial. *BMC Public Health* **2009**, *9*, 94. [CrossRef] [PubMed]

66. Léger, L.A.; Mercier, D.; Gadoury, C.; Lambert, J. The multistage 20 metre shuttle run test for aerobic fitness. *J. Sports Sci.* **1988**, *6*, 93–101. [CrossRef] [PubMed]

67. Pescatello, L.S. *ACSM's Guidelines for Exercise Testing and Prescription*; Wolters Kluwer: Amsterdam, The Netherlands, 2014.

68. Lezi, E.; Burns, J.M.; Swerdlow, R.H. Effect of high-intensity exercise on aged mouse brain mitochondria, neurogenesis, and inflammation. *Neurobiol. Aging* **2014**, *35*, 2574–2583. [CrossRef]

69. Li, J.; Ding, Y.H.; Rafols, J.A.; Lai, Q.; McAllister, J.P., 2nd; Ding, Y. Increased astrocyte proliferation in rats after running exercise. *Neurosci. Lett.* **2005**, *386*, 160–164. [CrossRef] [PubMed]

70. Saur, L.; Baptista, P.P.; de Senna, P.N.; Paim, M.F.; do Nascimento, P.; Ilha, J.; Bagatini, P.B.; Achaval, M.; Xavier, L.L. Physical exercise increases GFAP expression and induces morphological changes in hippocampal astrocytes. *Brain Struct. Funct.* **2014**, *219*, 293–302. [CrossRef]

71. Loprinzi, P.D. The role of astrocytes on the effects of exercise on episodic memory function. *Physiol. Int.* **2019**, *106*, 21–28. [CrossRef]

72. Chen, M.J.; Russo-Neustadt, A.A. Running exercise-induced up-regulation of hippocampal brain-derived neurotrophic factor is CREB-dependent. *Hippocampus* **2009**, *19*, 962–972. [CrossRef]

73. Molnar, E. Long-term potentiation in cultured hippocampal neurons. *Semin. Cell. Dev. Biol.* **2011**, *22*, 506–513. [CrossRef]

74. Horvath, S.; Pirazzini, C.; Bacalini, M.G.; Gentilini, D.; Di Blasio, A.M.; Delledonne, M.; Mari, D.; Arosio, B.; Monti, D.; Passarino, G.; et al. Decreased epigenetic age of PBMCs from Italian semi-supercentenarians and their offspring. *Aging* **2015**, *7*, 1159–1170. [CrossRef]

75. Christiansen, L.; Lenart, A.; Tan, Q.; Vaupel, J.W.; Aviv, A.; McGue, M.; Christensen, K. DNA methylation age is associated with mortality in a longitudinal Danish twin study. *Aging Cell* **2016**, *15*, 149–154. [CrossRef] [PubMed]

76. Woelfel, J.R.; Dudley-Javoroski, S.; Shields, R.K. Precision Physical Therapy: Exercise, the Epigenome, and the Heritability of Environmentally Modified Traits. *Phys. Ther.* **2018**, *98*, 946–952. [CrossRef] [PubMed]

77. Hunter, D.J.; James, L.; Hussey, B.; Wadley, A.J.; Lindley, M.R.; Mastana, S.S. Impact of aerobic exercise and fatty acid supplementation on global and gene-specific DNA methylation. *Epigenetics* **2019**, *14*, 294–309. [CrossRef] [PubMed]

78. Schenk, A.; Koliamitra, C.; Bauer, C.J.; Schier, R.; Schweiger, M.R.; Bloch, W.; Zimmer, P. Impact of Acute Aerobic Exercise on Genome-Wide DNA-Methylation in Natural Killer Cells-A Pilot Study. *Genes* **2019**, *10*, 380. [CrossRef] [PubMed]

79. Van Praag, H.; Kempermann, G.; Gage, F.H. Running increases cell proliferation and neurogenesis in the adult mouse dentate gyrus. *Nat. Neurosci.* **1999**, *2*, 266–270. [CrossRef] [PubMed]

80. van Praag, H. Neurogenesis and exercise: Past and future directions. *Neuromol. Med.* **2008**, *10*, 128–140. [CrossRef]

81. Lista, I.; Sorrentino, G. Biological mechanisms of physical activity in preventing cognitive decline. *Cell. Mol. Neurobiol.* **2010**, *30*, 493–503. [CrossRef]

82. Fernandes, J.; Arida, R.M.; Gomez-Pinilla, F. Physical exercise as an epigenetic modulator of brain plasticity and cognition. *Neurosci. Biobehav. Rev.* **2017**, *80*, 443–456. [CrossRef]

83. Denham, J. Exercise and epigenetic inheritance of disease risk. *Acta Physiol.* **2018**, *222*, e12881. [CrossRef]

84. Biterge, B.; Schneider, R. Histone variants: Key players of chromatin. *Cell Tissue Res.* **2014**, *356*, 457–466. [CrossRef]

85. Li, G.; Zhu, P. Structure and organization of chromatin fiber in the nucleus. *FEBS Lett.* **2015**, *589*, 2893–2904. [CrossRef] [PubMed]

86. Ramachandran, S.; Henikoff, S. Nucleosome dynamics during chromatin remodeling in vivo. *Nucleus* **2016**, *7*, 20–26. [CrossRef] [PubMed]

87. Di Liegro, C.M.; Schiera, G.; Di Liegro, I. H1.0 Linker Histone as an Epigenetic Regulator of Cell Proliferation and Differentiation. *Genes* **2018**, *9*, 310. [CrossRef] [PubMed]

88. Strahl, B.D.; Allis, C.D. The language of covalent histone modifications. *Nature* **2000**, *103*, 41–15. [CrossRef] [PubMed]

89. Yap, K.L.; Zhou, M.M. Structure and mechanisms of lysine methylation recognition by the chromodomain in gene transcription. *Biochemistry* **2011**, *50*, 1966–1980. [CrossRef] [PubMed]

90. Meier, K.; Brehm, A. Chromatin regulation: How complex does it get? *Epigenetics* **2014**, *9*, 1485–1495. [CrossRef]

91. Ausió, J.; Georgel, P.T. MeCP2 and CTCF: Enhancing the cross-talk of silencers. *Biochem. Cell Biol.* **2017**, *95*, 593–608. [CrossRef]

92. Jain, A.K.; Barton, M.C. Bromodomain Histone Readers and Cancer. *J. Mol. Biol.* **2017**, *429*, 2003–2010. [CrossRef]
93. Han, P.; Chang, C.P. Long non-coding RNA and chromatin remodeling. *RNA Biol.* **2015**, *12*, 1094–1098. [CrossRef]
94. Bartel, D.P. MicroRNAs: Genomics, biogenesis, mechanism, and function. *Cell* **2004**, *116*, 281–297. [CrossRef]
95. Mohr, A.M.; Mott, J.L. Overview of microRNA biology. *Semin. Liver Dis.* **2015**, *35*, 3–11. [CrossRef] [PubMed]
96. Uchida, S.; Dimmeler, S. Exercise controls non-coding RNAs. *Cell Metab.* **2015**, *21*, 511–512. [CrossRef] [PubMed]
97. Alibegovic, A.C.; Sonne, M.P.; Højbjerre, L.; Bork-Jensen, J.; Jacobsen, S.; Nilsson, E.; Faerch, K.; Hiscock, N.; Mortensen, B.; Friedrichsen, M.; et al. Insulin resistance induced by physical inactivity is associated with multiple transcriptional changes in skeletal muscle in young men. *Am. J. Physiol. Endocrinol. Metab.* **2010**, *299*, E752–E763. [CrossRef] [PubMed]
98. Pruunsild, P.; Kazantseva, A.; Aid, T.; Palm, K.; Timmusk, T. Dissecting the human BDNF locus: Bidirectional transcription, complex splicing, and multiple promoters. *Genomics* **2007**, *90*, 397–406. [CrossRef] [PubMed]
99. Chen, K.W.; Chen, L. Epigenetic Regulation of BDNF Gene during Development and Diseases. *Int. J. Mol. Sci.* **2017**, *18*, 571. [CrossRef] [PubMed]
100. De Assis, G.G.; Gasanov, E.V.; de Sousa, M.B.C.; Kozacz, A.; Murawska-Cialowicz, E. Brain derived neutrophic factor, a link of aerobic metabolism to neuroplasticity. *J. Physiol. Pharmacol.* **2018**, *69*, 351–358. [CrossRef]
101. Nofuji, Y.; Suwa, M.; Sasaki, H.; Ichimiya, A.; Nishichi, R.; Kumagai, S. Different circulating brain-derived neurotrophic factor responses to acute exercise between physically active and sedentary subjects. *J. Sports Sci. Med.* **2012**, *11*, 83–88. [PubMed]
102. Hung, C.L.; Tseng, J.W.; Chao, H.H.; Hung, T.M.; Wang, H.S. Effect of Acute Exercise Mode on Serum Brain-Derived Neurotrophic Factor (BDNF) and Task Switching Performance. *J. Clin. Med.* **2018**, *7*, 301. [CrossRef]
103. Saucedo Marquez, C.M.; Vanaudenaerde, B.; Troosters, T.; Wenderoth, N. High-intensity interval training evokes larger serum BDNF levels compared with intense continuous exercise. *J. Appl. Physiol. (1985)* **2015**, *119*, 1363–1373. [CrossRef]
104. Etnier, J.L.; Wideman, L.; Labban, J.D.; Piepmeier, A.T.; Pendleton, D.M.; Dvorak, K.K.; Becofsky, K. The Effects of Acute Exercise on Memory and Brain-Derived Neurotrophic Factor (BDNF). *J. Sport Exerc. Psychol.* **2016**, *38*, 331–340. [CrossRef]
105. Håkansson, K.; Ledreux, A.; Daffner, K.; Terjestam, Y.; Bergman, P.; Carlsson, R.; Kivipelto, M.; Winblad, B.; Granholm, A.C.; Mohammed, A.K. BDNF Responses in Healthy Older Persons to 35 Minutes of Physical Exercise, Cognitive Training, and Mindfulness: Associations with Working Memory Function. *J. Alzheimers Dis.* **2017**, *55*, 645–657. [CrossRef] [PubMed]
106. Kim, J.H.; Kim, D.-Y. Aquarobic exercises improve the serum blood irisin and brain-derived neurotrophic factor levels in elderly women. *Exp. Gerontol.* **2018**, *104*, 60–65. [CrossRef] [PubMed]
107. De la Rosa, A.; Solana, E.; Corpas, R.; Bartrés-Faz, D.; Pallàs, M.; Vina, J.; Sanfeliu, C.; Gomez-Cabrera, M.C. Long-term exercise training improves memory in middle-aged men and modulates peripheral levels of BDNF and Cathepsin, B. *Sci. Rep.* **2019**, *9*, 3337. [CrossRef] [PubMed]
108. De Azevedo, K.P.M.; de Oliveira Segundo, V.H.; de Medeiros, G.C.B.S.; de Sousa Mata, Á.N.; García, D.Á.; de Carvalho Leitão, J.C.G.; Knackfuss, M.I.; Piuvezam, G. Effects of exercise on the levels of BDNF and executive function in adolescents: A protocol for systematic review and meta-analysis. *Medicine* **2019**, *98*, e16445. [CrossRef] [PubMed]
109. Gomez-Pinilla, F.; Vaynman, S.; Ying, Z. Brain-derived neurotrophic factor functions as a metabotrophin to mediate the effects of exercise on cognition. *Eur. J. Neurosci.* **2008**, *28*, 2278–2287. [CrossRef] [PubMed]
110. Steiner, J.L.; Murphy, E.A.; McClellan, J.L.; Carmichael, M.D.; Davis, J.M. Exercise training increases mitochondrial biogenesis in the brain. *J. Appl. Physiol.* **2011**, *111*, 1066–1071. [CrossRef] [PubMed]
111. Tsankova, N.M.; Berton, O.; Renthal, W.; Kumar, A.; Neve, R.L.; Nestler, E.J. Sustained hippocampal chromatin regulation in a mouse model of depression and antidepressant action. *Nat. Neurosci.* **2006**, *9*, 519–525. [CrossRef] [PubMed]
112. Gomez-Pinilla, F.; Zhuang, Y.; Feng, J.; Ying, Z.; Fan, G. Exercise impacts brain-derived neurotrophic factor plasticity by engaging mechanisms of epigenetic regulation. *Eur. J. Neurosci.* **2011**, *33*, 383–390. [CrossRef] [PubMed]

113. Ieraci, A.; Mallei, A.; Musazzi, L.; Popoli, M. Physical exercise and acute restraint stress differentially modulate hippocampal brain-derived neurotrophic factor transcripts and epigenetic mechanisms in mice. *Hippocampus* **2015**, *25*, 1380–1392. [CrossRef]

114. Fernandes, A.; Li, Y.W. Focused microwave irradiation-assisted immunohistochemistry to study effects of ketamine on phospho-ERK expression in the mouse brain. *Brain Res.* **2017**, *1670*, 86–95. [CrossRef]

115. Gejl, A.K.; Enevold, C.; Bugge, A.; Andersen, M.S.; Nielsen, C.H.; Andersen, L.B. Associations between serum and plasma brain-derived neurotrophic factor and influence of storage time and centrifugation strategy. *Sci. Rep.* **2019**, *9*, 9655. [CrossRef] [PubMed]

116. Zhao, Y.; Zhang, A.; Wang, Y.; Hu, S.; Zhang, R.; Qian, S. Genome-wide identification of brain miRNAs in response to high-intensity intermittent swimming training in Rattus norvegicus by deep sequencing. *BMC Mol. Biol.* **2019**, *20*, 3. [CrossRef] [PubMed]

117. Beclin, C.; Follert, P.; Stappers, E.; Barral, S.; Coré, N.; de Chevigny, A.; Magnone, V.; Lebrigand, K.; Bissels, U.; Huylebroeck, D.; et al. miR-200 family controls late steps of postnatal forebrain neurogenesis via Zeb2 inhibition. *Sci. Rep.* **2016**, *6*, 35729. [CrossRef] [PubMed]

118. Pandey, A.; Singh, P.; Jauhari, A.; Singh, T.; Khan, F.; Pant, A.B.; Parmar, D.; Yadav, S. Critical role of the miR-200 family in regulating differentiation and proliferation of neurons. *J. Neurochem.* **2015**, *133*, 640–652. [CrossRef] [PubMed]

119. Jauhari, A.; Yadav, S. MiR-34 and MiR-200: Regulator of Cell Fate Plasticity and Neural Development. *Neuromol. Med.* **2019**. [CrossRef] [PubMed]

120. Choi, P.S.; Zakhary, L.; Choi, W.Y.; Caron, S.; Alvarez-Saavedra, E.; Miska, E.A.; McManus, M.; Harfe, B.; Giraldez, A.J.; Horvitz, H.R.; et al. Members of the miRNA-200 family regulate olfactory neurogenesis. *Neuron* **2008**, *57*, 41–55. [CrossRef] [PubMed]

121. Lee, S.T.; Chu, K.; Jung, K.H.; Yoon, H.J.; Jeon, D.; Kang, K.M.; Park, K.H.; Bae, E.K.; Kim, M.; Lee, S.K.; et al. MicroRNAs induced during ischemic preconditioning. *Stroke* **2010**, *41*, 1646–1651. [CrossRef]

122. Hu, T.; Zhou, F.J.; Chang, Y.F.; Li, Y.S.; Liu, G.C.; Hong, Y.; Chen, H.L.; Xiyang, Y.B.; Bao, T.H. miR21 is Associated with the Cognitive Improvement Following Voluntary Running Wheel Exercise in TBI Mice. *J. Mol. Neurosci.* **2015**, *57*, 114–122. [CrossRef]

123. Kou, X.; Li, J.; Liu, X.; Chang, J.; Zhao, Q.; Jia, S.; Fan, J.; Chen, N. Swimming attenuates d-galactose-induced brain aging via suppressing miR-34a-mediated autophagy impairment and abnormal mitochondrial dynamics. *J. Appl. Physiol.* **2017**, *122*, 1462–1469. [CrossRef]

124. Harman, D. Aging: A theory based on free radical and radiation chemistry. *J. Gerontol.* **1956**, *11*, 298–300. [CrossRef]

125. Frenzel, M.; Rommelspacher, H.; Sugawa, M.D.; Dencher, N.A. Ageing alters the supramolecular architecture of OxPhos complexes in rat brain cortex. *Exp. Gerontol.* **2010**, *45*, 563–572. [CrossRef] [PubMed]

126. Jones, T.T.; Brewer, G.J. Age-related deficiencies in complex I endogenous substrate availability and reserve capacity of complex IV in cortical neuron electron transport. *Biochim. Biophys. Acta* **2010**, *1797*, 167–176. [CrossRef] [PubMed]

127. Li, H.; Shen, L.; Hu, P.; Huang, R.; Cao, Y.; Deng, J.; Yuan, W.; Liu, D.; Yang, J.; Gu, H.; et al. Aging-associated mitochondrial DNA mutations alter oxidative phosphorylation machinery and cause mitochondrial dysfunctions. *Biochim. Biophys. Acta Mol. Basis Dis.* **2017**, *1863*, 2266–2273. [CrossRef] [PubMed]

128. Zhang, L.; Trushin, S.; Christensen, T.A.; Bachmeier, B.V.; Gateno, B.; Schroeder, A.; Yao, J.; Itoh, K.; Sesaki, H.; Poon, W.W.; et al. Altered brain energetics induces mitochondrial fission arrest in Alzheimer's Disease. *Sci. Rep.* **2016**, *6*, 18725. [CrossRef]

129. Hara, Y.; Yuk, F.; Puri, R.; Janssen, W.G.; Rapp, P.R.; Morrison, J.H. Presynaptic mitochondrial morphology in monkey prefrontal cortex correlates with working memory and is improved with estrogen treatment. *Proc. Natl. Acad. Sci. USA* **2014**, *111*, 486–491. [CrossRef] [PubMed]

130. Bubber, P.; Haroutunian, V.; Fisch, G.; Blass, J.P.; Gibson, G.E. Mitochondrial abnormalities in Alzheimer brain: Mechanistic implications. *Ann. Neurol.* **2005**, *57*, 695–703. [CrossRef] [PubMed]

131. Palomo, G.M.; Manfredi, G. Exploring new pathways of neurodegeneration in ALS: The role of mitochondria quality control. *Brain Res.* **2015**, *1607*, 36–46. [CrossRef]

132. Lee, K.S.; Huh, S.; Lee, S.; Wu, Z.; Kim, A.K.; Kang, H.Y.; Lu, B. Altered ER-mitochondria contact impacts mitochondria calcium homeostasis and contributes to neurodegeneration in vivo in disease models. *Proc. Natl. Acad. Sci. USA* **2018**, *115*, E8844–E8853. [CrossRef]

133. Marosi, K.; Mattson, M.P. BDNF mediates adaptive brain and body responses to energetic challenges. *Trends Endocrinol. Metab.* **2014**, *25*, 89–98. [CrossRef]

134. Gusdon, A.M.; Callio, J.; Distefano, G.; O'Doherty, R.M.; Goodpaster, B.H.; Coen, P.M.; Chu, C.T. Exercise increases mitochondrial complex I activity and DRP1 expression in the brains of aged mice. *Exp. Gerontol.* **2017**, *90*, 1–13. [CrossRef]

135. Parzych, K.R.; Klionsky, D.J. An overview of autophagy: Morphology, mechanism, and regulation. *Antioxid. Redox Signal.* **2014**, *20*, 460–473. [CrossRef] [PubMed]

136. Lira, V.A.; Okutsu, M.; Zhang, M.; Greene, N.P.; Laker, R.C.; Breen, D.S.; Hoehn, K.L.; Yan, Z. Autophagy is required for exercise training-induced skeletal muscle adaptation and improvement of physical performance. *FASEB J.* **2013**, *27*, 4184–4193. [CrossRef] [PubMed]

137. Sanchez, A.M.; Bernardi, H.; Py, G.; Candau, R.B. Autophagy is essential to support skeletal muscle plasticity in response to endurance exercise. *Am. J. Physiol. Regul. Integr. Comp. Physiol.* **2014**, *307*, R956–R969. [CrossRef] [PubMed]

138. Dethlefsen, M.M.; Halling, J.F.; Møller, H.D.; Plomgaard, P.; Regenberg, B.; Ringholm, S.; Pilegaard, H. Regulation of apoptosis and autophagy in mouse and human skeletal muscle with aging and lifelong exercise training. *Exp. Gerontol.* **2018**, *111*, 141–153. [CrossRef] [PubMed]

139. Huang, J.; Wang, X.; Zhu, Y.; Li, Z.; Zhu, Y.T.; Wu, J.C.; Qin, Z.H.; Xiang, M.; Lin, F. Exercise activates lysosomal function in the brain through AMPK-SIRT1-TFEB pathway. *CNS Neurosci. Ther.* **2019**. [CrossRef] [PubMed]

140. Palikaras, K.; Lionaki, E.; Tavernarakis, N. Coupling mitogenesis and mitophagy for longevity. *Autophagy* **2015**, *11*, 1428–1430. [CrossRef] [PubMed]

141. Moreira, O.C.; Estébanez, B.; Martínez-Florez, S.; de Paz, J.A.; Cuevas, M.J.; González-Gallego, J. Mitochondrial Function and Mitophagy in the Elderly: Effects of Exercise. *Oxid. Med. Cell. Longev.* **2017**, *2017*, 2012798. [CrossRef] [PubMed]

142. Dos Santos, J.M.; Moreli, M.L.; Tewari, S.; Benite-Ribeiro, S.A. The effect of exercise on skeletal muscle glucose uptake in type 2 diabetes: An epigenetic perspective. *Metabolism* **2015**, *64*, 1619–1628. [CrossRef]

143. Rezapour, S.; Shiravand, M.; Mardani, M. Epigenetic changes due to physical activity. *Biotechnol. Appl. Biochem.* **2018**, *65*, 761–767. [CrossRef]

144. Chaillou, T. Skeletal Muscle Fiber Type in Hypoxia: Adaptation to High-Altitude Exposure and Under Conditions of Pathological Hypoxia. *Front. Physiol.* **2018**, *9*, 1450. [CrossRef]

145. Gundersen, K. Excitation-transcription coupling in skeletal muscle: The molecular pathways of exercise. *Biol. Rev. Camb. Philos. Soc.* **2011**, *86*, 564–600. [CrossRef] [PubMed]

146. Masuzawa, R.; Konno, R.; Ohsawa, I.; Watanabe, A.; Kawano, F. Muscle type-specific RNA polymerase II recruitment during PGC-1α gene transcription after acute exercise in adult rats. *J. Appl. Physiol. (1985)* **2018**, *125*, 1238–1245. [CrossRef] [PubMed]

147. Pette, D.; Vrbová, G. What does chronic electrical stimulation teach us about muscle plasticity? *Muscle Nerve* **1999**, *22*, 666–677. [CrossRef]

148. Pette, D. Training effects on the contractile apparatus. *Acta Physiol. Scand.* **1998**, *162*, 367–376. [CrossRef]

149. Hughes, S.M.; Koishi, K.; Rudnicki, M.; Maggs, A.M. MyoD protein is differentially accumulated in fast and slow skeletal muscle fibres and required for normal fibre type balance in rodents. *Mech. Dev.* **1997**, *61*, 151–163. [CrossRef]

150. Macharia, R.; Otto, A.; Valasek, P.; Patel, K. Neuromuscular junction morphology, fiber-type proportions, and satellite-cell proliferation rates are altered in MyoD(-/-) mice. *Muscle Nerve* **2010**, *42*, 38–52. [CrossRef]

151. Parsons, S.A.; Millay, D.P.; Wilkins, B.J.; Bueno, O.F.; Tsika, G.L.; Neilson, J.R.; Liberatore, C.M.; Yutzey, K.E.; Crabtree, G.R.; Tsika, R.W.; et al. Genetic loss of calcineurin blocks mechanical overload-induced skeletal muscle fiber type switching but not hypertrophy. *J. Biol. Chem.* **2004**, *279*, 26192–26200. [CrossRef]

152. Oh, M.; Rybkin, I.I.; Copeland, V.; Czubryt, M.P.; Shelton, J.M.; van Rooij, E.; Richardson, J.A.; Hill, J.A.; De Windt, L.J.; Bassel-Duby, R.; et al. Calcineurin is necessary for the maintenance but not embryonic development of slow muscle fibers. *Mol. Cell. Biol.* **2005**, *25*, 6629–6638. [CrossRef]

153. Rana, Z.A.; Gundersen, K.; Buonanno, A. Activity-dependent repression of muscle genes by NFAT. *Proc. Natl. Acad. Sci. USA* **2008**, *105*, 5921–5926. [CrossRef]

154. Ehlers, M.L.; Celona, B.; Black, B.L. NFATc1 controls skeletal muscle fiber type and is a negative regulator of MyoD activity. *Cell Rep.* **2014**, *8*, 1639–1648. [CrossRef]

155. Gehlert, S.; Bloch, W.; Suhr, F. Ca^{2+}-dependent regulations and signaling in skeletal muscle: From electro-mechanical coupling to adaptation. *Int. J. Mol. Sci.* **2015**, *16*, 1066–1095. [CrossRef] [PubMed]

156. Cohen, P. The role of calcium ions, calmodulin and troponin in the regulation of phosphorylase kinase from rabbit skeletal muscle. *Eur. J. Biochem.* **1980**, *111*, 563–574. [CrossRef] [PubMed]

157. Dasgupta, M.; Honeycutt, T.; Blumenthal, D.K. The gamma-subunit of Skeletal Muscle Phosphorylase Kinase Contains Two Noncontiguous Domains That Act in Concert to Bind Calmodulin. *J. Biol. Chem.* **1989**, *264*, 17156–17163. [PubMed]

158. Sola-Penna, M.; Da, S.D.; Coelho, W.S.; Marinho-Carvalho, M.M.; Zancan, P. Regulation of mammalian muscle type 6-phosphofructo-1-kinase and its implication for the control of the metabolism. *IUBMB Life* **2010**, *62*, 791–796. [CrossRef] [PubMed]

159. Marxsen, J.H.; Stengel, P.; Doege, K.; Heikkinen, P.; Jokilehto, T.; Wagner, T.; Jelkmann, W.; Jaakkola, P.; Metzen, E. Hypoxia-inducible factor-1 (HIF-1) promotes its degradation by induction of HIF-α-prolyl-4-hydroxylases. *Biochem. J.* **2004**, *381 Pt 3*, 761–767. [CrossRef]

160. Lando, D.; Peet, D.J.; Whelan, D.A.; Gorman, J.J.; Whitelaw, M.L. Asparagine hydroxylation of the HIF transactivation domain: A hypoxic switch. *Science* **2002**, *295*, 858–861. [CrossRef] [PubMed]

161. Ke, Q.; Costa, M. Hypoxia-inducible factor-1 (HIF-1). *Mol. Pharmacol.* **2006**, *70*, 1469–1480. [CrossRef] [PubMed]

162. Zimna, A.; Kurpisz, M. Hypoxia-Inducible Factor-1 in Physiological and Pathophysiological Angiogenesis: Applications and Therapies. *BioMed Res. Int.* **2015**, *2015*, 549412. [CrossRef] [PubMed]

163. Fischer, M.; Richeit, P.; Knaus, P.; Coirault, C. YAP-Mediated Mechanotransduction in Skeletal Muscle. *Front. Physiol.* **2016**, *7*, 41. [CrossRef]

164. Panciera, T.; Azzolin, L.; Cordenonsi, M.; Piccolo, S. Mechanobiology of YAP and TAZ in physiology and disease. *Nat. Rev. Mol. Cell. Biol.* **2017**, *18*, 758–770. [CrossRef] [PubMed]

165. Imajo, M.; Miyatake, K.; Iimura, A.; Miyamoto, A.; Nishida, E. A molecular mechanism that links Hippo signalling to the inhibition of Wnt/b catenin signalling. *EMBO J.* **2012**, *31*, 1109–1122. [CrossRef] [PubMed]

166. Lessard, S.J.; MacDonald, T.L.; Pathak, P.; Han, M.S.; Coffey, V.G.; Edge, J.; Rivas, D.A.; Hirshman, M.F.; Davis, R.J.; Goodyear, L.J. JNK regulates muscle remodeling via myostatin/SMAD inhibition. *Nat. Commun.* **2018**, *9*, 3030. [CrossRef] [PubMed]

167. Aronson, D.; Boppart, M.D.; Dufresne, S.D.; Fielding, R.A.; Goodyear, L.J. Exercise stimulates c-Jun NH2 kinase activity and c-Jun transcriptional activity in human skeletal muscle. *Biochem. Biophys. Res. Commun.* **1998**, *251*, 106–110. [CrossRef] [PubMed]

168. Damas, F.; Ugrinowitsch, C.; Libardi, C.A.; Jannig, P.R.; Hector, A.J.; McGlory, C.; Lixandrão, M.E.; Vechin, F.C.; Montenegro, H.; Tricoli, V.; et al. Resistance training in young men induces muscle transcriptome-wide changes associated with muscle structure and metabolism refining the response to exercise-induced stress. *Eur. J. Appl. Physiol.* **2018**, *118*, 2607–2616. [CrossRef] [PubMed]

169. Wessner, B.; Liebensteiner, M.; Nachbauer, W.; Csapo, R. Age-specific response of skeletal muscle extracellular matrix to acute resistance exercise: A pilot study. *Eur. J. Sports Sci.* **2019**, *19*, 354–364. [CrossRef] [PubMed]

170. McGee, S.L.; Fairlie, E.; Garnham, A.P.; Hargreaves, M. Exercise-induced histone modifications in human skeletal muscle. *J. Physiol.* **2009**, *587 Pt 24*, 5951–5958. [CrossRef]

171. Barrès, R.; Yan, J.; Egan, B.; Treebak, J.T.; Rasmussen, M.; Fritz, T.; Caidahl, K.; Krook, A.; O'Gorman, D.J.; Zierath, J.R. Acute exercise remodels promoter methylation in human skeletal muscle. *Cell Metab.* **2012**, *15*, 405–411. [CrossRef]

172. Ling, C.; Rönn, T. Epigenetic adaptation to regular exercise in humans. *Drug Discov. Today* **2014**, *19*, 1015–1018. [CrossRef]

173. King-Himmelreich, T.S.; Schramm, S.; Wolters, M.C.; Schmelzer, J.; Möser, C.V.; Knothe, C.; Resch, E.; Peil, J.; Geisslinger, G.; Niederberger, E. The impact of endurance exercise on global and AMPK gene-specific DNA methylation. *Biochem. Biophys. Res. Commun.* **2016**, *474*, 284–290. [CrossRef]

174. Russell, A.P.; Lamon, S.; Boon, H.; Wada, S.; Güller, I.; Brown, E.L.; Chibalin, A.V.; Zierath, J.R.; Snow, R.J.; Stepto, N.; et al. Regulation of miRNAs in human skeletal muscle following acute endurance exercise and short-term endurance training. *J. Physiol.* **2013**, *591*, 4637–4653. [CrossRef]

175. Silva, G.J.J.; Bye, A.; El Azzouzi, H.; Wisløff, U. MicroRNAs as Important Regulators of Exercise Adaptation. *Prog. Cardiovasc. Dis.* **2017**, *60*, 130–151. [CrossRef] [PubMed]

176. Nielsen, S.; Scheele, C.; Yfanti, C.; Akerström, T.; Nielsen, A.R.; Pedersen, B.K.; Laye, M.J. Muscle specific microRNAs are regulated by endurance exercise in human skeletal muscle. *J. Physiol.* **2010**, *588 Pt 20*, 4029–4037. [CrossRef]

177. Pastore, A.; Piemonte, F. S-Glutathionylation signaling in cell biology: Progress and prospects. *Eur. J. Pharm. Sci.* **2012**, *46*, 279–292. [CrossRef] [PubMed]

178. Kramer, P.A.; Duan, J.; Gaffrey, M.J.; Shukla, A.K.; Wang, L.; Bammler, T.K.; Qian, W.J.; Marcinek, D.J. Fatiguing contractions increase protein S-glutathionylation occupancy in mouse skeletal muscle. *Redox Biol.* **2018**, *17*, 367–376. [CrossRef] [PubMed]

179. Iizuka, K.; Machida, T.; Hirafuji, M. Skeletal muscle is an endocrine organ. *J. Pharmacol. Sci.* **2014**, *125*, 125–131. [CrossRef] [PubMed]

180. Schnyder, S.; Handschin, C. Skeletal muscle as an endocrine organ: PGC-1a, myokines and exercise. *Bone* **2015**, *80*, 115125. [CrossRef] [PubMed]

181. Giudice, J.; Taylor, J.M. Muscle as a paracrine and endocrine organ. *Curr. Opin. Pharmacol.* **2017**, *34*, 49–55. [CrossRef] [PubMed]

182. Delezie, J.; Handschin, C. Endocrine Crosstalk Between Skeletal Muscle and the Brain. *Front. Neurol.* **2018**, *9*, 698. [CrossRef]

183. Matthews, V.B.; Astrom, M.B.; Chan, M.H.; Bruce, C.R.; Krabbe, K.S.; Prelovsek, O.; Akerström, T.; Yfanti, C.; Broholm, C.; Mortensen, O.H.; et al. Brain-derived neurotrophic factor is produced by skeletal muscle cells in response to contraction and enhances fat oxidation via activation of AMP-activated protein kinase. *Diabetologia* **2009**, *52*, 1409–1418. [CrossRef]

184. Ogborn, D.I.; Gardiner, P.F. Effects of exercise and muscle type on BDNF, NT-4/5, and TrKB expression in skeletal muscle. *Muscle Nerve* **2010**, *41*, 385–391. [CrossRef]

185. Pan, W.; Banks, W.A.; Fasold, M.B.; Bluth, J.; Kastin, A.J. Transport of brain-derived neurotrophic factor across the blood-brain barrier. *Neuropharmacology* **1998**, *37*, 1553–1561. [CrossRef]

186. Buck, M.R.; Karustis, D.G.; Day, N.A.; Honn, K.V.; Sloane, B.F. Degradation of extracellular-matrix proteins by human cathepsin B from normal and tumour tissues. *Biochem. J.* **1992**, *282*, 273–278. [CrossRef] [PubMed]

187. Linebaugh, B.E.; Sameni, M.; Day, N.A.; Sloane, B.F.; Keppler, D. Exocytosis of active cathepsin B enzyme activity at pH 7.0, inhibition and molecular mass. *Eur. J. Biochem.* **1999**, *264*, 100–109. [CrossRef] [PubMed]

188. Moon, H.Y.; Becke, A.; Berron, D.; Becker, B.; Sah, N.; Benoni, G.; Janke, E.; Lubejko, S.T.; Greig, N.H.; Mattison, J.A.; et al. Running-Induced Systemic Cathepsin B Secretion Is Associated with Memory Function. *Cell Metab.* **2016**, *24*, 332–340. [CrossRef] [PubMed]

189. Ji, L.L.; Gomez-Cabrera, M.C.; Vina, J. Exercise and hormesis: Activation of cellular antioxidant signaling pathway. *Ann. N. Y. Acad. Sci.* **2006**, *1067*, 425–435. [CrossRef] [PubMed]

190. Pereira, R.O.; Tadinada, S.M.; Zasadny, F.M.; Oliveira, K.J.; Pires, K.M.P.; Olvera, A.; Jeffers, J.; Souvenir, R.; Mcglauflin, R.; Seei, A.; et al. OPA1 deficiency promotes secretion of FGF21 from muscle that prevents obesity and insulin resistance. *EMBO J.* **2017**, *36*, 2126–2145. [CrossRef] [PubMed]

191. BonDurant, L.D.; Potthoff, M.J. Fibroblast Growth Factor 21: A Versatile Regulator of Metabolic Homeostasis. *Annu. Rev. Nutr.* **2018**, *38*, 173–196. [CrossRef] [PubMed]

192. Cuevas-Ramos, D.; Mehta, R.; Aguilar-Salinas, C.A. Fibroblast Growth Factor 21 and Browning of White Adipose Tissue. *Front. Physiol.* **2019**, *10*, 37. [CrossRef] [PubMed]

193. Hsuchou, H.; Pan, W.; Kastin, A.J. The fasting polypeptide FGF21 can enter brain from blood. *Peptides* **2007**, *28*, 2382–2386. [CrossRef]

194. Kuro-O, M. The Klotho proteins in health and disease. *Nat. Rev. Nephrol.* **2019**, *15*, 27–44. [CrossRef]

195. He, Z.; Tian, Y.; Valenzuela, P.L.; Huang, C.; Zhao, J.; Hong, P.; He, Z.; Yin, S.; Lucia, A. Myokine Response to High-Intensity Interval vs. Resistance Exercise: An Individual Approach. *Front. Physiol.* **2018**, *9*, 1735. [CrossRef] [PubMed]

196. Boström, P.; Wu, J.; Jedrychowski, M.P.; Korde, A.; Ye, L.; Lo, J.C.; Rasbach, K.A.; Boström, E.A.; Choi, J.H.; Long, J.Z.; et al. A PGC1-α-dependent myokine that drives brown-fat-like development of white fat and thermogenesis. *Nature* **2012**, *481*, 463–468. [CrossRef] [PubMed]

197. Wrann, C.D.; White, J.P.; Salogiannnis, J.; Laznik-Bogoslavski, D.; Wu, J.; Ma, D.; Lin, J.D.; Greenberg, M.E.; Spiegelman, B.M. Exercise induces hippocampal BDNF through a PGC-1α/FNDC5 pathway. *Cell Metab.* **2013**, *18*, 649–659. [CrossRef] [PubMed]

198. Phillips, C.; Baktir, M.A.; Srivatsan, M.; Salehi, A. Neuroprotective effects of physical activity on the brain: A closer look at trophic factor signaling. *Front. Cell. Neurosci.* **2014**, *8*, 170. [CrossRef] [PubMed]

199. Banks, W.A.; Kastin, A.J.; Broadwell, R.D. Passage of cytokines across the blood-brain barrier. *Neuroimmunomodulation* **1995**, *2*, 241–248. [CrossRef] [PubMed]

200. Rothaug, M.; Becker-Pauly, C.; Rose-John, S. The role of interleukin-6 signaling in nervous tissue. *Biochim. Biophys. Acta* **2016**, *1863*, 1218–1227. [CrossRef]

201. Semple, B.D.; Kossmann, T.; Morganti-Kossmann, M.C. Role of chemokines in CNS health and pathology: A focus on the CCL2/CCR2 and CXCL8/CXCR2 networks. *J. Cereb. Blood Flow Metab.* **2010**, *30*, 459–473. [CrossRef]

202. Pan, W.; Wu, X.; He, Y.; Hsuchou, H.; Huang, E.Y.; Mishra, P.K.; Kastin, A.J. Brain interleukin-15 in neuroinflammation and behavior. *Neurosci. Biobehav. Rev.* **2013**, *37*, 184–192. [CrossRef] [PubMed]

203. Proia, P.; Di Liegro, C.M.; Schiera, G.; Fricano, A.; Di Liegro, I. Lactate as a Metabolite and a Regulator in the Central Nervous System. *Int. J. Mol. Sci.* **2016**, *17*, 1450. [CrossRef]

204. Morland, C.; Andersson, K.A.; Haugen, Ø.P.; Hadzic, A.; Kleppa, L.; Gille, A.; Rinholm, J.E.; Palibrk, V.; Diget, E.H.; Kennedy, L.H.; et al. Exercise induces cerebral VEGF and angiogenesis via the lactate receptor HCAR1. *Nat. Commun.* **2017**, *8*, 15557. [CrossRef]

205. Tari, A.R.; Norevik, C.S.; Scrimgeour, N.R.; Kobro-Flatmoen, A.; Storm-Mathisen, J.; Bergersen, L.H.; Wrann, C.D.; Selbæk, G.; Kivipelto, M.; Moreira, J.B.N.; et al. Are the neuroprotective effects of exercise training systemically mediated? *Prog. Cardiovasc. Dis.* **2019**, *62*, 94–101. [CrossRef] [PubMed]

206. de Castro Abrantes, H.; Briquet, M.; Schmuziger, C.; Restivo, L.; Puyal, J.; Rosenberg, N.; Rocher, A.B.; Offermanns, S.; Chatton, J.Y. The lactate receptor HCAR1 modulates neuronal network activity through the activation of Gα and Gβ γ subunits. *J. Neurosci.* **2019**, *39*, 4422–4433. [CrossRef] [PubMed]

207. Iraci, N.; Leonardi, T.; Gessler, F.; Vega, B.; Pluchino, S. Focus on Extracellular Vesicles: Physiological Role and Signalling Properties of Extracellular Membrane Vesicles. *Int. J. Mol. Sci.* **2016**, *17*, 171. [CrossRef]

208. Di Liegro, C.M.; Schiera, G.; Di Liegro, I. Extracellular Vesicle-Associated RNA as a Carrier of Epigenetic Information. *Genes* **2017**, *8*, 240. [CrossRef]

209. Mateescu, B.; Kowal, E.J.; van Balkom, B.W.; Bartel, S.; Bhattacharyya, S.N.; Buzás, E.I.; Buck, A.H.; de Candia, P.; Chow, F.W.; Das, S.; et al. Obstacles and opportunities in the functional analysis of extracellular vesicle RNA—An ISEV position paper. *J. Extracell. Vesicles* **2017**, *6*, 1286095. [CrossRef] [PubMed]

210. Whitham, M.; Parker, B.L.; Friedrichsen, M.; Hingst, J.R.; Hjorth, M.; Hughes, W.E.; Egan, C.L.; Cron, L.; Watt, K.I.; Kuchel, R.P.; et al. Extracellular Vesicles Provide a Means for Tissue Crosstalk during Exercise. *Cell Metab.* **2018**, *27*, 237–251.e4. [CrossRef] [PubMed]

211. Choi, S.H.; Bylykbashi, E.; Chatila, Z.K.; Lee, S.W.; Pulli, B.; Clemenson, G.D.; Kim, E.; Rompala, A.; Oram, M.K.; Asselin, C.; et al. Combined adult neurogenesis and BDNF mimic exercise effects on cognition in an Alzheimer's mouse model. *Science* **2018**, *361*, eaan8821. [CrossRef] [PubMed]

212. Lourenco, M.V.; Frozza, R.L.; de Freitas, G.B.; Zhang, H.; Kincheski, G.C.; Ribeiro, F.C.; Gonçalves, R.A.; Clarke, J.R.; Beckman, D.; Staniszewski, A.; et al. Exercise-linked FNDC5/irisin rescues synaptic plasticity and memory defects in Alzheimer's models. *Nat. Med.* **2019**, *25*, 165–175. [CrossRef]

213. Young, M.F.; Valaris, S.; Wrann, C.D. A role for FNDC5/Irisin in the beneficial effects of exercise on the brain and in neurodegenerative diseases. *Prog. Cardiovasc. Dis.* **2019**, *62*, 172–178. [CrossRef]

214. Wang, K.; Li, H.; Wang, H.; Wang, J.H.; Song, F.; Sun, Y. Irisin Exerts Neuroprotective Effects on Cultured Neurons by Regulating Astrocytes. *Mediat. Inflamm.* **2018**, *2018*, 9070341. [CrossRef]

215. Jahangiri, Z.; Gholamnezhad, Z.; Hosseini, M. Neuroprotective effects of exercise in rodent models of memory deficit and Alzheimer's. *Metab. Brain Dis.* **2019**, *34*, 21–37. [CrossRef] [PubMed]

216. Carro, E.; Trejo, J.L.; Busiguina, S.; Torres-Aleman, I. Circulating insulin-like growth factor I mediates the protective effects of physical exercise against brain insults of different etiology and anatomy. *J. Neurosci.* **2001**, *21*, 5678–5684. [CrossRef] [PubMed]

217. Ding, Q.; Vaynman, S.; Akhavan, M.; Ying, Z.; Gomez-Pinilla, F. Insulin-like growth factor I interfaces with brain-derived neurotrophic factor-mediated synaptic plasticity to modulate aspects of exercise-induced cognitive function. *Neuroscience* **2006**, *140*, 823–833. [CrossRef] [PubMed]

218. Saab, B.J.; Mansuy, I.M. Neuroepigenetics of memory formation and impairment: The role of microRNAs. *Neuropharmacology* **2014**, *80*, 61–69. [CrossRef] [PubMed]

219. Schiera, G.; Contrò, V.; Sacco, A.; Macchiarella, A.; Cieszczyk, P.; Proia, P. From Epigenetics to Anti-Doping Application: A New Tool of Detection. *Hum. Mov.* **2017**, *18*, 3–10. [CrossRef]

220. Keifer, J.; Zheng, Z.; Ambigapathy, G. A MicroRNA-BDNF Negative Feedback Signaling Loop in Brain: Implications for Alzheimer's Disease. *Microrna* **2015**, *4*, 101–108. [CrossRef] [PubMed]

221. Wang, X.; Liu, P.; Zhu, H.; Xu, Y.; Ma, C.; Dai, X.; Huang, L.; Liu, Y.; Zhang, L.; Qin, C. miR-34a, a microRNA up-regulated in a double transgenic of Alzheimer's disease, inhibits bcl2 translation. *Brain Res. Bull.* **2009**, *80*, 268–273. [CrossRef] [PubMed]

222. Do, K.; Laing, B.T.; Landry, T.; Bunner, W.; Mersaud, N.; Matsubara, T.; Li, P.; Yuan, Y.; Lu, Q.; Huang, H. The effects of exercise on hypothalamic neurodegeneration of Alzheimer's disease mouse model. *PLoS ONE* **2018**, *13*, e0190205. [CrossRef] [PubMed]

223. Morris, J.K.; Vidoni, E.D.; Johnson, D.K.; Van Sciver, A.; Mahnken, J.D.; Honea, R.A.; Wilkins, H.M.; Brooks, W.M.; Billinger, S.A.; Swerdlow, R.H.; et al. Aerobic exercise for Alzheimer's disease: A randomized controlled pilot trial. *PLoS ONE* **2017**, *12*, e0170547. [CrossRef]

224. Hou, L.; Chen, W.; Liu, X.; Qiao, D.; Zhou, F.M. Exercise-induced neuroprotection of the nigrostriatal dopamine system in Parkinson's disease. *Front. Aging Neurosci.* **2017**, *9*, 358. [CrossRef]

225. Ibanez, P.; Bonnet, A.M.; Debarges, B.; Lohmann, E.; Tison, F.; Pollak, P.; Agid, Y.; Dürr, A.; Brice, A. Causal relation between α-synuclein gene duplication and familial Parkinson's disease. *Lancet* **2004**, *364*, 1169–1171. [CrossRef]

226. Nuytemans, K.; Theuns, J.; Cruts, M.; Van Broeckhoven, C. Genetic etiology of Parkinson disease associated with mutations in the SNCA, PARK2, PINK1, PARK7, and LRRK2 genes: A mutation update. *Hum. Mutat.* **2010**, *31*, 763–780. [CrossRef] [PubMed]

227. Lauzé, M.; Daneault, J.F.; Duval, C. The effects of physical activity in Parkinson's disease: A Review. *J. Parkinsons Dis.* **2016**, *6*, 685–698. [CrossRef] [PubMed]

228. Bhalsing, K.S.; Abbas, M.M.; Tan, L.C.S. Role of Physical Activity in Parkinson's Disease. *Ann. Indian Acad. Neurol.* **2018**, *21*, 242–249. [CrossRef] [PubMed]

229. Amara, A.W.; Memon, A.A. Effects of exercise on non-motor symptoms in Parkinson's disease. *Clin. Ther.* **2018**, *40*, 8–15. [CrossRef] [PubMed]

230. Ahlskog, J.E. Does vigorous exercise have a neuroprotective effect in Parkinson disease? *Neurology* **2011**, *77*, 288–294. [CrossRef]

231. Tajiri, N.; Yasuhara, T.; Shingo, T.; Kondo, A.; Yuan, W.; Kadota, T.; Wang, F.; Baba, T.; Tayra, J.T.; Morimoto, T.; et al. Exercise exerts neuroprotective effects on Parkinson's disease model of rats. *Brain Res.* **2010**, *1310*, 200–207. [CrossRef]

232. Collins, G.A.; Hill, L.E.; Chandramohan, Y.; Whitcomb, D.; Droste, S.K.; Reul, J.M. Exercise improves cognitive responses to psychological stress through enhancement of epigenetic mechanisms and gene expression in the dentate. *PLoS ONE* **2009**, *4*, e4330. [CrossRef]

233. Cohen, A.D.; Tillerson, J.L.; Smith, A.D.; Schallert, T.; Zigmond, M.J. Neuroprotective effects of prior limb use in 6-hydroxydopamine-treated rats: Possible role of GDNF. *J. Neurochem.* **2003**, *85*, 299–305. [CrossRef]

234. Aguiar, A.S., Jr.; Duzzioni, M.; Remor, A.P.; Tristão, F.S.; Matheus, F.C.; Raisman-Vozari, R.; Latini, A.; Prediger, R.D. Moderate-Intensity Physical Exercise Protects against Experimental 6-Hydroxydopamine-Induced Hemiparkinsonism Through Nrf2-Antioxidant Response Element Pathway. *Neurochem. Res.* **2016**, *41*, 64–72. [CrossRef]

235. Lee, J.M.; Kim, T.W.; Park, S.S.; Han, J.H.; Shin, M.S.; Lim, B.V.; Kim, S.H.; Baek, S.S.; Cho, Y.S.; Kim, K.H. Treadmill Exercise Improves Motor Function by Suppressing Purkinje Cell Loss in Parkinson Disease Rats. *Int. Neurourol. J.* **2018**, *22* (Suppl. 3), S147–S155. [CrossRef] [PubMed]

236. Putcha, D.; Ross, R.S.; Cronin-Golomb, A.; Janes, A.C.; Stern, C.E. Altered intrinsic functional coupling between core neurocognitive networks in Parkinson's disease. *NeuroImage Clin.* **2015**, *7*, 449–455. [CrossRef] [PubMed]

237. Fontanesi, C.; Kvint, S.; Frazzitta, G.; Bera, R.; Ferrazzoli, D.; Di Rocco, A.; Rebholz, H.; Friedman, E.; Pezzoli, G.; Quartarone, A.; et al. Intensive Rehabilitation Enhances Lymphocyte BDNF-TrkB Signaling in Patients with Parkinson's Disease. *Neurorehabil. Neural Repair.* **2015**, *30*, 411–418. [CrossRef] [PubMed]

238. Wang, H.Y.; Crupi, D.; Liu, J.; Stucky, A.; Cruciata, G.; Di Rocco, A.; Friedman, E.; Quartarone, A.; Ghilardi, M.F. Repetitive Transcranial Magnetic Stimulation Enhances BDNF-TrkB Signaling in Both Brain and Lymphocyte. *J. Neurosci.* **2011**, *31*, 11044–11054. [CrossRef] [PubMed]

239. Kelly, N.A.; Wood, K.H.; Allendorfer, J.B.; Ford, M.P.; Bickel, C.S.; Marstrander, J.; Amara, A.W.; Anthony, T.; Bamman, M.M.; Skidmore, F.M. High-Intensity Exercise Acutely Increases Substantia Nigra and Prefrontal Brain Activity in Parkinson's Disease. *Med. Sci. Monit.* **2017**, *23*, 6064–6071. [CrossRef]

240. Kim-Ha, J.; Kim, Y.J. Age-related epigenetic regulation in the brain and its role in neuronal diseases. *BMB Rep.* **2016**, *49*, 671–680. [CrossRef]

241. Paulsen, J.S.; Miller, A.C.; Hayes, T.; Shaw, E. Cognitive and behavioural changes in Huntington disease before diagnosis. *Handb. Clin. Neurol.* **2017**, *144*, 69–91. [CrossRef]

242. Trembath, M.K.; Horton, Z.A.; Tippett, L.; Collins, V.R.; Churchyard, A.; Roxburgh, R. A retrospective study of the impact of lifestyle on age at onset of Huntington disease. *Mov. Disord.* **2010**, *25*, 1444–1450. [CrossRef]

243. Mueller, S.M.; Petersen, J.A.; Jung, H.H. Exercise in Huntington's disease: Current state and clinical significance. *Tremor Other Hyperkinet. Mov.* **2019**, *9*, 601. [CrossRef]

244. Bohlen, S.; Ekwall, C.; Hellström, K.; Vesterlin, H.; Björnefur, M.; Wiklund, L.; Reilmann, R. Physical therapy in Huntington's disease—Toward objective assessments? *Eur. J. Neurol.* **2013**, *20*, 389–393. [CrossRef]

245. Cruickshank, T.M.; Thompson, J.A.; Domínguez, D.J.F.; Reyes, A.P.; Bynevelt, M.; Georgiou-Karistianis, N.; Barker, R.A.; Ziman, M.R. The effect of multidisciplinary rehabilitation on brain structure and cognition in Huntington's disease: An exploratory study. *Brain Behav.* **2015**, *5*, e00312. [CrossRef] [PubMed]

246. Quinn, L.; Hamana, K.; Kelson, M.; Dawes, H.; Collett, J.; Townson, J.; Roos, R.; van der Plas, A.A.; Reilmann, R.; Frich, J.C.; et al. A randomized, controlled trial of a multi-modal exercise intervention in Huntington's disease. *Parkinsonism Relat. Disord.* **2016**, *31*, 46–52. [CrossRef] [PubMed]

247. Wallace, M.; Downing, N.; Lourens, S.; Mills, J.; Kim, J.I.; Long, J.; Paulsen, J. Is There an Association of Physical Activity with Brain Volume, Behavior, and Day-to-day Functioning? A Cross Sectional Design in Prodromal and Early Huntington Disease. *PLoS Curr.* **2016**, *17*, 8. [CrossRef]

248. Mueller, S.M.; Gehrig, S.M.; Petersen, J.A.; Frese, S.; Mihaylova, V.; Ligon-Auer, M.; Khmara, N.; Nuoffer, J.M.; Schaller, A.; Lundby, C.; et al. Effects of endurance training on skeletal muscle mitochondrial function in Huntington disease patients. *Orph. J. Rare Dis.* **2017**, *12*, 184. [CrossRef] [PubMed]

249. Sølvsten, C.A.; de Paoli, F.; Christensen, J.H.; Nielsen, A.L. Voluntary physical exercise induces expression and epigenetic remodeling of VegfA in the rat hippocampus. *Mol. Neurobiol.* **2018**, *55*, 567–582. [CrossRef] [PubMed]

250. Contrò, V.; Schiera, G.; Macchiarella, A.; Sacco, A.; Lombardo, G.; Proia, P. Multiple sclerosis: Physical activity and well-being. *Trends Sport Sci.* **2017**, *2*, 53–58.

251. Mähler, A.; Balogh, A.; Csizmadia, I.; Klug, L.; Kleinewietfeld, M.; Steiniger, J.; Šušnjar, U.; Müller, D.N.; Boschmann, M.; Paul, F. Metabolic, Mental and Immunological Effects of Normoxic and Hypoxic Training in Multiple Sclerosis Patients: A Pilot Study. *Front. Immunol.* **2018**, *9*, 2819. [CrossRef] [PubMed]

252. Mulero, P.; Almansa, R.; Neri, M.J.; Bermejo-Martin, J.F.; Archanco, M.; Arenillas, J.F.; Téllez, N. Improvement of fatigue in multiple sclerosis by physical exercise is associated to modulation of systemic interferon response. *J. Neuroimmunol.* **2015**, *280*, 8–11. [CrossRef]

253. Naghibzadeh, M.; Ranjbar, R.; Tabandeh, M.R.; Habibi, A. Effects of Two Training Programs on Transcriptional Levels of Neurotrophins and Glial Cells Population in Hippocampus of Experimental Multiple Sclerosis. *Int. J. Sports Med.* **2018**, *39*, 604–612. [CrossRef]

254. Souza, P.S.; Goncalves, E.D.; Pedroso, G.S.; Farias, H.R.; Junqueira, S.C.; Marcon, R.; Tuon, T.; Cola, M.; Silveira, P.C.; Santos, A.R.; et al. Physical exercise attenuates experimental autoimmune encephalomyelitis by inhibiting peripheral immune response and blood-brain barrier disruption. *Mol. Neurobiol.* **2017**, *54*, 4723–4737. [CrossRef]

255. Houdebine, L.; Gallelli, C.A.; Rastelli, M.; Sampathkumar, N.K.; Grenier, J. Effect of physical exercise on brain and lipid metabolism in mouse models of multiple sclerosis. *Chem. Phys. Lipids* **2017**, *207 Pt B*, 127–134. [CrossRef]

256. Björklund, A.; Dunnett, S.B. Dopamine neuron systems in the brain: An update. *Trends Neurosci.* **2007**, *30*, 194–202. [CrossRef]

257. Arenas, E.; Denham, M.; Villaescusa, J.C. How to make a midbrain dopaminergic neuron. *Development* **2015**, *142*, 1918–1936. [CrossRef] [PubMed]

258. Poulin, J.F.; Zou, J.; Drouin-Ouellet, J.; Kim, K.Y.; Cicchetti, F.; Awatramani, R.B. Defining midbrain dopaminergic neuron diversity by single-cell gene expression profiling. *Cell Rep.* **2014**, *9*, 930–939. [CrossRef] [PubMed]

259. La Manno, G.; Gyllborg, D.; Codeluppi, S.; Nishimura, K.; Salto, C.; Zeisel, A.; Borm, L.E.; Stott, S.R.W.; Toledo, E.M.; Villaescusa, J.C.; et al. Molecular Diversity of Midbrain Development in Mouse, Human, and Stem Cells. *Cell* **2016**, *167*, 566–580. [CrossRef]

260. Poulin, J.F.; Caronia, G.; Hofer, C.; Cui, Q.; Helm, B.; Ramakrishnan, C.; Chan, C.S.; Dombeck, D.A.; Deisseroth, K.; Awatramani, R. Mapping projections of molecularly defined dopamine neuron subtypes using intersectional genetic approaches. *Nat. Neurosci.* **2018**, *21*, 1260–1271. [CrossRef] [PubMed]

261. Tiklová, K.; Björklund, Å.K.; Lahti, L.; Fiorenzano, A.; Nolbrant, S.; Gillberg, L.; Volakakis, N.; Yokota, C.; Hilscher, M.M.; Hauling, T.; et al. Single-cell RNA sequencing reveals midbrain dopamine neuron diversity emerging during mouse brain development. *Nat. Commun.* **2019**, *10*, 581. [CrossRef] [PubMed]

262. Bromberg-Martin, E.S.; Matsumoto, M.; Hikosaka, O. Dopamine in motivational control: Rewarding, aversive, and alerting. *Neuron* **2010**, *68*, 815–834. [CrossRef]

263. Palmiter, R.D. Dopamine signaling in the dorsal striatum is essential for motivated behaviors: Lessons from dopamine-deficient mice. *Ann. N. Y. Acad. Sci.* **2008**, *1129*, 35–46. [CrossRef]

264. Berke, J.D. What does dopamine mean? *Nat. Neurosci.* **2018**, *21*, 787–793. [CrossRef]

265. Waterson, M.J.; Horvath, T.L. Neuronal Regulation of Energy Homeostasis: Beyond the Hypothalamus and Feeding. *Cell Metab.* **2015**, *22*, 962–970. [CrossRef] [PubMed]

266. Shohamy, D.; Adcock, R.A. Dopamine and adaptive memory. *Trends Cogn. Sci.* **2010**, *14*, 464–472. [CrossRef] [PubMed]

267. Perreault, M.L.; Jones-Tabah, J.; O'Dowd, B.F.; George, S.R. A physiological role for the dopamine D5 receptor as a regulator of BDNF and Akt signalling in rodent prefrontal cortex. *Int. J. Neuropsychopharmacol.* **2013**, *16*, 477–483. [CrossRef] [PubMed]

268. Pristerà, A.; Blomeley, C.; Lopes, E.; Threlfell, S.; Merlini, E.; Burdakov, D.; Cragg, S.; Guillemot, F.; Ang, S.L. Dopamine neuron-derived IGF-1 controls dopamine neuron firing, skill learning, and exploration. *Proc. Natl. Acad. Sci. USA* **2019**, *116*, 3817–3826. [CrossRef] [PubMed]

269. Hötting, K.; Röder, B. Beneficial effects of physical exercise on neuroplasticity and cognition. *Neurosci. Biobehav. Rev.* **2013**, *37 Pt B*, 2243–2257. [CrossRef]

270. Kang, S.S.; Jeraldo, P.R.; Kurti, A.; Miller, M.E.; Cook, M.D.; Whitlock, K.; Goldenfeld, N.; Woods, J.A.; White, B.A.; Chia, N.; et al. Diet and exercise orthogonally alter the gut microbiome and reveal independent associations with anxiety and cognition. *Mol. Neurodegener.* **2014**, *9*, 36. [CrossRef] [PubMed]

271. Herrera, J.J.; Fedynska, S.; Ghasem, P.R.; Wieman, T.; Clark, P.J.; Gray, N.; Loetz, E.; Campeau, S.; Fleshner, M.; Greenwood, B.N. Neurochemical and behavioural indices of exercise reward are independent of exercise controllability. *Eur. J. Neurosci.* **2016**, *43*, 1190–1202. [CrossRef]

272. Zhu, X.; Ottenheimer, D.; DiLeone, R.J. Activity of D1/2 Receptor Expressing Neurons in the Nucleus Accumbens Regulates Running, Locomotion, and Food Intake. *Front. Behav. Neurosci.* **2016**, *10*, 66. [CrossRef]

273. Nock, N.L.; Minnes, S.; Alberts, J.L. Neurobiology of substance use in adolescents and potential therapeutic effects of exercise for prevention and treatment of substance use disorders. *Birth Defects Res.* **2017**, *109*, 1711–1729. [CrossRef]

274. Matta Mello, P.E.; Cevada, T.; Sobral Monteiro-Junior, R.; Teixeira, G.T.; da Cruz, R.E.; Lattari, E.; Blois, C.; Camaz Deslandes, A. Neuroscience of exercise: From neurobiology mechanisms to mental health. *Neuropsychobiology* **2013**, *68*, 1–14. [CrossRef]

275. Crush, E.A.; Frith, E.; Loprinzi, P.D. Experimental effects of acute exercise duration and exercise recovery on mood state. *J. Affect. Dis.* **2018**, *229*, 282–287. [CrossRef] [PubMed]

276. Greenwood, B.N. The role of dopamine in overcoming aversion with exercise. *Brain Res.* **2018**, *1713*, 102–107. [CrossRef] [PubMed]

277. Knab, A.M.; Lightfoot, J.T. Does the difference between physically active and couch potato lie in the dopamine system? *Int. J. Biol. Sci.* **2010**, *6*, 133–150. [CrossRef] [PubMed]

278. Ruegsegger, G.N.; Booth, F.W. Running from Disease: Molecular Mechanisms Associating Dopamine and Leptin Signaling in the Brain with Physical Inactivity, Obesity, and Type 2 Diabetes. *Front. Endocrinol.* **2017**, *8*, 109. [CrossRef] [PubMed]

279. Kami, K.; Tajima, F.; Senba, E. Activation of mesolimbic reward system via laterodorsal tegmental nucleus and hypothalamus in exercise-induced hypoalgesia. *Sci. Rep.* **2018**, *8*, 11540. [CrossRef] [PubMed]

280. El Mestikawy, S.; Glowinski, J.; Hamon, M. Tyrosine hydroxylase activation in depolarized dopaminergic terminals–involvement of Ca2+-dependent phosphorylation. *Nature* **1983**, *302*, 830–832. [CrossRef] [PubMed]

281. Itagaki, C.; Isobe, T.; Taoka, M.; Natsume, T.; Nomura, N.; Horigome, T.; Omata, S.; Ichinose, H.; Nagatsu, T.; Greene, L.A.; et al. Stimulus-coupled interaction of tyrosine hydroxylase with 14-3-3 proteins. *Biochemistry* **1999**, *38*, 15673–15680. [CrossRef]

282. Greenwood, B.N.; Foley, T.E.; Le, T.V.; Strong, P.V.; Loughridge, A.B.; Day, H.E.; Fleshner, M. Long-term voluntary wheel running is rewarding and produces plasticity in the mesolimbic reward pathway. *Behav. Brain Res.* **2011**, *217*, 354–362. [CrossRef]

283. Foley, T.E.; Fleshner, M. Neuroplasticity of dopamine circuits after exercise: Implications for central fatigue. *Neuromol. Med.* **2008**, *10*, 67–80. [CrossRef]

284. Droste, S.K.; Schweizer, M.C.; Ulbricht, S.; Reul, J.M. Long-term voluntary exercise and the mouse hypothalamic-pituitary-adrenocortical axis: Impact of concurrent treatment with the antidepressant drug tianeptine. *J. Neuroendocrinol.* **2006**, *18*, 915–925. [CrossRef]

285. Chen, C.; Nakagawa, S.; An, Y.; Ito, K.; Kitaichi, Y.; Kusumi, I. The exercise-glucocorticoid paradox: How exercise is beneficial to cognition, mood, and the brain while increasing glucocorticoid levels. *Front. Neuroendocrinol.* **2017**, *44*, 83–102. [CrossRef] [PubMed]

286. Sutoo, D.; Akiyama, K. Regulation of brain function by exercise. *Neurobiol. Dis.* **2003**, *13*, 1–14. [CrossRef]

287. Finberg, J.P.M. Inhibitors of MAO-B and COMT: Their effects on brain dopamine levels and uses in Parkinson's disease. *J. Neural. Transm.* **2019**, *126*, 433–448. [CrossRef] [PubMed]

288. Chen, H.; Pine, D.S.; Ernst, M.; Gorodetsky, E.; Kasen, S.; Gordon, K.; Goldman, D.; Cohen, P. The MAOA gene predicts happiness in women. *Prog. Neuropsychopharmacol. Biol. Psychiatry* **2013**, *40*, 122–125. [CrossRef] [PubMed]

289. Hill, L.D.; Lorenzetti, M.S.; Lyle, S.M.; Fins, A.I.; Tartar, A.; Tartar, J.L. Catechol-O-methyltransferase Val158Met polymorphism associates with affect and cortisol levels in women. *Brain Behav.* **2018**, *8*, e00883. [CrossRef] [PubMed]

290. Boekhoudt, L.; Omrani, A.; Luijendijk, M.C.; Wolterink-Donselaar, I.G.; Wijbrans, E.C.; van der Plasse, G.; Adan, R.A. Chemogenetic activation of dopamine neurons in the ventral tegmental area, but not substantia nigra, induces hyperactivity in rats. *Eur. Neuropsychopharmacol.* **2016**, *26*, 1784–1793. [CrossRef] [PubMed]

291. Watkins, B.A. Endocannabinoids, exercise, pain, and a path to health with aging. *Mol. Asp. Med.* **2018**, *64*, 68–78. [CrossRef] [PubMed]

292. Hillard, C.J. Circulating Endocannabinoids: From Whence Do They Come and Where are They Going? *Neuropsychopharmacology* **2018**, *43*, 155–172. [CrossRef] [PubMed]

293. Raichlen, D.A.; Foster, A.D.; Seillier, A.; Giuffrida, A.; Gerdeman, G.L. Exercise-induced endocannabinoid signaling is modulated by intensity. *Eur. J. Appl. Physiol.* **2013**, *113*, 869–875. [CrossRef] [PubMed]

294. Brellenthin, A.G.; Crombie, K.M.; Hillard, C.J.; Koltyn, K.F. Endocannabinoid and Mood Responses to Exercise in Adults with Varying Activity Levels. *Med. Sci. Sports Exerc.* **2017**, *49*, 1688–1696. [CrossRef] [PubMed]

295. Cohen, K.; Abraham, W.; Aviv, W. Modulatory effects of cannabinoids on brain neurotransmission. *Eur. J. Neurosci.* **2019**. [CrossRef] [PubMed]

296. King-Himmelreich, T.S.; Möser, C.V.; Wolters, M.C.; Schmetzer, J.; Schreiber, Y.; Ferreirós, N.; Russe, O.Q.; Geisslinger, G.; Niederberger, E. AMPK contributes to aerobic exercise-induced antinociception downstream of endocannabinoids. *Neuropharmacology* **2017**, *124*, 134–142. [CrossRef] [PubMed]

297. Fuss, J.; Steinle, J.; Bindila, L.; Auer, M.K.; Kirchherr, H.; Lutz, B.; Gass, P. A runner's high depends on cannabinoid receptors in mice. *Proc. Natl. Acad. Sci. USA* **2015**, *112*, 13105–13108. [CrossRef] [PubMed]

298. Sonnenblick, Y.; Taler, M.; Bachner, Y.G.; Strous, R.D. Exercise, Dehydroepiandrosterone (DHEA), and Mood Change: A Rationale for the "Runners High"? *Isr. Med. Assoc. J.* **2018**, *20*, 335–339. [PubMed]

299. Prough, R.A.; Clark, B.J.; Klinge, C.M. Novel mechanisms for DHEA action. *J. Mol. Endocrinol.* **2016**, *56*, R139–R155. [CrossRef] [PubMed]

300. Kimura, T.; Kaneko, F.; Iwamoto, E.; Saitoh, S.; Yamada, T. Neuromuscular electrical stimulation increases serum brain-derived neurotrophic factor in humans. *Exp. Brain Res.* **2019**, *237*, 47–56. [CrossRef]

301. Costa, M.T.S.; Vieira, L.P.; Barbosa, E.O.; Mendes Oliveira, L.; Maillot, P.; Ottero Vaghetti, C.A.; Giovani Carta, M.; Machado, S.; Gatica-Rojas, V.; Monteiro-Junior, R.S. Virtual Reality-Based Exercise with Exergames

as Medicine in Different Contexts: A Short Review. *Clin. Pract. Epidemiol. Ment. Health* **2019**, *15*, 15–20. [CrossRef]

302. Wouters, M.; Evenhuis, H.M.; Hilgenkamp, T.I.M. Physical fitness of children and adolescent with moderate to severe intellectual disabilities. *Disabil. Rehabil.* **2019**, 1–11. [CrossRef]

303. Del Coso, J.; Moreno, V.; Gutiérrez-Hellín, J.; Baltazar-Martins, G.; Ruíz-Moreno, C.; Aguilar-Navarro, M.; Lara, B.; Lucía, A. ACTN3 R577X Genotype and Exercise Phenotypes in Recreational Marathon Runners. *Genes* **2019**, *10*, 413. [CrossRef]

304. Kandel, E. *The Disordered Mind: What Unusual Brains Tell Us about Ourselves*; Little Brown Book Group, Ed.; Hachette: London, UK, 2018.

 © 2019 by the authors. Licensee MDPI, Basel, Switzerland. This article is an open access article distributed under the terms and conditions of the Creative Commons Attribution (CC BY) license (http://creativecommons.org/licenses/by/4.0/).

MDPI

St. Alban-Anlage 66

4052 Basel

Switzerland

Tel. +41 61 683 77 34

Fax +41 61 302 89 18

www.mdpi.com

Genes Editorial Office

E-mail: genes@mdpi.com

www.mdpi.com/journal/genes

www.ingramcontent.com/pod-product-compliance
Lightning Source LLC
LaVergne TN
LVHW071520180726
843512LV00014B/1129

9 783039 284801